RENEWALS 691-4574

DATE DUE

FEB 0 4			
MAY 0 2			
MAR - 6			

DIELECTRIC SPECTROSCOPY OF POLYMERS

DIELECTRIC SPECTROSCOPY OF POLYMERS

PÉTER HEDVIG

Research Institute for Plastics
Budapest, Hungary

A HALSTED PRESS BOOK

JOHN WILEY & SONS
NEW YORK

Published in the U.S.A.
by Halsted Press, a
Division of John Wiley & Sons, Inc.,
NEW YORK

Library of Congress Catalog Card №.: 75-9653
Hedvig, Péter.

Dielectric Spectroscopy of Polymers
"A Halsted Press book."
Bibliography: p.
Includes index.
1. Polymers and polymerization — Spectra.
2. Plastics — Electric properties. I. Title.
QC 463. P5H4 548'.8432 75-9653

ISBN 0-470-36747-4

© Akadémiai Kiadó, Budapest 1977

Co-edition of
Adam Hilger Ltd, Bristol
Halsted Press, New York

and of

Akadémiai Kiadó, Budapest

Printed in Hungary

CONTENTS

SYMBOLS

c	velocity of light
c_p	specific heat at constant pressure
c_r	specific heat at constant volume
$\mathscr{D}$	electric displacement vector
d	distance
e	charge
exp	exponential function
E	energy
E^*	complex tensional modulus (Young's modulus)
E'	tensional storage modulus
E''	tensional loss modulus
$\mathscr{E}$	electric field vector
J	bond charge
f	distribution function
F	free energy
$\mathscr{F}$	force
g_r	Kirkwood reduction factor, steric factor
G^*	complex mechanical torsional modulus
G'	torsional storage modulus
G''	torsional loss modulus
h	Planck constant
$\hbar = \dfrac{h}{2\pi}$	
H	hamiltonian function
$\hat{H}$	hamiltonian operator
$\mathscr{H}$	magnetic field vector
$\mathscr{I}$	Coulomb integral
i	imaginary unit
I	current
I	ionicity of bond
j	current density vector
J^*	complex mechanical compliance
J'	real part of complex torsional compliance, storage compliance
J''	unreal part of complex torsional compliance, loss compliance
k	rate constant of reactions

k	Boltzmann's constant
k	propagation wave vector
$\mathscr{H}$	exchange integral
l	distance
$\mathscr{L}$	Laplace transform
$\hat{L}$	Liouville operator
m^*	effective mass
m	mass
m	molar fraction
n	particle number
n	index of refraction
N_A	Avogadro's number
p	linear momentum, vector
p	linear momentum, absolute value
p	pressure
P	probability
$\mathscr{P}$	electric polarization vector
q	general coordinate
Q	quality factor
r	radius vector
R	gas constant
R	bond distance
R	electric resistance
S	entropy
t	time
T	temperature
T_g	glass–rubber transition temperature
T_{gg}	glass–glass transition temperature
T_{ll}	liquid–liquid transition temperature
T_m	crystalline melting temperature
T_1	nuclear magnetic spin–lattice relaxation time
T_2	nuclear magnetic spin–spin relaxation time
$T_{1\varrho}$	rotating frame spin–lattice relaxation time
X	Pauling electronegativity
X	Flory's interaction parameter
Y	admittance
v	velocity, absolute value
$\mathbf{v}$	velocity, vector
v	volume
V	voltage, potential
w_f	non-equilibrium free volume
w	weight fraction
α	thermal volume expansion coefficient
γ	mechanical strain
γ_I	nuclear magnetogyric ratio
Γ	gamma function
δ	inductive parameter
δ	Dirac delta function

10

ε^*	complex permittivity tensor
ε'	real part of complex permittivity
ε''	unreal part of complex permittivity
ϵ	extinction coefficient
η	viscosity
θ	moment of inertia
θ	modulus of a thermodynamical distribution function
ν	frequency
μ	chemical potential
$\boldsymbol{\mu}$	dipole moment, vector
μ	dipole moment, absolute value
μ_{eff}	effective dipole moment, vector
λ	wavelength
Λ	logarithmic decrement of mechanical damping
ϕ	dipole autocorrelation function
φ_i	atomic orbitals
ψ_i	molecular orbitals
χ^*	complex susceptibility
χ'	real part of complex susceptibility
χ''	unreal part of complex susceptibility
σ	mechanical stress
σ	electrical conductivity
ϱ	specific resistivity
ϱ	density
ϱ_e	electron charge density
ω	angular frequency
Ω	ohm, unit of resistivity

11

CHAPTER 1

BASIC PRINCIPLES OF DIELECTRIC SPECTROSCOPY

Dielectric spectroscopy is based on the interaction of electromagnetic radiation with the electric dipole moments of the material under test. The frequency range of the radiation is between 10^{-6} Hz and about 10^{10} Hz. Above 10^{10} Hz, in the infrared optical and ultraviolet region, the absorption and emission of radiation is due to changes in the induced dipole moments, which are dependent on the polarizability of the atoms or molecules. At lower frequencies the contribution of the induced dipole moments becomes small in comparison with that of the permanent dipole moments of the system. Consequently dielectric spectroscopy is useful for studying polar molecules in the gaseous phase or in solution, where absorption of radiation is mainly due to the reorientation of permanent dipoles. This method, introduced by Debye in 1931, has been used since then to determine molecular dipole moments and to study liquid and solid structures.

In the condensed phase the situation is rather complicated because the electronic states of the system cannot be described in terms of molecular orbitals; collective crystal states (excitons) are to be considered. It has been shown that the dielectric behaviour of a solid can be rigorously interpreted only in terms of exciton states (Rice and Jortner 1967). As a consequence of this, induced dipole moments may contribute to the absorption of radiation appreciably even at the low frequency range. Moreover the dipole moments of molecules or groups are strongly influenced by intermolecular interactions which can also be described in terms of exciton states.

In polymeric solids and visco-elastic liquids the contribution of the exciton states to the permanent dipole moments is not very large; at least it is weak enough to consider certain groups of atoms or bonds individually. In the ground state, correspondingly, one can regard a polymeric solid containing certain polar groups at first approximation as a system of not very strongly interacting electrical dipoles. This way the methods of dielectric spectroscopy developed originally for gases and solutions can be more or less accurately applied to polymeric solids too. The work done in this field in the past 25–30 years has been based on this approximation. Only recently is attention being paid to the effects of induced dipole polarization in polymers, which is directly connected with the exciton states and correspondingly with the physical structure of the solid.

The response of a material to an external electric field is essentially a statistical effect. It is not possible to observe the orientation of the individual moments; only the bulk polarization of the assembly can be measured. This means that dielectric spectroscopy is based on statistical thermodynamical considerations; therefore this chapter, besides dealing with the origin of the molecular dipole moments, also briefly discusses the response of a system of dipole moments to external electric fields.

1.1 The origin of dipole moments

Dipole moments originate from the asymmetry of the positive and negative charge densities of the system. Positive charges arise from the nuclei; they are localized. Changes in the positive charge density result from structural transformations of the molecule (isomeric transitions, rotation of groups or that of the solid. e.g. recrystallization, local motions, rotation of groups, vibration, etc.). Negative charge densities arise from the electronic system, which is delocalized. The extent of delocalization depends on the chemical structure.

The total dipole moment of a molecule is

$$\mu = \int \mathbf{r} \left[\varrho_e(\mathbf{r}) + \varrho_n(\mathbf{r}) \right] d\mathbf{r} \qquad (1.1)$$

where $\varrho_e(r) \, \varrho_n(r)$ are the electronic and nuclear charge densities respectively.

In molecules the electron densities can be calculated by constructing the molecular wavefunction as a linear combination of atomic orbitals (see, for example, Pilar, 1968)

$$\psi = \sum_i \lambda_i \varphi_i$$

where the λ_i are weight factors and the φ_i are the atomic orbitals.

The electron charge density is expressed as

$$\varrho_e = -e\,\psi^*\psi$$

where e is the charge of the electron, ψ^* is the complex conjugate of ψ.
The total electronic charge in a volume v is

$$e(v) = -e \int \psi^*\psi\,dv$$

In principle, it is possible to calculate the fractional electronic charge distribution of a molecule. Knowing the equilibrium position of the nuclei, the positive charge distribution is also known and thus the dipole moment of the molecule can be calculated. In practice such calculations are extremely difficult. This is why the total dipole moment of a molecule is more often interpreted as being a vectorial sum of bond moments

$$\mu = \sum_{i=1}^{N} \mu_i$$

The absolute values of bond moments are approximately expressed as

$$\mu_i = 4.8\,R\,I \qquad (1.2)$$

where R is the bond distance, I is the ionicity of the bond. The ionicity of a bond between molecules A and B can be expressed in terms of the Pauling-electronegativity values (Pauling, 1960) X_A and X_B as follows

$$I(AB) = 1 - \exp\left[-\frac{(X_A - X_B)^2}{4}\right] \qquad (1.3)$$

The Pauling electronegativities are defined as

$$X_A - X_B = [0.18\,E_{AB} - (E_{AA}\,E_{BB})^{1/2}]^{1/2} \qquad (1.4)$$

where E_{AB} is the total binding energy of the bond, E_{AA}, E_{BB} respectively are the binding energies of the corresponding like-atoms.
According to the type of the electron configuration, the following three main kinds of bond-dipole moments are distinguished (see, for example, Hedvig, 1975):

(1) *σ-moments*. These are defined by the ionic parts of the σ-bond eigenfunctions. The molecular orbital of the σ-bond formed by molecules A and B is

$$\psi_{AB} = \lambda_A \varphi_A + \lambda_B \varphi_B$$

where φ_A and φ_B are the atomic orbitals, λ_A and λ_B are weight factors.

The σ-bond moment is defined as

$$\mu_\sigma = (\lambda_A^2 - \lambda_B^2)\, eR + 4e\, \lambda_A \lambda_B\, r_{AB} \tag{1.5}$$

where $\lambda_A^2 - \lambda_B^2$ is referred to as Mullikan's bond charge, R is the bond distance. The bond charge represents the ionic character of the bond. The second term of Eqn. (1.5) is the contribution of the overlap charge, where

$$r_{AB} = \int r \varphi_A \varphi_B \, dv$$

is the overlap distance for the electron densities of A and B.

The definition of Eqn. (1.5) is only approximate because it assumes that the overlap charge is equally divided between atoms A and B, which evidently cannot be true generally. For a more refined discussion of electron charge densities see Politzer and Harris (1970).

(2) *Mesomeric or π-moments* arise from the deformation of the π-electron density due to the inductive effect of a substituent. The main effect of the substituent is to change the coulombic integral of the π-electron system. For substituted benzenes, for example, the ring carbon coulombic integral is expressed as

$$\mathcal{I} = \mathcal{I}(C) + \delta\,\mathcal{K}(C) \tag{1.6}$$

where $\mathcal{I}(C)$ ist the coulombic integral of the ring carbon in the unsubstituted compound

$$\mathcal{I}(C) = \langle \varphi_i | \hat{H} | \varphi_i \rangle$$

where φ_i is the atomic orbital, $\hat{H}$ is the hamiltonian operator of the system, $\mathcal{K}(C)$ is the exchange integral of the unsubstituted compound

$$\mathcal{K}(C) = \langle \varphi_i | \hat{H} | \varphi_k \rangle$$

δ is the parameter representing the inductive effect of the substituent; its value depends on the electronegativity of the substituent and on the polarizability of the bonds.

As an illustrative example let us consider the effect of chlorine substitution on the π-electron densities of benzene (Higasi *et al.*, 1965).

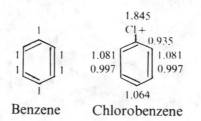

Benzene   Chlorobenzene

It is seen that the uniform π-electron density is deformed by the chlorine substitution resulting in a mesomeric dipole moment of 2.3 d. The negative charge center is at the ring, the positive is at the chlorine atom.

(3) *Lone pair moments* arise from non-bonding electrons; they are different from zero only for hybridized orbitals. The lone pair moment is expressed as

$$\mu_{lp} = \frac{4e}{1+\lambda^2} \, r_{sp} \tag{1.7}$$

where λ is the mixing parameter of the hybrid orbital built on s- and p-orbitals

$$\varphi_i = \frac{s_i + \lambda p_i}{(1+\lambda^2)^{1/2}} \tag{1.8}$$

r_{sp} is the overlap distance between the s and p orbitals

$$r_{sp} = \int r \, s p \, dv \tag{1.9}$$

For the hybrid orbitals of carbon the lone pair moments are the following

$$\lambda = 1, \quad \mu_{lp}(sp) = 4.5 \text{ d}$$
$$\lambda = 2, \quad \mu_{lp}(sp^2) = 4.2 \text{ d}$$
$$\lambda = 3, \quad \mu_{lp}(sp^3) = 3.9 \text{ d}$$

In Table 1.1 the values of some bond moments are shown. These moments have been calculated from the measured values of the corresponding methyl-compounds by assuming $C^- - H^+$ polarization of the CH bonds and tetrahedral configuration. For chloromethane, for example, the tetrahedral angle is 109.5°. Correspondingly, the component

Table 1.1 Bond dipole moments (Minkin *et al.*, 1968)

Bond	Dipole moment in debye units (10^{-18} e.s.u)	Bond	Dipole moment in debye units
$\geqslant \overset{+}{C}-\overset{-}{F}$	1.39	$>\overset{+}{C}=\overset{-}{N}-$	1.4
$\geqslant \overset{+}{C}-\overset{-}{Cl}$	1.47	$>\overset{+}{C}=\overset{-}{O}$	2.4
$\geqslant \overset{+}{C}-\overset{-}{N}<$	0.45	$>\overset{+}{C}=\overset{-}{S}$	2.0
$\geqslant \overset{+}{C}-\overset{-}{O}-$	0.7	$-\overset{+}{C}\equiv\overset{-}{N}$	3.1
$\geqslant \overset{+}{C}-\overset{-}{S}-$	0.9	$\overset{+}{H}-\overset{-}{O}-$	1.5
		$\overset{+}{H}-\overset{-}{N}<$	1.3
		$\overset{+}{H}-\overset{-}{S}-$	0.7
$\overset{+}{C(sp^3)}-\overset{-}{C(sp^2)}$	0.69	$>Si-C\leqslant$	1.2
$\overset{+}{C(sp^3)}-\overset{-}{C(sp)}$	1.48	$>Si-H$	1.0
$C(sp^2)-C(sp)$	1.15	$>Si-N<$	1.55

of the C^--H^+ dipole moment along the $C-Cl$ axis is $\mu(C-H) \cos 180° - 109.5° = \frac{1}{3}\mu(C-H)$. As there are three $C-H$ bonds to contribute to the total moment along the $C-Cl$ axis and $\mu(C-H) = 0.4$ debye, the total $C-Cl$ bond moment is

$$\mu(C-Cl) = \mu(CH_3Cl) - \mu(C-H) = 1.87 - 0.4 = 1.47 \text{ debye}$$

Other configurations can be calculated similarly.

The bond dipole moment values of Table 1.1 are given in debye units, 1 debye being equal to 10^{-18} e.s.u. As the charge of the electron (unit charge) is 4.8×10^{-10} e.s.u., in order to produce a dipole moment of 1 debye the positive and negative unit charge-centers should be separated by 0.28 Å ($\approx 3 \times 10^{-9}$ cm). The direction of the dipole moment is taken conventionally from the negative to the positive charge-center. This is also indicated in Table 1.1.

For dipole moment values see Minkin *et al.* (1968), Osipov and Minkin (1965), Smith (1955), McClellan (1963).

To assume that the total dipole moment is a vectorial sum of the bond-moments is, of course, an oversimplification; the charge density distribution of a bond is evidently influenced by the neighbours through inductive effects and also by overlap of the electron densities. This is illustrated in Table 1.2 where the $C-H$, $C-Cl$ and $C-F$ bond

Table 1.2 Inductive effect of halogen substitution on the bond dipole moments (Smith, 1955)

| Compound | Bond moment debye | | Total moment of the molecule debye |
	C – H − +	C – Cl + −	
CH_3Cl	0	1.86	1.87
CH_2Cl_2	0.16	1.25	1.62
$CHCl_3$	0.19	0.92	1.2
	C – H − +	C – F + −	Total
CH_3F	0	1.81	1.79
CH_2F_2	0.23	1.45	1.96
CHF_3	0.32	1.22	1.64

moments are shown by different halogen substitutions for chloro- and fluoro-methanes. It is seen that the $C-H$ bond moment is considerably increased by the halogen substitution, while the carbon-halogen bond moment is decreased.

Recent studies have shown (Nagy, 1973) that the classical view of a molecule being built on atoms bound by chemical bonds is not valid. In such a simple molecule as ethylene, for example, calculations show that the bond order between the hydrogen atoms is not negligible; it is almost one-third of the $C-H$ bond order. The molecular diagram of ethylene according to these calculations is

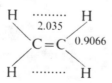

This implies that the dipole moment of a group or a molecule should be evaluated directly from the nuclear and electronic charge distributions, which can be calculated without assuming chemical bond structure. Such calculations are, however, rather cómplicated and expensive because long computer-times are needed. The semiclassical view based on the Valence Bond Theory (Pauling, 1960) is still useful, although quantitative agreement between theory and experiment is usually poor.

The effective dipole moment

In dielectric spectroscopy of gases and solutions the units oriented by the electric field are the molecules as a whole. This process is based on the thermal motion which makes all possible dipole orientations approximately equally probable. This probability distribution of the orientation is influenced by the external electric field to result in macroscopic polarization (see, for example, Fröhlich, 1958). In this process the molecules are considered as being rigid configurations; the molecular dipole moment itself does not change during thermal motion. This approximation is evidently not true for macromolecular systems, not even in dilute solution. The shape of the molecule and, correspondingly, the total molecular dipole moment change during thermal motion; the molecules evidently cannot be treated as rigid configurations. Correspondingly, by considering dipole orientation in macromolecular systems, only parts of the molecule are to be considered, the configuration of which is unchanged during thermal motion.

Even for small molecules there are some cases when one part of the molecule is rigid and another part is flexible. By considering the effective dipole moment the bond-moments of the rigid part are to be added vectorially. If the molecule contains N polar groups of configuration which is unchanged during thermal motion, the effective dipole moment is expressed as

$$\mu_{eff} = \left[\sum_{i=1}^{N} \mu_{xi}^2 + \sum_{i=1}^{N} \mu_{yi}^2 + \sum_{i=1}^{N} \mu_{zi}^2 \right]^{1/2} \tag{1.10}$$

where μ_{xi}, μ_{yi}, μ_{zi} are the bond moment components along the coordinate axes x, y, z.

If the molecule contains N polar groups which rotate freely the effective average moment is

$$\mu_{eff} = \sum_{i=1}^{N} \mu_i^2 + 2 \sum_{i=1}^{N} \sum_{s<i} \prod_{k=1}^{s+1} \cos\theta_k\, \mu_i\, \mu_s \qquad (1.11)$$

where θ_k is the angle between momenta i and s.

As a particular example for this consider the α-bromo-ω-cyanodecane molecule

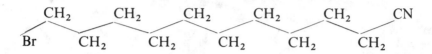

This molecule has two polar groups $C-Br$ and $C-CN$, the bond moments are $\mu(Br) = 1.42$ debye, $\mu(CN) = 3.1$ debye. Assuming free rotation about the $C-C$ bonds and considering the angle between the two polar groups $\theta = 70 \cdot 5°$ Eqn. (1.11) becomes

$$\mu_{eff}^2 = \mu^2(Br) + \mu^2(CN) - 2\mu(Br)\,\mu(CN) \cos 70.5°$$

The effective dipole moment is

$$\mu_{eff} = 3.92 \text{ debye}$$

The experimental value is 3.95 debye in benzene solution (Huisgen, 1957), so the agreement is good. This means that in solution the molecule exhibits thermal motion as a whole while the polar end-groups rotate freely. If, however, the solution is frozen and rotation of the end-groups is hindered, the dielectric behaviour of the system will be determined only by the moments $\mu(Br)$ and $\mu(CN)$. This also means that even for small molecules, and always for macromolecules, the effective dipole moment may depend on the temperature.

In considering effective dipole moments it is helpful to use group moments instead of bond moments. Group moments are the dipole moments of such chemical groups which are known as stable configurations during thermal motion. The values of some group moments interesting in the chemistry of macromolecules are collected in Table 1.3. Two main types are considered—one is when the polar group is connected to a phenyl ring, the other is when it is connected to a methyl

21

Table 1.3 Group dipole moments (Minkin et al., 1968)

Group Ph − X	Direction of the moment	Angle of the group moment with Ph − X	Dipole moment debye	Group CH$_3$ − X	Angle of the group moment with CH$_3$ − X	Dipole moment debye
Ph − CH$_3$	←	0	0.37	H$_3$C − CH$_3$	0	0
Ph − CF$_3$	→	0	− 2.54	H$_3$C − CF$_3$	0	− 2.32
Ph − CCl$_3$	→	0	− 2.04	H$_3$C − CCl$_3$ + ← −	0	− 1.57
Ph − CN	→	0	− 4.05	H$_3$C − CN + ← −	0	− 3.47
Ph − CHO	146°↗	146	− 2.96	H$_3$C − CHO + ← −	125	− 2.49
Ph − COCH$_3$	132°↗	132	− 2.96	H$_3$C − COCH$_3$ + ← −	120	− 2.75
−	−	−	−	H$_3$C − COOH + ← −	106	− 1.63
Ph − COOCH$_3$ + −	120°↗	110	− 1.83	−	−	−
Ph − COOC$_2$H$_5$	118°↗	118	− 1.9	H$_3$C − COOC$_2$H$_5$ − → +	89	1.8
Ph − F	→	0	− 1.47	H$_3$C − F + ← −	0	− 1.79
Ph − Cl	→	0	− 1.59	H$_3$C − Cl + ← −	0	− 1.87
Ph − OH	↑	90	1.55	−	−	−
Ph − OCH$_3$ − → +	72°↘	72	1.28	H$_3$C − OCH$_3$ − → +	124	1.28
Ph − OCF$_3$ − → +	160°↗	160	− 2.36	−	−	−
Ph − OCOCH$_3$ − → +	66°↘	66	1.69	−	−	−
Ph − NO + ← −	149°↗	149	− 3.09	−	−	−
Ph − NO$_2$ + ← −	→	0	− 4.01	H$_3$C − NO$_2$ + ← −	0	− 3.10

group. Besides the group moment values, the direction is also shown. The angles listed represent the angle of the dipole moment with the Ph − X and the H$_3$C − X bond.

Effective dipole moments are evidently dependent on the stereo-chemical structure of the molecule or group. As a simple example, let us consider the *cis* and *trans* isomers of difluoroethylene. In the *trans*-form the effective dipole moment is zero because the oppositely directed C−F dipoles cancel each other. In the *cis*-form the effective dipole moment is rather high, as the C−F dipoles are oriented the same way:

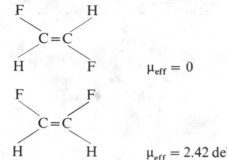

$$\mu_{eff} = 0$$

$$\mu_{eff} = 2.42 \text{ debye (Laurie, 1961)}$$

Similarly

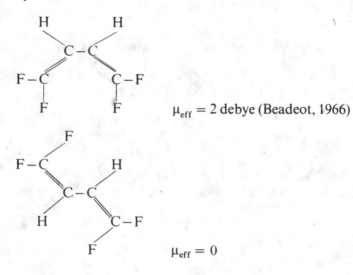

$$\mu_{eff} = 2 \text{ debye (Beadeot, 1966)}$$

$$\mu_{eff} = 0$$

Evidently the effective dipole moment of macromolecules should also depend on their stereochemical structure. One would expect the highest effective dipole moments for isotactic polymers, where the monomer units, and correspondingly the same moments, are oriented regularly the same way.

In atactic polymers where the substituent groups are arranged at random the individual bond moments are more likely to be cancelled, resulting in smaller effective dipole moment. In Fig. 1.1 the three main

23

stereochemical forms of a vinyl polymer are illustrated. The main chain exhibits a zigzag configuration, the polar substituent groups are shown by circles.

In Table 1.4 the effective dipole moments of some polymers of different tacticities are compared on the basis of dielectric measure-

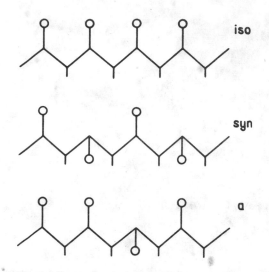

Fig. 1.1 Stereochemical forms of a vinyl polymer

Table 1.4 The effect of stereoregularity on the effective dipole moments of some macromolecules in solution. (Effective dipole moment in debye units.)

Polymer	Isotactic	Atactic	Syndiotactic	
Polystyrene in toluene 311 K	0.44	0.36	—	Kriegbaum, Roig (1959)
Polymethyl methacrylate in benzene 303 K	1.42	1.28	1.26	Mikhailov, Krasnev (1967)
Polybutyl methacrylate in benzene 293 K	1.52	1.45	1.45	Mikhailov, Krasnev (1967) Borisova *et al.* (1962)
Polyvinyl isobutyrate in benzene 298 K	1.16	1.07	—	Takeda *et al.* (1960)
Poly-tert-butyl-methacrylate in benzene 293 K	1.73	1.54	—	Mikhailov, Burstein (1964)

24

ments in dilute solution; the isotactic forms always exhibit higher effective dipole moments than the atactic or syndiotactic forms, as is expected (Mikhailov, Krasnev, 1967).

In the condensed phase or in not-too-diluted solutions the dipole moments of the molecules or groups would strongly interact. This interaction results in a decrease of the total dipole moment which is effective in interaction with the electric field. This reduction factor is defined as (Kirkwood, 1939)

$$g_r = \frac{\mu_{eff}^2 \text{ (condensed)}}{\mu_0^2 \text{ (gas)}} \qquad (1.12)$$

where μ_0 is the total dipole moment of the group or molecule, μ_{eff} is the effective dipole moment measured in the condensed state. The value of the reduction factor is determined by the number and distance of the nearest polar neighbours to the group considered. For polymers the Kirkwood reduction factor g_t is significantly different from unity, even for dilute solutions, as a result of the intramolecular interactions of the polar units of a macromolecule. The effective dipole moments and the reduction factors for some polymers in dilute solution are collected in Table 1.5. The calculated dipole moments for the monomer units are also shown.

In bulk polymers in the solid and in the viscoelastic fluid state the intramolecular interactions are evidently significant too. The corresponding reduction factors, however, have not been determined; the configuration of the polymer molecules in bulk polymers is not yet known, and it is also difficult to determine what part of the molecule behaves rigidly during thermal motion in a given temperature range.

Effective dipole moments of polymers are usually determined from the measured values of the static (relaxed) permittivity ε_0 and the high-frequency optical (unrelaxed) one ε_∞. For evaluation of the dipole moment the Fröhlich (1949) theory is usually used. According to this theory, the macroscopical polarization of a sphere is calculated by averaging the dipoles inside the sphere and by regarding the outside region as a continuum. For polymer molecules in solution, according to the Fröhlich-Kirkwood theory, the dielectric permittivity is expressed as

$$\varepsilon_0 - \varepsilon_\infty = \frac{3\,\varepsilon_0}{2\,\varepsilon_0 + \varepsilon_\infty}\,\frac{4\,\pi\,N_r}{3\,kT}\left(\frac{\varepsilon_\infty + 2}{3}\right)^2 g_r\,\mu_0^2 \qquad (1.13)$$

where N_r is the concentration of the repeat units, μ_0 is their dipole moment, g_r is the Kirkwood reduction factor which is expressed

Table 1.5 Dipole moments and reduction factors of some polymers

	μ_{eff}	$\mu_0\sqrt{g_r}$	μ_0	g_r
Polymethyl methacrylate	1.6—1.7	1.33	1.79	0.55
atactic	1.6			
isotactic	1.7			
Polypropyl methacrylate		1.41	1.89	0.56
Polyisopropyl methacrylate		1.45	1.87	0.61
Polybutyl methacrylate	1.4—1.5	1.38	1.88	0.55
atactic	1.4			
isotactic	1.5			
Poly tert-butyl methacrylate				
atactic	1.5			
isotactic	1.7			
Polyphenyl methacrylate				
atactic	1.4			
isotactic	1.6			
Polymethyl acrylate		1.37	1.75	0.68
Polyethyl acrylate		1.58	1.82	0.76
Polypropyl acrylate		1.58	1.83	0.75
Polybutyl acrylate		1.52	1.85	0.67
Polyvinyl acetate		1.66	1.83	0.82
Polyvinyl propionate		1.71	1.82	0.82
Polyvinyl butyrate		1.63	1.77	0.81
Polyethyl methacrylate		1.35	1.85	0.53

as an average of the spatial orientations within the sphere. In dilute solutions when the polymer molecules are not entangled the reduction factor is expressed as

$$g_r = 1 + \sum_{i=1}^{N} \langle \cos\gamma_{oi} \rangle_{av} \tag{1.14}$$

where γ_{oi} is the angle between one repeat unit of the chain o and another one within the same chain i. In concentrated solutions or in bulk when entanglement also occurs

$$g_r = 1 + \sum_{i} \sum_{k} \cos\gamma_{oi}^k \tag{1.15}$$

where k refers to a different molecule. It can be assumed that the interactions affecting the reduction factor g_r are extended to a few repeat units within the same chain and between chains too. This means

26

that the static dielectric constant, according to this model, should not depend on the molecular weight of the polymer. This is experimentally observed. For a detailed discussion of the structural factors determining the Kirkwood reduction factor in polymeric solution see Volkenstein (1963).

Defect dipole moments

In solids, besides permanent dipole moments of the polar bonds, moments associated with structural inhomogeneities are also effective in determining the dielectric behaviour. This is connected with the fact that some electrons in the solid are shared by many molecules. The electronic state of such a system cannot be described in terms of individual molecular orbitals; combinations of molecular orbitals, i.e. crystal orbitals, should be used. Crystal orbitals can be constructed on the basis of molecular orbitals by assuming that intermolecular interactions are weak. This is referred to as the weak-binding or Frenkel approximation (Frenkel, 1931). In this approach the crystal corresponding to N units of like-molecules unit cells or aggregates containing h sites each is expressed as

$$\psi_{k,\mu}^j = \frac{1}{\sqrt{N}} \sum_{\mu=1}^h \sum_{n=1}^N B_\mu^g \, \phi_{n\mu}^j \, F(k, R_n, t) \qquad (1.16)$$

where $\phi_{n,\mu}^j$ is the molecular orbital excited to the j-th state at the μ-th site in the n-th unit cell, B_μ^g are weight-factors, $g = 1, 2, \ldots h$, $F(k, R_n, t)$ is a function representing the symmetry of the crystal or aggregate, k is a vector representing the transformation properties of the group (unit cell, aggregate), R_n is the position vector of the n-th unit cell.

For crystals the $F(k, R_n, t)$ function is periodic

$$F(k, R_n, t) = \exp\left\{ i \, k \, R_n - \frac{iE_j}{\hbar} t \right\} \qquad (1.17)$$

and $\psi_{k,\mu}^j$ is a wave, k is a wave-vector, E_j is the excitation energy.

The corresponding state is referred to as the exciton state. It means that excitation travels through the crystal as a wave over all μ-sites.

When the system is not crystalline the $F(k, R_n, t)$ function is not periodic but can be constructed as a Fourier series of periodic functions.

In the absence of intermolecular interactions the exciton state would be hN-fold degenerate. This degeneration is lifted by the interactions to result in an exciton energy-band. The energy eigenvalues are generally expressed as

$$E^{(j)} = E_0^{(j)} + E_D + E_M^{(j)}(\mathbf{k}) \tag{1.18}$$

where $E_0^{(j)}$ is the energy of the unperturbed molecule, E_D is a constant which arises from the translationally equivalent sites, $E_M^{(j)}(\mathbf{k})$ is an energy-splitting caused by the translationally non-equivalent sites. $E_M^{(j)}(\mathbf{k})$ has h components, where h is the number of sites in the unit cell. The corresponding splitting of the spectrum lines have been experimentally observed, the effect is referred to as Davydov splitting (Davydov, 1948).

Exciton states are not only effective in the excited state of the solid but in the ground state too. The ground state of the solid cannot be constructed purely by a combination of the ground-state molecular orbitals; some excited states will be mixed up by configurational interaction. Such an intermixing of higher energy orbitals to the ground state is often encountered in molecules.

In another approximation the intermolecular interactions are considered strong; charge transfer between subsequent sites is regarded possible. This approach is referred to as tight binding or Wannier approximation (Wannier, 1937). Wannier constructed crystal orbitals in terms of electron-hole pair eigenfunctions. The situation is similar to that of considering the ionicity of bonds; in this case transfer of an electron from one atom to another is considered. In the Wannier case, electron-hole pairs can move correlated, resulting in transfer of excitation energy as

$$
\begin{array}{ll}
M_1^{\pm} M_2 M_3 \ldots M_n & \psi_1^* \psi_2 \ldots \psi_n \\
M_1 M_2^{\pm} M_3 \ldots M_n & \psi_1 \psi_2^* \ldots \psi_n \\
\qquad \vdots & \qquad \vdots
\end{array}
$$

where M means a molecule and ψ is the molecular orbital; the star signifies excitation.

The other possibility is uncorrelated motion of electrons and holes, i.e. charge transfer:

$$
\begin{array}{ll}
M_1^+ M_2^- M_3 M_4 \ldots M_n & \psi_1^+ \psi_2^- \psi_3 \ldots \psi_n \\
M_1^+ M_2 M_3^- M_4 \ldots M_n & \psi_1^+ \psi_2 \psi_3^+ \ldots \psi_n
\end{array}
$$

The Wannier exciton energy is expressed as

$$E_{exc} = E_g + \frac{m_{exc}^* [e^2/\varepsilon]^2}{2\,h^2 n^2} + \frac{h^2 k^2}{2(m_e^* + m_h^*)} \qquad (1.19)$$

where E_g is the energy-gap between the conduction band and the valence band, m_{exc}^*, m_e^*, m_h^* are the effective masses of the exciton, electron and hole respectively, k is the exciton wave vector, n is a quantum number, ε is the dielectric permittivity of the medium.

The Wannier exciton energies expressed by Eqn. (1.19) are only valid for semiconductors, where the intermolecular interactions are very strong. In this case the radius of the electron-hole combination may be as large as 100 Å. For usual macromolecular solids (not organic polymeric semiconductors) the situation is in between the Frenkel and Wannier case. In the Frenkel approximation the electron-hole distance is in the order of 1 Å.

The radius of the Wannier exciton is

$$R_0 = \frac{n^2 h^2}{m_{exc}^* (e^2/\varepsilon)} \qquad (1.20)$$

For further details of exciton theory see Knox (1963) and Davydov (1968).

From the exciton picture some important conclusions about the polarization of a bulk polymer can be drawn:

Case I. The binding energy and correspondingly the radius of the electron-hole pair depend on the regularity of the structure. At structural defects the binding energy is decreased and the radius is increased. In the presence of an external electric field, a dipole is therefore formed at structural defects. This will be referred to as defect dipole. A structural defect can be a missing lattice site, a dislocation or an interface of a crystallite. A defect dipole formed by the polarization of a Wannier exciton state at a missing lattice point is schematically illustrated in Fig. 1.2. It is noted that for formation of a defect dipole it is not necessary to have a periodical crystalline structure.

Case II. As the radius and binding energy of the exciton is dependent on the dielectric permittivity of the medium, any inhomogeneity in the permittivity should result in a change and correspondingly in the formation of dipoles in the presence of an external field. This phenomenon has long been known as the Maxwell-Wagner-Sillars (MWS) effect. If the sample contains regions of different permittivities, in an external electric field interfacial dipole polarization builds up. The build-up of dipole polarization at an interface of two media of different permittivities is schematically illustrated in Fig. 1.3. The MWS polarization in heterogeneous systems will be discussed in Chapter 5.

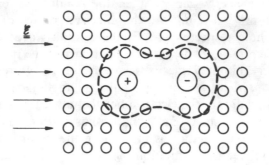

Fig. 1.2 Defect dipole formed at a missing lattice site in a solid

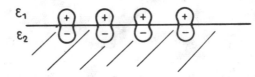

Fig. 1.3 Defect dipoles formed at the interface of two media

It is seen that, as well as the permanent dipole moments of the molecules or groups, induced dipole moments can be present either from the physical (Case I) or chemical (Case II) inhomogeneities or defects of the solid. The defect dipole moments formed this way can be extremely large. For an electron-hole formation 5 Å apart the dipole moment would be about 20 debye, which is by an order of magnitude higher than the bond moments. It will be shown later that such a high dipole polarization is indeed observed in polymeric solids. A technique based on this phenomenon is referred to as dielectric depolarization spectroscopy; it will be discussed in Chapter 3.

The exciton picture makes it possible to describe the dielectric polarization in solids without introducing Lorentz's internal field. For a general treatment see Rice and Jortner (1967).

1.2 Statistical theory

Polymers in bulk are practically never in thermodynamical equilibrium. As dielectric spectroscopy is based on the statistical thermal motion of some parts of the polymer molecules in the presence of an external electric field, the statistical thermodynamical aspects are of basic importance. The response of an assembly of dipoles to an external electric field is a thermodynamically non-equilibrium process itself. In an assembly of macromolecules the thermal energy is never equally distributed among the degrees of motional freedom; it is not possible to define a single temperature for the system as it is usual for gases. Thus a macromolecular system is approximated as a series of subsystems which might exhibit local equilibria characterized by local partial temperatures. The concept of temperature in non-equilibrium systems is discussed in detail by Zoubarov (1971). Here the linear response of a system to external perturbation will be discussed briefly, with special emphasis on the effect of external electric field.

Linear reaction of a system to an external perturbation

The simplest approach to the study of non-equilibrium processes is to assume that the external perturbation energy is small with respect to the total energy of the system. The perturbation is described by the time-dependent hamiltonian function H_t' which represents the interaction energy of the system with the external force field. It is assumed that

$$H_t'(p, q)_{t = -\infty} = 0$$

where p and q are the general momenta and coordinates respectively. The perturbation hamiltonian can be generally expressed as

$$H_t' = -\sum_i B_i(p, q) \, \mathscr{F}_i(t) \qquad (1.21)$$

where $B_i(p, q)$ are dynamical variables affected by the external forces $\mathscr{F}_i(t)$. In the case of a system of dipoles in an electric field $\mathscr{F}_i(t) \equiv \mathscr{E}(t)$ is the external electric field, $B \equiv \mathscr{P}$, the polarization, i.e. the dipole moment concentration of the system.

When the perturbation is periodic

$$H_t'(p, q) = \sum_\omega B_\omega(p, q) \exp(-i\omega t) \qquad (1.22)$$

where ω is the angular frequency of the force field.

31

In the presence of the external perturbation the response of the system is described by the Liouville equation (see e.g. in Prigogine, 1962)

$$\frac{df}{dt} = \{H_0 + H'_t, f\} = i\hat{L}f \qquad (1.23)$$

where f is the distribution function, $\hat{L}$ is the Liouville operator,

$$\{H_0 + H'_t, f\} = \frac{\partial(H_0 + H'_t)}{\partial q}\frac{\partial f}{\partial p} - \frac{\partial(H_0 + H'_t)}{\partial p}\frac{\partial f}{\partial q}$$

is the classical Poisson expression; H_0 is the hamiltonian function of the unperturbed system.

It is assumed that at $t = -\infty$ the system was in equilibrium. If there is no possibility for mass exchange with the reservoir the distribution function is

$$f(-\infty) = f_0 = Q^{-1} \exp\left(-\frac{H_0}{\theta}\right) \qquad (1.24)$$

where

$$Q = \int \exp\left(-\frac{H_0}{\theta}\right) dv$$

f_0 in Eqn. (1.24) is the distribution function of the Gibbs canonic ensemble and θ is the modulus, which is equal to kT only at thermal equilibrium; v is the phase volume.

On the basis of the Liouville equation the non-equilibrium distribution function can be determined as

$$f(t) = f_0 + \int_{-\infty}^{t} \exp[i(t - t')\hat{L}]\{H'_{t'}, f(t')\} dt' \qquad (1.25)$$

When the perturbation is small, $H'_{t'} \ll H_0$, this integral equation can be solved by iteration. At first approximation one obtains

$$f(t) = f_0 + \int_{-\infty}^{t} \{H'_{t'}(t' - t), f_0\} dt' \qquad (1.26)$$

By using this distribution function, it is possible to express the expectational value of any dynamical variable A as

$$\langle A \rangle = \langle A \rangle_0 + \int_{-\infty}^{+\infty} \langle\langle A \ H'_{t'}(t'-t)\rangle\rangle \ dt' \qquad (1.27)$$

where

$$\langle\langle A \ H'_{t'}(t'-t)\rangle\rangle = \vartheta(t-t') \langle \{A, H'_{t'}(t'-t)\}\rangle_0 =$$
$$= \vartheta(t-t') \langle \{A(t), H'_{t'}(t')\}\rangle_0$$

Here $\langle \ \rangle_0$ means averaging with respect to the equilibrium distribution function, $\vartheta(t-t')$ is a function defined as

$$\vartheta(t) = \begin{array}{l} 1 \text{ if } t > 0 \\ 0 \text{ if } t < 0 \end{array}$$

$\langle\langle A(t) \ H'_{t'}(t')\rangle\rangle$ is referred to as the Green function. It describes the linear response of the system to an external perturbation. If

$$H'_{t'} = B \ \delta(t-t')$$

i.e. an instantaneous perturbation is applied to the system at t', the response of any dynamical variable A is expressed as

$$\langle A \rangle = \langle A \rangle_0 + \langle\langle A(t) B(t')\rangle\rangle$$

Thus the Green function describes the time dependence of the deviation of the average value of a variable from the equilibrium value after the application of an infinitesimal perturbation pulse. $\delta(t-t')$ is the Dirac function.

In the general case when the perturbation has the form of Eqn. (1.21) the linear reaction of a variable A is described by the Kubo formula (Kubo, 1958)

$$\langle A \rangle = \langle A \rangle_0 - \sum_j \int_{-\infty}^{+\infty} \langle\langle A(t) B_j(t')\rangle\rangle \ \mathscr{F}_j(t') \ dt' \qquad (1.28)$$

$$\langle A \rangle = \langle A \rangle_0 + \sum_j \frac{1}{\theta} \int_{-\infty}^{t} \langle A(t) \dot{B}_j(t')\rangle_0 \ \mathscr{F}_j(t') \ dt' \qquad (1.29)$$

where $\dot{\underline{B}}$ means the time derivative of B. The Kubo formula can also be expressed by

$$\langle A \rangle = \langle A \rangle_0 + \sum_j \int_{-\infty}^{t} \phi_{AB_j}(t-t')\, \mathscr{F}_j(t')\, dt' \qquad (1.30)$$

where

$$\phi_{AB_j}(t-t') = \frac{1}{\theta} \langle \dot{B}_j(t')\, A(t) \rangle_0 \qquad (1.31)$$

is referred to as the time correlation function. It is related directly to the Green function as

$$\langle\langle A(t)\ B(t') \rangle\rangle = -\vartheta(t-t')\, \phi_{AB}(t-t') \qquad (1.32)$$

In the case of periodic perturbation

$$\langle A \rangle = \langle A \rangle_0 + \sum_\omega \exp(-i\omega t) \langle\langle AB_\omega \rangle\rangle_\omega \qquad (1.33)$$

$$\langle\langle A\ B_\omega \rangle\rangle_\omega = \int_{-\infty}^{+\infty} \langle\langle AB_\omega(t) \rangle\rangle \exp(-i\omega t)\, dt$$

For a single-frequency periodic perturbation

$$H_t' = -\mathscr{F}_0 \cos(\omega t)\, B(p, q)$$

$$\langle A \rangle = \langle A \rangle_0 + \mathrm{Re}\,[\chi^*(\omega)\, \mathscr{F}_0 \exp(-i\omega t)] \qquad (1.34)$$

where

$$\chi^*(\omega) = -\langle\langle AB \rangle\rangle_\omega = -\int_{-\infty}^{+\infty} \exp(-i\omega t) \langle\langle AB(t) \rangle\rangle\, dt$$

is referred to as the general complex susceptibility of the system. Re [ ] means the real part of the expression in brackets, $\mathscr{F}_0$ is the amplitude of the external force field, $B(p, q)$ is the dynamical variable which is affected by the force field.

It is seen that the linear reaction of a system to an external periodic field is described by the general complex susceptibility, which is related to the Green function. The overall response, correspondingly, has the same character in any form of external perturbation: mechanical, electric or magnetic.

34

The linear reaction of a statistical system to an external force $\mathscr{F}(t)$ can be formulated also by considering the deviation from the equilibrium value of a variable as a response.

$$\Delta A(t) = \langle A(t) \rangle - \langle A \rangle_0$$

$$= \int_{-\infty}^{t} \phi_{AB}(t-t') \mathscr{F}(t') \, dt' \equiv \int_{-\infty}^{t} \frac{1}{\theta} \langle A(t') \dot{B}(t) \rangle_0 \mathscr{F}(t') \, dt'$$

$$\equiv \int_{-\infty}^{+\infty} \langle\langle A(t') B(t) \rangle\rangle \mathscr{F}(t') \, dt'$$

(1.35)

$\mathscr{F}(t)$ can be considered as an input signal to the system, ΔA as the output. In this respect the response of a statistical system to external perturbation is described the same way as that of any macroscopic system, electronic, mechanical or pneumatic, used in the regulation technique. In the theory of regulation, in considering the response of an electric or magnetic network to an external signal the same general formalism is used.

A system of charges in a periodic external electric field

As a particular example of the general treatment discussed above let us consider a periodic external electric field

$$\mathscr{E} = \mathscr{E}_0 \cos(\omega t) \equiv \mathrm{Re}\left[\mathscr{E}_0 \exp(-i\omega t)\right]$$

(1.36)

acting to a system of electric charges.

The perturbation hamiltonian is the dipole interaction energy

$$H'_t = -\sum_j e_j \mathscr{D} r_j \cos(\omega t) = -\mathscr{D}\mathscr{P} \cos(\omega t)$$

(1.37)

where $\mathscr{P} = \sum e_j r_j$ the polarization vector, i.e. the net dipole moment concentration, $\mathscr{D} = \varepsilon^* \mathscr{E}$ is the displacement vector, r_j is the radius vector of charge e_j. In this expression the amplitude of the external force field $\mathscr{F} \equiv \mathscr{D}$, the dynamical variable which is affected by the field is the polarization $B \equiv \mathscr{P}$, i.e. the macroscopic dipole moment density. Let us choose the current as a dynamical variable $A \equiv I$. According to Eqn. (1.33)

$$\langle I_\alpha \rangle = \int_{-\infty}^{+\infty} \langle\langle I_\alpha(t) \mathscr{P}_\beta(t') \rangle\rangle \mathscr{D}_\beta \cos\omega t' \, dt'$$

(1.38)

as $\langle I_\alpha \rangle_0 = 0$ and $I_\alpha(t) = \sum e_j \dot{r}_{j\alpha}(t) = \dot{\mathcal{P}}_\alpha(t)$. α and β indicate directions as the current is generally not directed the same way as the displacement $\mathcal{D}$.

By introducing the susceptibility the average current is

$$\langle I_\alpha \rangle = \sum_\beta \mathrm{Re}\,[\chi^*_{\alpha\beta}(\omega) \exp(-i\omega t)\,\mathcal{D}_\beta] \qquad (1.39)$$

where

$$\chi^*_{\alpha\beta} = -\int_{-\infty}^{+\infty} \exp(-i\omega t)\,\langle\langle I_\alpha \dot{\mathcal{P}}_\beta(t) \rangle\rangle\,dt \qquad (1.40)$$

In this calculation the displacement is regarded as an external field, which is connected with the main internal field $\mathcal{E}$ acting on the charges as

$$\mathcal{D} = \varepsilon^* \mathcal{E}$$

where ε^* is the complex dielectric permittivity. This means that the coulombic interaction between charges is taken into account directly, this interaction is not introduced in the hamiltonian of the system.

In another approach the external field is taken to be equal to the mean internal field. The coulomb interaction effects are taken into account in the hamiltonian of the system. By this approach the electrical conductivity is expressed as

$$\sigma_{\alpha\beta} = -\int_{-\infty}^{+\infty} \exp(-i\omega t)\,\langle\langle I_\alpha\ \dot{\mathcal{P}}_\beta(t) \rangle\rangle\,dt \qquad (1.41)$$

The mean current

$$\langle I_\alpha \rangle = \sum_\beta \mathrm{Re}\,[\sigma_{\alpha\beta}(\omega) \exp(-i\omega t)]\,\mathcal{E}_\beta \qquad (1.42)$$

The relations between the conductivity $\sigma_{\alpha\beta}$ and the susceptibility $\chi_{\alpha j}$ components are

$$\sigma_{\alpha\beta} = \sum_\gamma \chi_{\alpha\gamma} \varepsilon_{\gamma\beta} \qquad (1.43)$$

or in the tensor form

$$\sigma = \chi \varepsilon \qquad (1.44)$$

36

As

$$\mathscr{D} = \mathscr{E} + 4\pi \langle \mathscr{P} \rangle = \mathscr{E} + \frac{4\pi}{i\omega} \langle I \rangle = \varepsilon^* \mathscr{E}$$

the complex permittivity tensor is expressed as

$$\varepsilon^*(\omega) = 1 + \frac{4\pi \chi^*(\omega)}{i\omega} \qquad (1.45)$$

the complex conductivity tensor

$$\sigma^*(\omega) = \chi^*(\omega) \left[1 + \frac{4\pi \chi^*(\omega)}{i\omega} \right]^{-1} \qquad (1.46)$$

It is seen that the complex permittivity and resistivity are related to the complex susceptibility which, in turn, is connected with the Green function of the system, or using the representation of Eqn. (1.30) with the time correlation function.

The electrical conductivity tensor components can also be expressed as

$$\sigma_{\alpha\beta}(\omega) = \frac{1}{\theta} \int_0^\infty \exp(i\omega t) \langle I_\alpha I_\beta(t) \rangle \, dt \qquad (1.47)$$

Similarly for the susceptibility

$$\chi_{\alpha\beta}(\omega) = \frac{1}{\theta} \int_0^\infty \exp(i\omega t) \langle I_\beta I_\alpha(t) \rangle \, dt \qquad (1.48)$$

These expressions follow directly from the general Kubo formulae (1.28) and (1.29); they are referred to as the Kubo formulae for the electrical conductivity and susceptibility respectively.

The dipole autocorrelation function

Cole (1965) calculated the dielectric permittivity of a spherical molecule in solution on the basis of the Kubo formalism outlined above. The average polarization, i.e. the macroscopic dipole moment density of the system, is expressed as

$$\langle \mathscr{P} \rangle = \frac{N\alpha\mathscr{E}_0}{v} - \frac{N\langle \mu(0)\,\mu(0)\rangle_0}{3\,kTv} \int_{-\infty}^t \mathscr{E}_0(t')\,\dot{\phi}(t-t')\,dt' \qquad (1.49)$$

where

$$\phi(t) = \frac{\langle \mu(0)\,\mu(t) \rangle}{\langle \mu(0)\,\mu(0) \rangle} \tag{1.50}$$

Here N is the total number of molecules, v is the volume, averaging is to be performed with respect to the equilibrium distribution function, μ is the dipole moment of a single unit, α is the polarizability.

By applying an external periodic field and considering the relationship of the external field $\mathscr{E}$ with the internal (Lorentz) field $\mathscr{E}_0$ as

$$\mathscr{E}\exp(i\omega t) = \frac{3\,\mathscr{E}_0}{\varepsilon^* + 2}\exp(i\omega t) \tag{1.51}$$

where by definition

$$\varepsilon^* - 1 = \frac{4\pi\,\langle \mathscr{P} \rangle}{\mathscr{E}\,\exp(i\omega t)}$$

according to Eqns (1.49) and (1.51) one obtains

$$\frac{\varepsilon^* - \varepsilon_\infty}{\varepsilon_0 - \varepsilon_\infty} = \left\{ 1 + \left(\frac{\varepsilon_0 + 2}{\varepsilon_\infty + 2} \right) \left[\mathscr{L}(-\dot{\phi})^{-1} - 1 \right] \right\}^{-1} \tag{1.52}$$

where

$$\varepsilon_0 - \varepsilon_\infty = 4\pi N \frac{\varepsilon_0 + 2}{3} \frac{\varepsilon_\infty + 2}{3} \frac{\langle \mu(0)\,\mu(0) \rangle}{3\,kTv} \tag{1.53}$$

and $\mathscr{L}(-\dot{\phi})$ is the Laplace transform of ϕ

$$\mathscr{L}(-\dot{\phi}) = -\int_0^\infty \exp(-i\omega t)\,\dot{\phi}(t)\,dt$$

In the particular case when all the interactions of $\mu(0)$ with its neighbours are neglected

$$\phi(t) = \frac{\langle \mu(0)\,\mu(t) \rangle}{\langle \mu(0)\,\mu(0) \rangle} = \exp(-t/\tau) \tag{1.54}$$

38

where τ is the relaxation time of the response, one obtains

$$\frac{\varepsilon^* - \varepsilon_\infty}{\varepsilon_0 - \varepsilon_\infty} = \frac{1}{1 + i\omega \dfrac{\varepsilon_0 + 2}{\varepsilon_\infty + 2}\tau} = \frac{1}{1 + i\omega\bar{\tau}} \qquad (1.55)$$

$\bar{\tau} = \dfrac{\varepsilon_0 + 2}{\varepsilon_\infty + 2}\tau$ is referred to as the macroscopic relaxation time.

Equation (1.54) is referred to as the single relaxation time approximation. Equation (1.55) is the Debye equation for spherical dipole in the single relaxation approximation; it can be also derived classically (Debye, 1921). As shown above, this formula follows directly from the general Kubo formulae.

$\phi(t) = \dfrac{\langle \mu(0)\,\mu(t) \rangle}{\langle \mu(0)\,\mu(0) \rangle}$ is referred to as the dipole-moment autocorrela-

tion function. It is useful to formulate the linear response of a system of dipoles to external electric field in a fairly general way. The dipole-moment autocorrelation function describes the decrease of the projection of the average dipole moment to an original direction as time develops. For detailed calculations see Boon and Rice (1967) and Résibois (1964). The method has been recently reviewed by Williams (1972).

In the more general case when the system contains N dipole moments which are not necessarily like ones, the macroscopic dipole moment autocorrelation function is expressed as

$$\phi(t) = \frac{\displaystyle\sum_{i=1}^{N} \langle \mu_i(0)\,\mu_i(t) \rangle + 2 \sum_{i=2}^{N}\sum_{k=1}^{i-1} \langle \mu_i(0)\,\mu_k(t) \rangle}{\displaystyle\sum_{i=1}^{N} \mu_i^2 + \sum_{i=2}^{N}\sum_{k=1}^{i-1} \langle \mu_i(0)\,\mu_k(0) \rangle} \qquad (1.56)$$

where μ_i is the dipole moment of the i-th unit. The terms $\mu_i(0)\,\mu_k(0)$ are referred to as cross-correlation terms, they represent the time-dependent orientation correlations between dipoles i and k. The limiting values of the autocorrelation function are

$$\phi_{(0)} = 1$$
$$\phi_{(\infty)} = 0$$

In terms of the dipole autocorrelation function the complex permittivity ε^* in a system containing one kind of rigid dipoles is expressed as

$$\frac{\varepsilon^*(i\omega)-\varepsilon_\infty}{\varepsilon_0-\varepsilon_\infty} = \left\{1+\frac{3\,\varepsilon_0}{2\,\varepsilon_0+\varepsilon_\infty}\left[\mathscr{L}(-\dot{\phi})\right]^{-1}-1\right\}^{-1} \qquad (1.57)$$

where

$$\mathscr{L}(-\dot{\phi}) = \int\limits_0^\infty \exp(-i\omega t)\,[-\dot{\phi}(t)]\,dt$$

The factor $3\,\varepsilon_0/(2\,\varepsilon_0+\varepsilon_\infty)$ appearing in Eqn. (1.57) is due to the local field effect.

If the local field effect is neglected, i. e.

$$\frac{3\,\varepsilon_0}{2\,\varepsilon_0+\varepsilon_\infty} \approx 1 \qquad (1.58)$$

$$\frac{\varepsilon^*(i\omega)-\varepsilon_\infty}{\varepsilon_0-\varepsilon_\infty} = \int\limits_0^\infty \exp(-i\omega t)\,[-\dot{\phi}(t)]\,dt \qquad (1.59)$$

For the real and unreal parts

$$\frac{\varepsilon'(\omega)-\varepsilon_\infty}{\varepsilon_0-\varepsilon_\infty} = \int\limits_0^\infty\left[-\dot{\phi}(t)\right]\cos\omega t\,dt \qquad (1.60)$$

$$\frac{\varepsilon''(\omega)}{\varepsilon_0-\varepsilon_\infty} = \int\limits_0^\infty\left[-\dot{\phi}(t)\right]\sin\omega t\,dt \qquad (1.61)$$

In the further simplified case when

$$\phi(t) = \exp\left(-\frac{t}{\tau}\right)$$

the single relaxation time equation (1.55) is obtained.

The actual form of the dipole correlation function is determined by considering the degrees of motional freedom of the dipoles. For evaluating $\phi(t)$ some models should be used. If, for example, the dipole relaxation is determined by rotation of groups the energy barrier model developed by Fröhlich (1958), Hoffman (1952, 1955) can be used. The system is in this case approximated by a series of potential barriers. The occupation probability of the sites' potential minima is generally expressed as

$$P(t) = \exp(Tt) \tag{1.62}$$

where T is the transition matrix composed of the rate constants of transition between subsequent sites.

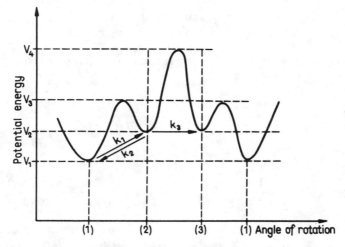

Fig. 1.4 A 3-site potential well for calculation of hindered rotation of a polar group

The dipole correlation function is

$$\phi(t) = \sum_i P_i^0 \xi_i(t) \tag{1.63}$$

where

$$\xi_i(t) = \sum_j P_{ji}(t) \mu_i \mu_j$$

is the decay function for the dipoles located at site i at $t = 0$, P_i^0 is the equilibrium occupational probability of site i.

As a particular example a three-site potential barrier is shown in Fig. 1.4 after Williams (1972). The dipole correlation function is

$$\phi(t) = \frac{1}{(1+2x)^2} (1+2x \cos\theta_{12})^2 + 2x(1-\cos\theta_{12})^2$$

$$\psi_2(t) + 2x(1+2x)(1-\cos^2\theta_{12}) \psi_3(t)$$

41

where θ_{12} is the angle between the dipole orientations in sites (1) and (2), $x = \dfrac{k_1}{k_2}$ the ratio of the rate constants

$$\psi_2(t) = \exp\left[-(k_2 + 2k_1)\, t\right]$$
$$\psi_3(t) = \exp\left[-(k_2 + 2k_3)\, t\right]$$

The transition matrix has the following form

$$T = \begin{pmatrix} -2k_1 & k_2 & k_2 \\ k_1 & -(k_2+k_3) & k_3 \\ k_1 & k_3 & -(k_2+k_3) \end{pmatrix} \qquad (1.64)$$

The complex permittivity due to the rotational polarization represented by the potential of Fig. 1.4 is

$$\frac{\varepsilon^* - \varepsilon_\infty}{\varepsilon_0 - \varepsilon_\infty} = \frac{C_2}{1 + i\omega\tau_2} + \frac{C_3}{1 + i\omega\tau_3} \qquad (1.65)$$

where $\tau_2 = (k_2 + 2k_1)^{-1}$ $\tau_3 = (k_2 + 2k_3)^{-1}$

$$C_2 = \frac{1}{2}\,\frac{1 - \cos\theta_{12}}{1 + x\ (1 + \cos\theta_{12})} \qquad C_3 = \frac{(1 + 2x)\,(1 + \cos\theta_{12})}{2\ 1 + x\ (1 + \cos\theta_{12})}$$

It is seen that the model predicts a response that can be described by two relaxation times τ_2 and τ_3 determined by the rate constants of transition through the corresponding potential barriers. By increasing the number of potential valleys the model evidently would lead to several relaxation times. In real systems the motion of a subsystem, i.e. a rotating group attached to the polymer chain is to be always represented by a series of potential valleys, which means that a series of relaxation times is needed to describe the response. This problem will be discussed in the next section.

Dispersion relations

From the general mathematical properties of the Green functions the following dispersion relations can be derived for the general susceptibility $\chi^* = \chi' - i\chi''$

$$\chi'(\omega) = \frac{1}{\pi} \int_{-\infty}^{+\infty} \chi''(\omega') \frac{2\omega'}{\omega'^2 - \omega^2}\, d\omega' \qquad (1.66)$$

$$\chi''(\omega) = -\frac{1}{\pi} \int_{-\infty}^{+\infty} \chi'(\omega') \frac{2\omega'}{\omega'^2 - \omega^2}\, d\omega' \qquad (1.67)$$

These relations are referred to as the Kramers Kronig relations. These equations imply that the dielectric susceptibilities χ' and χ'' or permittivities ε' and ε'' are related to each other; they express two sides of the same information.

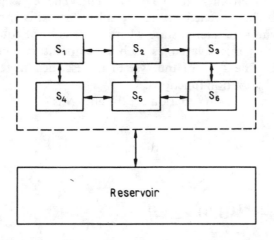

Fig. 1.5 Illustration of interaction among thermodynamical sub-systems

Local equilibrium

Polymers cannot be approximated as single statistical systems, but rather as a series of subsystems representing different kinds of units and of motions. The thermal motion of the main chain is evidently different from that of the side groups; they represent two different statistical thermodynamical assemblies. When such a system is subjected to an external perturbing field the new equilibrium is reached first within the subsystems, then between subsystems, and finally with the heat reservoir. This is schematically shown in Fig. 1.5. Evidently temperature can be defined only within a subsystem at local equilibrium. Thus, in a polymeric system, generally, a series of partial temperatures is to be defined to describe the thermodynamical state. Complete equilibrium with the thermostate is practically never reached in macromolecular systems.

The previous considerations are valid only at first approximation, as by considering the response of a system to external perturbation only the linear response has been taken into account and the modifying effect of the system to the external force (feedback) is neglected. Irreversible processes, such as diffusion and chemical reaction, evidently cannot be treated this way. For a more general theory see Zoubarov (1971).

One of the general treatments is based on the conservation of energy, mass and momentum within a subsystem. The conception is that equilibrium is first reached within the subsystem; this is referred to as local equilibrium. In such a state where there is local equilibrium, the main thermodynamical variables such as temperature, entropy should vary in space. The state of the system is described in terms of local equilibrium distribution functions.

The local conservation equation for the energy is

$$\frac{\partial H_i(r)}{\partial t} + \text{div } j_{H_i}(r) = I_{H_i}(r) \tag{1.68}$$

where

$$H_i(r) = \sum_i \frac{p_i^2}{2m} + \frac{1}{2} \sum_i V(r_i - r) \tag{1.69}$$

is the energy density, $V(r_i - r)$ is the interaction potential between the units at r and r_i, $I_{H_i}(r)$ is the energy current, and $j_{H_i}(r)$ is the energy-current density, p_i is the linear momentum. The Fourier components of the energy density are

$$H_k = \int \exp(-ikr) H(r) \, dr \tag{1.70}$$

The local conservation of mass is expressed as

$$\frac{\partial n_i}{\partial t} + \text{div } j_{n_i}(r) = I_i(r) \tag{1.71}$$

where

$$n_i = \sum_{i=1}^{N} \delta(r_i - r) \tag{1.72}$$

is the particle density, $j_{n_i}(r)$ is the mass current density, $j_{n_i}(r) = \sum_i \frac{P_i}{m} \delta(r_i - r)$, $I_i(r)$ is the mass current. The Fourier components of the particle density are

$$n(k) = \int \exp(-ikr) n(r) \, dr \tag{1.73}$$

44

The mass current $I_i(r)$ represents the rate of formation of particles in the subsystem i. Evidently

$$\sum_{i=1}^{N} I_i(r) = 0 \qquad (1.74)$$

where N is the total number of subsystems.

The local conservation of momentum is expressed as

$$\frac{\partial p_i(r)}{\partial t} + \text{div } T_i(r) = f_i(r) \qquad (1.75)$$

where $T_i(r)$ is the stress tensor, $f_i(r)$ is the interaction force density of the subsystem i with all the other subsystems.

For the whole system the total energy, mass and momentum densities are respectively

$$H(r) = \sum_{i=1}^{N} H_i(r) \qquad (1.76)$$

$$\varrho(r) = \sum_{i=1}^{N} m_i n_i(r) \qquad (1.77)$$

$$p(r) = \sum_{i=1}^{N} p_i(r) \qquad (1.78)$$

where N is the total number of subsystems.

For the total system the conservation equations are

$$\frac{\partial H(r)}{\partial t} + \text{div } j_H(r) = 0 \qquad (1.79)$$

$$\frac{\partial \varrho(r)}{\partial t} + \text{div } j_\rho(r) = 0 \qquad (1.80)$$

$$\frac{\partial p(r)}{\partial t} + \text{div } T(r) = 0 \qquad (1.81)$$

Using the conservation equations, it is possible to determine the distribution function corresponding to local equilibrium by seeking maximum value of the informational entropy

$$\delta S = -\langle \ln f \rangle = \text{maximum} \qquad (1.82)$$

The result is

$$f_{local} = Q^{-1} \exp \left\{ - \int \frac{1}{\theta(r)} [H(r) - \mu(r)] \, n(r) \, dr \right\} \qquad (1.83)$$

where

$$Q = \int_v \exp \left\{ - \int_r \frac{1}{\theta(r)} [H(r) - \mu(r)] \, n(r) \, dr \right\} dv$$

$\mu(r)$ is the thermodynamic potential.

The local partial temperature at position r is defined as

$$\frac{1}{kT(r)} \equiv \frac{1}{\theta(r)} = \frac{\delta S}{\delta \langle H(r) \rangle_{local}} \qquad (1.84)$$

where the subscript 'local' means averaging with respect to the local equilibrium distribution function f_{local}.

For the thermodynamic potential

$$\frac{\mu(r)}{\theta(r)} = - \frac{\delta S}{\delta \langle n(r) \rangle_{local}} \qquad (1.85)$$

The spatial dependence for θ and μ can be expressed in a Fourier series form if the structure is periodic. In this case

$$\frac{1}{\theta(r)} = \sum_k \frac{1}{\theta_k} \exp(ikr) \qquad (1.86)$$

$$\frac{\mu(r)}{\theta(r)} = \sum_k v_k \exp(ikr) \qquad (1.87)$$

$$\frac{1}{\theta_k} = \frac{\partial S}{\partial \langle H_k \rangle_{local}} \qquad (1.88)$$

$$v_k = - \frac{\partial S}{\partial \langle n_k \rangle_{local}} \qquad (1.89)$$

From the local conservation equations the following equations for the total energy and mass of the subsystems can be derived

$$\frac{\partial H_i}{\partial t} = E_i I_{H_i} \qquad \sum_{i=1}^N I_{H_i} = 0 \qquad (1.90)$$

$$\frac{\partial N_i}{\partial t} = I_{N_i} \qquad \sum_{i=1}^N I_{N_i} = 0 \qquad (1.91)$$

These are referred to as relaxation equations; they generally describe the time dependence of a variable in a subsystem as a result of the interaction with the other subsystems. This interaction is described by the functions $I_{H_i}(t) I_{N_i}(t)$.

1.3 Phenomenological theory

In the previous section the linear response of an assembly to an external mechanical or electrical perturbation has been discussed on the basis of non-equilibrium statistical thermodynamics. It has been shown that by application of an external force field the system is transferred to a non-equilibrium state which tends to reach a new equilibrium by a time-dependent relaxation process. If the external force field is not too strong, the non-equilibrium state can be described in terms of perturbed equilibrium distribution functions. The behaviour of a constant of motion is generally described in terms of Green functions or time-correlation functions which are connected with the general complex susceptibility of the system. It appears that the statistical thermo-dynamical treatment can be applied to any constant of motion; it is possible to describe the response of the system to mechanical, electric or magnetic force fields by using the same formalism.

The statistical thermodynamical approach, although it is very sophisticated, is very difficult to use for practical calculations because of computational difficulties. In practice, very often illustrative phenomenological models are used which are easy to handle. Some of them will be discussed in this section very briefly. For a more detailed discussion see Holzmüller and Altenburg (1961) and McKelvey (1962).

Basic models for a viscoelastic system

The linear elastic behaviour of an ideal elastic Hookean material is described as

$$\sigma = G\gamma \tag{1.92}$$

or

$$\gamma = J\sigma \tag{1.93}$$

where σ is the stress, γ is the strain, G is the hookean modulus and J is the compliance; $J = \dfrac{1}{G}$. The response of such a material to an external force field or to a deformation is that of an ideal spring.

The electrical analogue of Eqn. (1.92) is charging a capacitor:

$$V_0 = \frac{e_0}{C_0} = \frac{1}{C_0} \int I(t)\, dt \qquad (1.94)$$

where V_0 is the voltage, e_0 is the charge of the capacitor, $I(t)$ is the charging current, $C_0 = A/4\pi d$ is the capacity, where A is the cross-sectional area and d is the distance between the plates of the capacitor.

By inserting a dielectric material between the plates of the capacitor the capacity increases:

$$V = \frac{e_0}{C_0} = \frac{e}{C} \qquad (1.95)$$

where e_0 is referred to as the free charge, e $(e_0 < e)$ is the true charge. By the difference of the true and free charges the dielectric polarization is defined as

$$\mathscr{P} = \frac{1}{A} \cdot (e - e_0) \qquad (1.96)$$

The dielectric permittivity is defined as

$$\varepsilon = \frac{C}{C_0} = \frac{e}{e_0} \qquad (1.97)$$

Correspondingly

$$\mathscr{P} = \frac{e}{A}\left(1 - \frac{1}{\varepsilon}\right) \qquad (1.98)$$

By definition the electric field is generated by the free charges

$$\mathscr{E} = 4\pi\,\frac{e_0}{A} \qquad (1.99)$$

The electric displacement is generated by the true charges

$$\mathscr{D} = 4\pi\,\frac{e}{A} \qquad (1.100)$$

From Eqns (1.98), (1.99), (1.100)

$$\mathscr{D} = \varepsilon\,\mathscr{E} \qquad (1.101)$$
$$\mathscr{P} = \chi\,\mathscr{E} \qquad (1.102)$$

where χ is the electrical susceptibility.

48

It is seen that the response of an ideal lossless capacitor to an external electric field $\mathscr{E}$ is analogous to the response of an elastic body to a mechanical stress. The resulting deformation in the electric case is the polarization $\mathscr{P}$ or the displacement $\mathscr{D}$. According to the analogous mechanical equation (1.93), the mechanical compliance is analogous with the dielectric permittivity.

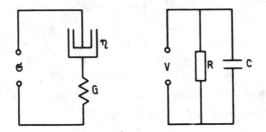

Fig. 1.6 The electrical and mechanical Maxwell models for a viscoelastic fluid

The other extreme case is when the material exhibits viscous flow without elasticity. The corresponding mechanical equation is

$$\sigma = \eta \frac{d\gamma}{dt} \tag{1.103}$$

where η is the viscosity.

The corresponding electrical equation is Ohm's law,

$$V = RI \tag{1.104}$$

as the current $I = \dfrac{de}{dt}$; R is resistivity. The general form of Eqn (1.104) is

$$\mathscr{E} = \varrho j \tag{1.105}$$

where $\mathscr{E}$ is the field strength, j is the current density, ϱ is the specific resistivity, which is in general a tensor.

The response of a viscoelastic system can be described in terms of models that consist of combination of dissipative viscous elements (dashpots or resistors) and elastic elements (springs or capacitors). A simple combination is shown in Fig. 1.6, where a spring and a dashpot are connected in series and a resistor and capacitor are connected in parallel. This combination is referred to as the Maxwell model. The corresponding equations are: *for the mechanical case*

$$\frac{d\gamma}{dt} = \frac{1}{G}\frac{d\sigma}{dt} + \frac{1}{\eta}\sigma$$
(1.106)

and *for the electric case*

$$\frac{de}{dt} = C\frac{dV}{dt} + \frac{1}{R}V$$
(1.107)

The solution of these equations in the case when a constant deformation is applied to the sample, and the time dependence of the stress as measured by a stress relaxation experiment is

$$\sigma(t) = \gamma_0 G \exp\left(-\frac{t}{\eta/G}\right)$$
(1.108)

As by definition $G = \sigma/\gamma$

$$G(t) = G \exp\left(-\frac{t}{\tau_\gamma}\right)$$
(1.109)

$\tau_\gamma = \eta/G$ is referred to as the stress relaxation time which characterizes the decay of the stress by application of a constant strain. In the electric case

$$V(t) = V_0 \exp\left(-\frac{t}{RC}\right)$$
(1.110)

This equation describes the discharge of the capacitor through the resistor R; the corresponding relaxation time is $\tau_e = RC$.

In practice, instead of the voltage of the capacitor the discharge current is measured

$$I(t) = I(0) \exp\left(-\frac{t}{\tau_e}\right)$$
(1.111)

Another basic model of a viscoelastic material is referred to as the Kelvin–Voigt model. It is illustrated in Fig. 1.7. As it consists of parallel coupled spring and dashpot, i.e. series coupled resistor-

50

capacitor systems, the corresponding equations are

$$\sigma = G\gamma + \frac{\eta d\gamma}{dt} \qquad (1.112)$$

$$V = \frac{1}{C}e + RI \qquad (1.113)$$

The solution is in the case of a constant stress σ_0 (creep experiment)

$$\gamma(t) = \frac{\sigma_0}{G}[1 - \exp(-t/\tau_\sigma)] \qquad (1.114)$$

where $\tau_\sigma = \eta/G_1$ is the retardation time of the strain exhibited by the material at constant stress.

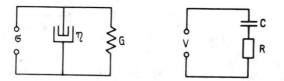

Fig. 1.7 The electrical and mechanical Kelvin–Voigt models for a viscoelastic fluid

By dividing Eqn. (1.114) by the constant stress σ_0, one obtains

$$J(t) = \frac{1}{G}[1 - \exp(-t/\tau_\sigma)] \qquad (1.115)$$

In the electrical case the solution of Eqn. (1.113) is

$$I(t) = \frac{1}{G}[1 - \exp(-t/\tau_\sigma)] \qquad (1.116)$$

Charge is seldom measured directly; it is more usual to measure the current at constant voltage (charge experiment)

$$I(t) = \frac{de(t)}{dt} = \frac{V_0}{R}[\exp(-t/\tau_V)] \qquad (1.117)$$

The Maxwell or Kelvin–Voigt models alone are not sufficient to describe the general response of a viscoelastic solid, not even at first approximation. Stress relaxation can be interpreted on the basis of the Maxwell model, but not creep. On the other hand the Kelvin–Voigt

4*

model is useful to describe creep but not stress relaxation. Evidently a combination of these models should be used. The simplest one is shown in Fig. 1.8; it is referred to as a combined model for linear viscoelastic

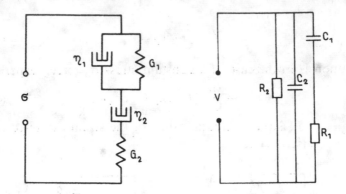

Fig. 1.8 The combined model for a viscoelastic fluid

solids. As it contains two elastic and one viscous components, the equation is

$$\sigma + \left(\frac{\eta_1 + \eta_2}{G_1} + \frac{\eta_2}{G_2}\right)\frac{d\sigma}{dt} + \frac{\eta_1\eta_2}{G_1G_2}\frac{d^2\sigma}{dt^2} = \eta_2\frac{d\gamma}{dt} + \frac{\eta_1\eta_2}{G_1}\frac{d^2\gamma}{dt^2}$$

(1.118)

In the electric case:

$$V + [(R_1 + R_2)C_1 + R_2C_2]\frac{dV}{dt} + R_1R_2C_1C_2\frac{d^2V}{dt^2} = R_2I + R_1R_2C_1\frac{dI}{dt}$$

(1.119)

The solutions for three basic cases follow:

(a) *Stress relaxation, i.e. electric discharge experiment when* $\eta_2 = \infty$

$$\gamma = \gamma_0 = \text{const}$$
$$e = e_0 = \text{const}$$

$$\sigma(t) = G_0\gamma_0 + (G_\infty - G_0)\gamma_0 \exp(-t/\tau_\gamma)$$

(1.120)

where $G_0 = \dfrac{\sigma(\infty)}{\gamma_0}$ is the relaxed modulus, $G_\infty = \dfrac{\sigma(0)}{\gamma_0}$ is the unrelaxed modulus, $\tau_\gamma = \eta_1/G_1 + G_2$ is the relaxation time of the stress decay.

$$V(t) = V_0 - (V_\infty - V_0) \exp(-t/\tau_e)$$

(1.121)

where $\tau_e = \dfrac{R_1 C_1 C_2}{C_1 + C_2} C_2$; the subscript e refers to constant charge (discharge experiment). $V_\infty = e_0 / C_2$, $V_0 = e_0 / C_1 + C_2$

(b) *Creep, i.e. electric-charge experiment*

$$\sigma = \sigma_0 = \text{const}$$
$$V = V_0 = \text{const}$$

$$\gamma(t) = \frac{\sigma_0}{G_2} + \frac{\sigma_0}{G_1} [1 - \exp(-t/\tau_\sigma)] + \frac{\sigma_0}{\eta_2} t \qquad (1.122)$$

where $\tau_\sigma = \eta_1 / G_1$ is the creep retardation time. By introducing the relaxed compliance $J_0 = \dfrac{1}{G_0}$ and $J_\infty = \dfrac{1}{G_\infty}$

$$\gamma(t) = J_0 \sigma_0 + (J_0 - J_\infty) \sigma_0 \exp(-t/\tau_\sigma) + \frac{\sigma_0}{\eta_2} t \qquad (1.123)$$

It is seen that the stress relaxation time τ_γ is different from the creep retardation time τ_σ Similarly in the electric case the relaxation time measured at constant charge (discharge experiment) is different from that measured at constant voltage (charge experiment).

(c) *Application of sinusoidal stress*

$$\sigma^* = \sigma_0 \exp i\omega t, \quad \mathcal{E} = \mathcal{E}_0 \exp i\omega t \qquad (1.124)$$

In this case the response of the material is expressed in terms of complex moduli

$$G^* = G' + iG''$$

or complex compliances

$$J^* = J' + iJ''$$

$$J^* = \frac{1}{G^*}$$

$$G' = \frac{J'}{|J^*|^2}$$

$$G'' = \frac{J''}{|J^*|^2}$$

The stress–strain relationship is

$$\sigma^* = G^* \gamma^*$$
$$\gamma^* = J^* \sigma^*$$

(1.125)

It is also usual to define the mechanical loss tangent as

$$\tan \delta_m = \frac{J''}{J'} = \frac{G''}{G'}$$

(1.126)

It can be shown that the unreal components of the moduli are related to the dissipation of the mechanical energy.

The response of the material to a periodic electric field $\mathscr{E} = \mathscr{E}_0 \exp i\omega t$ is generally expressed in terms of complex polarizability or susceptibility or in terms of complex permittivity.

$$\chi^* = \frac{\mathscr{P}^*}{\mathscr{E}^*} = \chi' - i\chi''$$

(1.127)

$$\varepsilon^* = \frac{\mathscr{D}^*}{\mathscr{E}^*} = \varepsilon' - i\varepsilon''$$

(1.128)

where $\mathscr{D}$ is the displacement vector, $\mathscr{P}$ is the polarization vector. The electrical loss tangent is

$$\tan \delta_e = \frac{\varepsilon''}{\varepsilon'}$$

(1.129)

The unreal part of the dielectric permittivity is related to the electric energy dissipation by the material (dielectric loss).

The response of a viscoelastic material to a periodic mechanical force field is similar to that of an electric field when the J^*, ε^* representations are used:

$$\frac{J'(\omega) - J_\infty}{J_0 - J_\infty} = \frac{1}{1 + \omega^2 \tau_\sigma^2}$$

(1.130)

$$\frac{J''(\omega)}{J_0 - J_\infty} = \frac{\omega \tau_\sigma}{1 + \omega^2 \tau_\sigma^2}$$

(1.131)

$$\tan \delta_m = \frac{(J_0 - J_\infty) \omega \tau_m}{J_0 + J_\infty \omega^2 \tau_m^2}$$

(1.132)

$$\frac{G' - G_0}{G_\infty - G_0} = \frac{\omega^2 \tau_\gamma^2}{1 + \omega^2 \tau_\gamma^2} \qquad (1.133)$$

$$\frac{G''}{G_\infty - G_0} = \frac{\omega \tau_\gamma}{1 + \omega^2 \tau_\gamma^2} \qquad (1.134)$$

$$\tan \delta_m = \frac{G_\infty - G_0}{G_0 + G_\infty \omega^2 \tau_m^2} \qquad (1.135)$$

where J_0, J_∞ are the relaxed and unrelaxed compliances respectively and $\tau_m = (\tau_\sigma \tau_\gamma)^{1/2}$ is the mechanical relaxation time. The actual forms of $G'(\omega)$, $G''(\omega)$, $J'(\omega)$, $J''(\omega)$ are shown in Fig. 1.9. The electric response is similarly

$$\frac{\varepsilon'(\omega) - \varepsilon_\infty}{\varepsilon_0 - \varepsilon_\infty} = \frac{1}{1 + \omega^2 \tau_e^2} \qquad (1.136)$$

$$\frac{\varepsilon''(\omega)}{\varepsilon_0 - \varepsilon_\infty} = \frac{\omega \tau_e}{1 + \omega^2 \tau_e^2} \qquad (1.137)$$

$$\tan \delta_e = \frac{(\varepsilon_0 - \varepsilon_\infty) \omega \tau_e}{\varepsilon_0 + \varepsilon_\infty \omega^2 \tau_e^2} \qquad (1.138)$$

where ε_0, ε_∞ are the relaxed and unrelaxed permittivities; τ_e is the dielectric relaxation time. $\tau_e = (\varepsilon_\infty / \varepsilon_0)^{1/2}$

The mechanical relaxation time according to the model of Fig. 1.8 is

$$\tau_m = (\tau_\sigma \tau_\gamma)^{1/2} = \left[\frac{\eta_1}{G_1} \frac{\eta_1}{G_1 + G_2} \right]^{1/2} \qquad (1.139)$$

The dielectric relaxation time is

$$\tau_e = \left[R_1 C_1 + \frac{R_1 R_2}{R_1 + R_2} C_1 \right]^{1/2} \qquad (1.140)$$

The actual forms of $\varepsilon'(\omega)$, $\varepsilon''(\omega)$ and $\tan \delta(\omega)$ are shown in Fig. 1.10. It is seen that $\varepsilon''(\omega)$ exhibits a maximum at $\omega_0 \tau_e = 1$. For convenience the frequency is plotted in a logarithmic scale. $\varepsilon'(\ln \omega)$ is referred to as the dielectric dispersion-spectrum band, $\varepsilon''(\ln \omega)$ is the absorption-spectrum band. According to the general Kramers–Kronig Eqns (1.66)

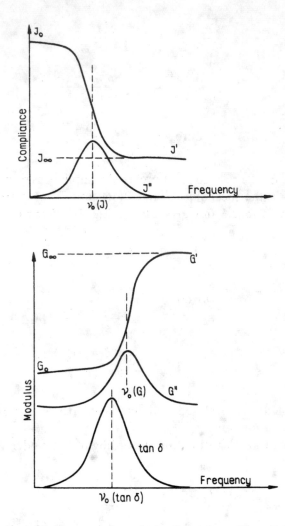

Fig. 1.9 Illustration of the frequency dependence of modulus and compliance components. Single-relaxation-time approximation

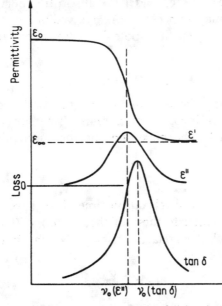
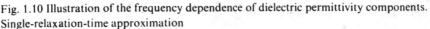

Fig. 1.10 Illustration of the frequency dependence of dielectric permittivity components. Single-relaxation-time approximation

and (1.67), ε' and ε'' are not independent

$$\varepsilon'(\omega) = \varepsilon_\infty + \frac{2}{\pi} \int_0^\infty \frac{\varepsilon''(\omega')\,\omega'}{(\omega')^2 - \omega^2}\, d\omega' \qquad (1.141)$$

$$\varepsilon''(\omega) = \frac{2}{\pi} \int_0^\infty \frac{\omega[\varepsilon'(\omega') - \varepsilon_\infty]}{(\omega')^2 - \omega^2}\, d\omega' \qquad (1.142)$$

The total area under the absorption curve is correspondingly

$$A = \int_{-\infty}^{+\infty} \varepsilon''(\ln\omega)\, d\ln\omega = \frac{\pi}{2}\,(\varepsilon_0 - \varepsilon_\infty) \qquad (1.143)$$

$\varepsilon_0 - \varepsilon_\infty$ is referred to as the oscillator strength of the transition, or the dielectric increment.

It is seen that the dielectric dispersion and absorption spectra are related to each other; they supply the same information about the transition in the case of the single relaxation time approximation. According to the Kirkwood–Fröhlich Eqn. (1.13), the oscillator strength $\varepsilon_0 - \varepsilon_\infty$, i.e. the area under the absorption curve, is related to the total dipole-moment concentration involved in the transition. From the transition frequency ω_0 the dipole relaxation time can be calculated from

$$\omega_0 \tau_\varepsilon = 1$$

Similarly, by plotting the moduli G' and G'' or the compliances J' and J'' as functions of $\ln \omega$, the mechanical dispersion and absorption spectra are obtained. The maximum of the mechanical loss curve corresponds to

$$\omega_0 \tau_m = 1$$

where τ_m is the mechanical relaxation time. The shape of the $J'(\ln\omega)$, $J''(\ln\omega)$ curves are similar to those of the $\varepsilon'(\ln\omega)$, $\varepsilon''(\ln\omega)$ curves. The mechanical and dielectric relaxation times so far have been found experimentally to be equal

$$\tau_\varepsilon = \tau_m$$

Distribution of relaxation times

By comparing Eqns (1.136) and (1.137) with Eqn. (1.55) of the previous section it is seen that the $\varepsilon'(\omega)$ and $\varepsilon''(\omega)$ dependences obtained from the Kelvin–Voigt model correspond to those obtained from the Kubo formulae by using single exponential decay of the correlation function. In the statistical picture it corresponds to the first approximation when the effect of the motion of the neighbouring dipoles is neglected.

For considering the general case it is possible to use a model consisting of a large number of Kelvin–Voigt units. This results in the expressions

$$\frac{\varepsilon' - \varepsilon_\infty}{\varepsilon_0 - \varepsilon_\infty} = \sum_{i=1}^{N} \frac{\lambda_i}{1 + \omega^2 \tau_i^2} \tag{1.144}$$

$$\frac{\varepsilon''}{\varepsilon_0 - \varepsilon_\infty} = \sum_{i=1}^{N} \frac{\lambda_i \omega \tau_i}{1 + \omega^2 \tau_i^2} \tag{1.145}$$

where τ_i is the relaxation time of the i-th unit and λ_i is the corresponding weight factor.

By increasing the number of units it is possible to define a relaxation-time distribution function. A more sophisticated treatment is possible by considering the linearity of the response of the system in general. The corresponding phenomenological theory is based on the superposition principle.

The superposition principle, discovered by Hopkinson (1877), is a consequence of the general statistical theory outlined in the previous section considering linear response of a system to small external perturbations. The principle states that a small increase in the external perturbation, e. g. stress or electric field, produces a small increment of a property of the system, e. g. strain or electric displacement, and that these increments can be added to produce the total effect.

For example, according to the single-relaxation-time model the strain produced by application of a constant stress σ_0 is

$$\gamma(t) = \sigma_0 J_\infty + \sigma_0 (J_0 - J_\infty)\, \varphi_\sigma(t) \tag{1.146}$$

where $\varphi_\sigma(t)$ is referred to as the normalized creep function, for the single relaxation time approach

$$\varphi_\sigma(t) = 1 - \exp(-t/\tau_\sigma) \tag{1.147}$$

By increasing the stress by $d\sigma$ at time t the change in strain is

$$d\gamma = J_\infty\, d\sigma + (J_0 - J_\infty)\, \varphi_\sigma(t - t')\, d\sigma \tag{1.148}$$

According to the superposition principle the total strain at time t is

$$\gamma(t) = J_\infty\, \sigma(t) + (J_0 - J_\infty) \int_{-\infty}^{t} \dot\sigma(t')\, \varphi_\sigma(t - t')\, dt' \tag{1.149}$$

By partial integration

$$\gamma(t) = J_\infty\, \sigma(t) + (J_0 - J_\infty) \int_{-\infty}^{t} \sigma(t')\, \dot\varphi_\sigma(t - t')\, dt' \tag{1.150}$$

The integral in this expression is essentially the same as that in the Kubo formula (1.30) so that the superposition principle is a phenomenological formulation of the Kubo theory.

The dielectric behaviour can be similarly formulated as

$$\mathscr{D}(t) = \varepsilon_\infty \mathscr{E}(t) + (\varepsilon_0 - \varepsilon_\infty) \int_{-\infty}^{t} \mathscr{E}(t')\, \dot\varphi_e(t - t')\, dt' \tag{1.151}$$

59

where φ_e is the normalized dielectric decay function at constant field. In the case of single relaxation time approximation

$$\varphi_e(t) = 1 - \exp(-t/\tau_e) \tag{1.152}$$

From Eqns (1.151) and (1.152) the single-relaxation time equations (1.136), (1.137) and (1.138) can be derived (see e. g. in McCrum *et al*. 1967).

The extension of the single-relaxation time approximation to the general case may be done by introducing such a function instead of the decay function $\varphi(t)$ which contains the relaxation time as a variable. This is referred to as the relaxation time distribution function $F_J(\ln\tau)$, which is normalized as

$$\int_{-\infty}^{+\infty} F_J(\ln\tau)\, d\tau = 1 \tag{1.153}$$

The total decay function is thus expressed as

$$\dot\phi_\sigma(t) = \int_{-\infty}^{+\infty} \frac{F_J(\ln\tau)}{\tau} \exp(-t/\tau)\, d\ln\tau \tag{1.154}$$

Similarly, when the strain is constant

$$\dot\phi_\gamma(t) = \int_{-\infty}^{+\infty} \frac{F_G(\ln\tau)}{\tau} \exp(-t/\tau)\, d\ln\tau \tag{1.155}$$

At constant stress the compliance is expressed as

$$\frac{J(t) - J_\infty}{J_0 - J_\infty} = \int_{-\infty}^{+\infty} F_J(\ln\tau)\,[1 - \exp(-t/\tau)]\, d\ln\tau \tag{1.156}$$

The real and unreal parts of the compliances in the case of sinusoidal stress

$$\frac{J'(\omega) - J_\infty}{J_0 - J_\infty} = \int_{-\infty}^{+\infty} \frac{F_J(\ln\tau)}{1 + \omega^2\tau^2}\, d\ln\tau \tag{1.157}$$

$$\frac{J''(\omega)}{J_0 - J_\infty} = \int_{-\infty}^{+\infty} \frac{F_J(\ln\tau)\,\omega\tau}{1 + \omega^2\tau^2} d\ln\tau \tag{1.158}$$

Similarly for the electric case at constant field

$$\frac{\varepsilon'(t)-\varepsilon_\infty}{\varepsilon_0-\varepsilon_\infty} = \int_{-\infty}^{+\infty} F_e(\ln\tau)\,[1-\exp(-t/\tau)]\,d\ln\tau \qquad (1.159)$$

In the case of sinusoidal field

$$\frac{\varepsilon'(\omega)-\varepsilon_\infty}{\varepsilon_0-\varepsilon_\infty} = \int_{-\infty}^{+\infty} \frac{F_e(\ln\tau)}{1+\omega^2\tau^2}\,d\ln\tau \qquad (1.160)$$

$$\frac{\varepsilon''(\omega)}{\varepsilon_0-\varepsilon_\infty} = \int_{-\infty}^{+\infty} \frac{F_e(\ln\tau)\omega\tau}{1+\omega^2\tau^2}\,d\ln\tau \qquad (1.161)$$

Here $F_e(\ln\tau)$ is the normalized distribution function of the dielectric relaxation times.

It is seen that in the representation using distribution of relaxation times the response of the system to a mechanical perturbation appears to be the same as that to an electric perturbation. The difference is represented by the difference between the distribution functions F_J or F_G and F_e. As the distribution functions are very difficult to determine experimentally, so far no essential difference between the response of polymers to electric and mechanical perturbations could be detected. Havriliak and Negami (1967) showed that by minor modification of the mechanical and electric equations the two distribution functions appear to be identical. This makes it possible, at least at first approximation, to derive information about the mechanical properties of polymers from dielectric data.

Arc diagrams

When ε' and ε'' are plotted against frequency the resulting curve shapes are only modified by the presence of distribution of relaxation times. This effect is more apparent when ε' is plotted against ε'' as shown in Fig. 1.11 for the single relaxation time approximation (dashed curve) and for a real case of polyvinyl acetate (solid curve) after Havriliak and Negami (1967). In the case of single relaxation time a semicircle should be obtained because, according to Eqns (1.136) and (1.137),

$$[\varepsilon'(\omega)-\varepsilon'(\omega_0)]^2 + [\varepsilon''(\omega)]^2 = [\varepsilon''(\omega_0)]^2 \qquad (1.162)$$

where ω_0 is the frequency corresponding to the maximum of ε''.

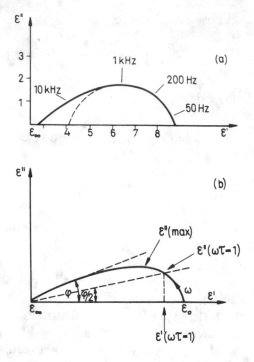

Fig. 1.11 (a) The Cole–Cole arc plot in the case of single-relaxation-time approximation (dashed curve) and for a real case: polyvinyl acetate (solid curve) (after Havriliak and Negami, 1967); (b) evaluation of relaxation parameters from a Cole–Cole plot

It is seen that there is a very significant deviation from the semicircle in the experimental plot, which means that the single-relaxation-time approximation is not valid. As evaluation of the distribution function from the measured spectra is extremely difficult, empirical corrections have been introduced in the single-relaxation-time equations in order to fit the empirical arc plots. Such corrections have been introduced by Cole and Cole (1941), Fuoss and Kirkwood (1941), Davidson and Cole (1950) and Scaife (1963). In these methods parameters are introduced which should characterize roughly the distribution function.

Scaife (1963), for example, introduced two parameters to the single relaxation time expression as

$$\frac{\varepsilon^*(\omega) - \varepsilon_\infty}{\varepsilon_0 - \varepsilon_\infty} = [1 + (i\omega\tau_0)^{1-a}]^b \qquad (1.163)$$

where the parameters

$$0 \leq a \leq 1$$
$$0 \leq b \leq 1$$

are determined from the actual arc plot and τ_0 is the mean relaxation time. For polyvinyl acetate, the arc plot of which is shown in Fig. 1.11b, $a = 0.09$ and $b = 0.45$.

The mean relaxation time is obtained from the experimental plot by using the relation

$$\tan(\varphi/2) = \frac{\varepsilon''(\omega\tau = 1)}{\varepsilon'(\omega\tau = 1) - \varepsilon_\infty} \tag{1.164}$$

where

$$\varphi = (1 - a)\frac{b}{2}$$

is the angle indicated in Fig. 1.11b. The mean relaxation time is obtained as

$$\tau_0 = \frac{1}{\omega_0}$$

where ω_0 is the frequency which corresponds to the ε' and ε'' value of the intersection of the arc with the $\varphi/2$ line.

The correction used by Scaife (1963) is a generalized one of that of Cole and Cole (1941), where $b = 0$, and that of Davidson and Cole (1950), where $a = 0$.

Temperature dependence of the relaxation time. Temperature as a variable

It has been shown by a simplified example in the previous section (see Fig. 1.4) that the relaxation times are connected with the transition probabilities between subsequent minimum-energy configurations of the dipoles, i.e. between minima of the potential valleys. From this it follows that according to Eyring's general theory of rate-processes (Glasstone *et al.*, 1941) the temperature dependence should be

$$\tau(T) = \tau_0 \exp\left(\frac{E}{kT}\right) \tag{1.165}$$

where E is the activation energy of the transition.

Equation (1.165) has been found valid for very many transitions in polymers, especially for those in which rotation of side groups or local motions of the main chain are involved. These problems will be discussed in detail in Chapter 2.

For another type of transition in polymers, the glass–rubber transition, Eqn. (1.165) proved not to be valid. The temperature dependence of the relaxation time is described in the case by the semi-empirical Williams–Landel–Ferry (WLF) equation (Williams *et al.*, 1955):

$$\ln\left(\frac{\tau}{\tau_0}\right) = -\frac{A(T-T_0)}{B+T-T_0} \tag{1.166}$$

where T_0 is the transition temperature, A and B are constants.

The WLF equation is a consequence of a general principle known as the time–temperature superposition. This principle was discovered by Leaderman (1943) by observing that the creep curves of polymers measured at different temperatures could be shifted to produce a master curve. The shift is given by

$$\ln a_T = -\frac{A(T-T_0)}{B+T-T_0} \tag{1.167}$$

The time–temperature superposition principle implies that the relaxation-time distribution function at a temperature T_0 is shifted in the $\ln \tau$ axis by $\ln a_T$ but the shape is unchanged.

In the case of the Arrhenius equation the shift is

$$\ln a_T = \frac{E}{k}\left(\frac{1}{T}-\frac{1}{T_0}\right) \tag{1.168}$$

The WLF equation (1.167) has been found valid in the high-temperature side of the glass–rubber transition. At low temperatures a significant deviation from this equation is observed. This means that the time–temperature superposition principle is not valid by passing through the glass–rubber transition. This problem will be discussed in detail in Chapter 2, Section 2.2. It will be shown there that the validity of the WLF equation can be extended to temperatures below T_g down to about $T_g - 100\,^\circ C$ by introducing effective temperature to account for the non-equilibrium state of the glass (Rusch, 1968).

In the WLF equation the temperature dependence of the relaxed and unrelaxed moduli, i.e. permittivities, are not taken into account.

McCrum and Morris (1964) calculated the effect of temperature on the relaxed and unrelaxed compliances, i.e. moduli, by introducing the following parameters:

$$J_0(T) - J_\infty(T) = b_T[J_0(T_0) - J_\infty(T_0)] \qquad (1.169)$$

$$J_\infty(T) = c_T J_\infty(T_0)$$
$$J_0(T) = d_T J_0(T_0) \qquad (1.170)$$

The parameters b_T, c_T and d_T can be estimated from the experimental results.

For the glass transition of amorphous polymers the temperature dependence of the unrelaxed modulus J_∞ is small in comparison with that of the relaxed one. This is also true for the dielectric case. The shift equation is then for the creep compliance

$$J\left(T_0, \frac{t}{a_T}\right) = \frac{1}{d_T}\left[J(T, t) - c_T J_\infty(T_0)\right] + J_\infty(T_0) \qquad (1.171)$$

For the dynamic-compliance

$$J(T_0, a_T, \omega) = \frac{1}{d_T}\left[J(T, \omega) - c_T I_\infty(T_0)\right] + J_\infty(T_0) \qquad (1.172)$$

Equations (1.171) and (1.172) are referred to as the McCrum–Morris (MM) equations. Equation (1.171) indicates that, by choosing the parameters correctly, a master curve can be constructed from the $I(t)$ or $J(\omega)$ curves measured at different temperatures. This has been verified for several polymeric systems.

According to the Kirkwood–Fröhlich equation (1.13) the temperature dependence of the oscillator strength of the dielectric transition is determined by the temperature dependence of ε_0, ε_∞, the effective dipole moment μ_{eff} and that of the reduction factor g_r. The unrelaxed permittivity according to the Clausius–Mosotti formula is independent of the temperature

$$\frac{\varepsilon_\infty - 1}{\varepsilon_\infty + 2} \frac{M}{\varrho} = \frac{4\pi N_0}{3}\alpha \qquad (1.173)$$

where M is the molecular weight, ϱ is the density N_0 is the concentration of the units and α is the polarizability of the unit.

The temperature dependence of the relaxed permittivity in the case of freely rotating dipoles is

$$\frac{\varepsilon_0 - 1}{\varepsilon_0 + 2} \frac{M}{\varrho} = \frac{4\pi N_0}{3} \left(\alpha + \frac{\mu_{\text{eff}}^2}{3\,kT} \right) \tag{1.174}$$

The temperature dependence of the oscillator strength is mainly determined by the $1/T$ factor occurring in Eqn. (1.13) and by the temperature dependence of the steric factor g_r

$$\varepsilon_0 - \varepsilon_\infty = \frac{C}{T}\, g_r(T)\, \mu_{\text{eff}}^2(T) \tag{1.175}$$

where the factor C is independent of or only slightly dependent on the temperature.

The effective dipole moment μ_{eff} depends on the temperature only if the configuration of the molecule is changed. The steric factor is temperature dependent because it is a measure of the intramolecular interactions; by definition

$$g_r = 1 - \frac{\langle \mu_{\text{eff}}^2 \rangle_{\text{av}}}{\mu_0^2} \tag{1.176}$$

From the temperature dependence of the oscillator strength $\varepsilon_0 - \varepsilon_\infty$ correspondingly important information about intramolecular interactions can be gained.

From Eqn. (1.175) it is seen that when the molecular configuration μ_{eff} and the intramolecular interactions g_r are not changed in the temperature range, $\varepsilon_0 - \varepsilon_\infty$ should decrease with increasing temperature. This is generally observed in dilute solution or sometimes in the molten state. The oscillator strength of polymers is more generally increased by increasing temperature, indicating that the dependence is mainly governed by the change in the intramolecular interactions $g_r(T)$. This problem will be discussed in some detail in Chapter 2. In any case the temperature dependence of the oscillator strength introduces a certain complication when dielectric spectra measured as a function of the temperature at constant frequency are interpreted.

From the time–temperature superposition principle, i.e. from the assumption that the shape of the relaxation time distribution function is

66

not changed as a function of the temperature, it follows that in principle it is possible to use the temperature as a variable instead of frequency in mechanical or dielectric relaxation experiments.

In the simplest case of the single-relaxation time approximation when the relaxation time depends on the temperature, according to Eqn. (1.165) the dielectric loss factor is

$$\varepsilon''(T) = (\varepsilon_0 - \varepsilon_\infty)_T \frac{\exp\left\{\frac{E}{k}\left(\frac{1}{T} - \frac{1}{T_m}\right)\right\}}{1 + \exp\left\{\frac{2E}{k}\left(\frac{1}{T} - \frac{1}{T_m}\right)\right\}} \qquad (1.177)$$

This function exhibits maximum at a temperature T_m, the corresponding relaxation time is

$$\tau(T_m) = \tau_0 \exp\left(\frac{E}{kT_m}\right) = \frac{1}{\omega_0} \qquad (1.178)$$

where ω_0 is the fixed angular frequency.

The activation energy E can also be expressed in the case of $\omega\tau_0 \ll 1$ as follows (Schallamach, 1946)

$$E = \left[\varepsilon_0(T_m) - \varepsilon_\infty(T_m)\right] \frac{k\pi}{2\int_0^\infty \varepsilon''(1/T)\,d(1/T)} \qquad (1.179)$$

Considering distribution of relaxation times, the average activation energy E is also given by Eqn. (1.179).

Similarly for the mechanical case

$$\langle E\rangle_{av} = [J_0(T_m) - J_\infty(T_m)] \frac{k\pi}{2} \frac{1}{\int_0^\infty J''(1/T)\,d(1/T)} \qquad (1.180)$$

$$\langle E\rangle_{av} = [G_\infty(T_m) - G_0(T_m)] \frac{k\pi}{2} \frac{1}{\int_0^\infty G''(1/T)\,d(1/T)} \qquad (1.181)$$

The average activation energy determined this way is independent of the frequency and of the shape of the relaxation-time distribution function.

When the dielectric spectrum is measured as a function of the temperature it is possible to define the limiting values as

$$\varepsilon_0 = \varepsilon'(T \to \infty) \equiv \varepsilon'(1/T \to 0)$$

$$\varepsilon_\infty = \varepsilon'(T \to 0) \equiv \varepsilon'(1/T \to \infty)$$

(1.182)

These limiting values are not necessarily the same as those measured when the frequency is chosen as a variable.

The limiting values ε_0 and ε_∞ can be most easily determined from the arc plots of ε' against ε'' by extrapolation.

CHAPTER 2

STRUCTURAL TRANSITIONS AND MOLECULAR MOBILITIES IN POLYMERS

As discussed in Chapter 1, the dielectric behaviour of a polymer is determined by the charge distribution and also by the statistical thermal motion of the polar groups. Evidently the chemical structure is the basic factor; in polymers which have polar bonds in their main chains or in side groups one should expect *ab initio* high dielectric permittivity. This is, however, only a necessary condition. Polytetrafluoroethylene for example has rather polar C-F groups (1.39 debye); nevertheless the change of its permittivity as a function of temperature or as a function of frequency is very small ($\varepsilon' \approx 2.2 - 2.3$, $\varepsilon'' \approx 10^{-4} - 10^{-3}$). The reason for this lies in the molecular configuration and physical structure of the polymer which makes the effective dipole moment small for the units which behave rigidly during thermal motion.

This means that in polymers in the solid or in the viscoelastic liquid state the physical structure, i.e. the way of packing the molecules, is of great importance in determining the dielectric behaviour. In condensed systems the overall response of the material to the application of an external electric field is also determined by the local inhomogeneities (defects) of the structure where, even in nonpolar materials, dipoles are formed. By considering these factors one can understand the experimental fact that structural transitions in polymers are generally accompanied by changes in dielectric permittivity. This is why, along with the other methods (mechanical relaxation spectroscopy, broad-line nuclear magnetic resonance, differential scanning calorimetry, dilatometry, infrared spectroscopy, X-ray and neutron scattering), dielectric spectroscopy has become an important tool in studying structural transitions and molecular mobilities in polymers.

In this chapter the concepts of structural transitions and molecular mobilities are discussed briefly and the effect of these transitions on dielectric behaviour is analysed. By illustrative examples it is shown what kind of transitions can be detected by dielectric spectroscopy and the results are compared with those obtained by various other methods.

2.1 Transition types in polymers

A polymer may exist in a solid state (amorphous and crystalline, usually mixed), in a viscoelastic fluid (rubber) and in a viscous fluid state. In some polymers, e.g. in crosslinked resins, there are no viscoelastic and viscous fluid states; the polymer does not melt at all. Polymers do not exist in the gaseous state because they would decompose before evaporation.

Polymers usually form very poor crystals. Although it is possible to grow single crystals of many polymers, their X-ray diffraction spectra always show the existence of a considerable amorphous background. A real polymeric solid is usually a mixture of crystalline and amorphous phases, i.e. its physical structure is heterogeneous. Even in purely amorphous polymers, structural heterogeneity has been discovered by electron microscopy and by the electron diffraction technique (Yeh, 1972). Polymer molecules are found to form aggregates of different form and size depending on the preparation and on the thermal history of the material. This aggregate structure is also referred to as supermolecular structure (Kargin, Slonimsky, Kitaigorodsky, 1957). Even in the fluid state and in solution, aggregate structure is often found to be present.

Structural inhomogeneities in polymeric materials are formed as a consequence of the difference of the thermodynamic behaviour of macromolecules with respect to that of small molecules. Statistical thermodynamics of polymeric systems, especially of solutions, has been discussed in detail by Flory (1953) and by Volkenstein (1963). From the peculiar thermodynamical behaviour of macromolecular systems it follows that they can exist in different crystal forms or in different aggregate forms simultaneously. This phenomenon is known in the physics of low-molecular-weight organic compounds as polymorphism (see, e.g., McCrone, 1965).

A generalized concept of polymorphism is illustrated schematically in Fig. 2.1. In this figure the free energy of the system is plotted schematically against a variable which represents the symmetry of the configuration. For simple spherical molecules (cubic symmetry) the energetically most favourable configuration is evidently the cubic lattice; all other possible configurations would represent much higher energy, so the system would arrange into a cubic lattice in a reasonably short time under suitable thermodynamical conditions.

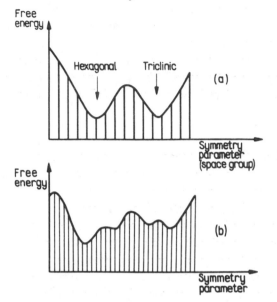

Fig. 2.1 Schematic illustration of polymorphism

In the case of molecules of lower symmetry the most favourable crystal lattice will be that which is near to the symmetry of the molecular electron configuration. Generally, the lower is the symmetry the shallower the corresponding potential valley in Fig. 2.1 will be. This means that in a given temperature range two or more crystal forms may coexist at certain temperatures; one form may be transformed into another.

An aggregate also has a definite symmetry as a crystal unit cell does, but this symmetry is not repeated all over the solid, i. e. the structure does not exhibit long-range order; no sharp X-ray patterns are observed. The symmetry of the aggregate is repeated in these systems only in a statistical sense. It is possible to broaden the concept of polymorphism to an amorphous polymeric solid exhibiting aggregate structure. Aggregates of definite size- and shape-distribution can be considered to form a definite structure having a definite mean free energy. Plotting a short-range symmetry parameter instead of the long-range symmetry (Fig. 2.1 b), the corresponding energy curve will be approximately

71

continuous, exhibiting minima which are less sharp than those of the corresponding crystal as a result of the statistical spread of the symmetry. This way it is possible to define a structural transition from one amorphous aggregate state into another. This view is essentially an extension of the paracrystalline model introduced originally by Hosemann (1950) for highly crystalline polymers. The concept is also implicitly used in the free-volume theory of glassy substances by considering that the system has several states of different configuration corresponding to approximately the same energy. The free-volume concept will be discussed in detail in section 2.2 of the present chapter.

Structural transitions

We shall define structural transitions in polymeric systems on the basis of the generalized concept of polymorphism illustrated in Fig. 2.1. Transition from one crystal form to another is evidently a structural transition; it is often encountered in polymers. The crystalline melting is also simply regarded as a structural transition in which the ordered system becomes disordered or less ordered. It is possible, however, to regard the transition from one aggregate form in an amorphous polymer into another as a structural transition because it involves large scale rearrangement of the structure. Such a transition is the glass–rubber transition in amorphous polymers, where the rigid glass which has a specific supermolecular structure is transformed to the viscoelastic fluid state which has another. A peculiarity of the glass transition is that it strongly depends on the direction and speed of the temperature variation. Experimental evidence and a more detailed discussion of this problem will be given in section 2.2 of the present chapter. The disappearance of the aggregate structure observed well above the glass–rubber transition is also regarded as a structural transition; it is from this point of view similar to melting of crystalline polymer: i.e. an order–disorder process.

Structural transitions will be considered here as being characterized by the following macroscopic features:

(a) The specific volume of the material changes abruptly at the transition. This is observed by measuring the thermal dilatation at constant pressure as a function of the temperature (Kovács, 1958a, b).

(b) Differential calorimetry shows enthalpy change at the transition— Wunderlich and Bodily (1964). This can also be explained on the basis of extension of polymorphism to amorphous systems (Fig. 2.1).

(c) The temperature dependence of the mechanical or dielectric relaxation time cannot be described by the simple Arrhenius equation (1.165) as the activation energy (enthalpy) is not constant. This means that by plotting the logarithm of the relaxation times against reciprocal frequency no straight line is obtained.

(d) The oscillator strength of the dielectric spectrum band $\varepsilon_0 - \varepsilon_\infty$ corresponding to dipoles attached to the main chain is increased as a function of the temperature to reach a maximum value above T_g (Ishida, 1969); the $1/T$ dependence which would follow from the Kirkwood–Fröhlich equation (1.13) is not obeyed. The reason for this is that the units which behave as rigid configurations during thermal motion change at the transition, resulting in changes in the effective dipole moment concentration.

(e) Structural transitions are especially sensitive to thermal pretreatments.

Transitions involving local molecular motion

Besides structural transitions in polymers, transitions may occur which do not involve large-scale structural rearrangement; just the local motion of some parts of the molecule is changed. Such a transition is, for example, liberation, i.e. freezing, of the rotation of side groups attached to the main polymer chain. The thermal motion of side groups is evidently different from that of the main chain; they represent a separate subsystem in the sense of statistical thermodynamics discussed in Chapter 1. This implies that the system of side groups is characterized by a specific partial temperature and specific relaxation time, i.e. distribution of relaxation times. If the side group contains polar bonds, freezing, i.e. liberation of this rotation, is represented by a significant change in the dielectric permittivity ε' and loss factor ε''. The corresponding relaxation process is sometimes referred to as dipole-group relaxation (Mikhailov, Borisova, 1964a). It is illustrated in Fig. 2.2a.

Another possibility of rotation of short segments without involving large-scale rearrangement of the structure is the crankshaft-type rotation of groups in the main chain (Schatzki, 1962, 1965; Boyer, 1963). Such a motion is illustrated in Fig. 2.2b. It is a conformational isomerisation of the main-chain segments with an estimated activation energy of 13 kcal/mole for linear hydrocarbon polymers. Similar conformational transitions in the main chain have been discussed in detail by Volken-stein (1963) and by Helfand (1971).

Local mode transitions also result from vibrations of short-chain segments about their equilibrium positions. Such a motion is termed local mode process (Yamafuji, Ishida, 1962; Gotlib, Salikhov, 1962; Saito, 1963).

Transitions involving local motion of groups are characterized by the following main features: (a) The specific volume of the material is not

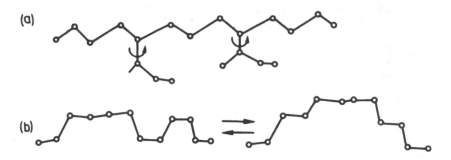

Fig. 2.2 Main types of local motion of polymer chains in bulk: (a) rotation of a side group (b) local isomerization of short main-chain segments: crankshaft-type rotation

significantly changed at the transition; no abrupt change of the thermal dilatation versus temperature curve is observed. (b) Differential calorimetry shows no significant enthalpy change; only changes of specific heat may be detected at such transitions. (c) The temperature dependence of the mechanical and dielectric relaxation time is satis-factorily described by the Arrhenius equation (1.165); it is possible to define a temperature-independent activation energy (enthalpy) for the process. (d) The oscillator strength of the dielectric spectrum band $\varepsilon_0 - \varepsilon_\infty$ is a monotonous function of the temperature; at the transition no maximum is exhibited. (e) Transitions involving local motion are not very sensitive to thermal pretreatments.

This specification of the transitions in polymeric systems into two main groups is evidently not a strict one. The local motion of the chains might involve some structural rearrangement and, on the other hand, in some cases the structural transition might run parallel with local motions. According to a fairly large body of experimental evidence, however (see, for example, McCrum 1967; Sazhin, 1970 and Hedvig 1969), the transitions involving local motions are well separated from

74

the structural transitions. Correspondingly side group or short-chain segments can be treated as individual thermodynamic subsystems. The situation is somewhat similar to the problem of nuclear and electronic spin relaxation. The nuclear or electronic spin systems are regarded as separate assemblies exhibiting their own partial temperatures, which are quite different from that of the lattice (see, for example, Hedvig, 1969). In a real polymeric system one should consider a series of assemblies formed by identical units of the structure. Each assembly has its own statistics and its own way of establishing equilibrium with the surroundings, i.e. with the other assemblies. Classification of the system into two parts is a simplification of this general view based on the experimental evidence.

Example of transitions in an amorphous polymer: polymethyl methacrylate

For illustrating the different types of transitions in amorphous polymers, poly-methyl methacrylate (PMMA) will be discussed here in some detail. The repeat unit of this polymer is

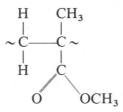

'Conventional' PMMA is polymerized at elevated temperatures by a free radical initiation. This polymer exhibits about 60% syndiotacticity (Bovey and Tiers, 1960); it is amorphous. By polymerization at low temperatures a higher degree of syndiotacticity can be achieved; such a polymer is crystallizable. Isotactic, partially crystalline, PMMA can be prepared by using stereo-specific catalysts.

The main transition temperatures in conventional (amorphous) PMMA which can be detected by mechanical relaxation spectroscopy or by broad-line NMR are shown schematically in Fig. 2.3. For labelling the transitions, two conventional notations are useful; one way is to denote the transitions by Greek letters α, β, γ, etc. starting usually from the highest observed transition temperature, or starting from the glass-transition temperature. The latter notation is used in Fig. 2.3, where the glass–rubber transition at about 110°C is denoted by α.

75

There is a liquid–liquid transition at about 210°C; it is denoted by α'. The glass–glass type of transitions which are observed in this polymer at −50°C, −170°C and near absolute zero temperature are denoted by β, γ and δ respectively. An alternative notation suggested by Boyer (1968) is also useful. According to this the glass–rubber transition is denoted by T_g, the glass–glass transitions by $T_{gg}^{(1)}$, $T_{gg}^{(2)}$... and the liquid-liquid transitions by $T_{ll}^{(1)}$, $T_{ll}^{(2)}$... The transitions indicated in

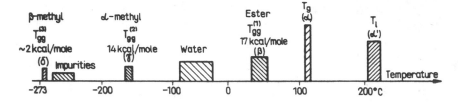

Fig. 2.3 Illustration of transitions in amorphous polymethyl methacrylate

Fig. 2.3 correspond to low frequencies (∼ 1 Hz). At higher frequencies the peaks are shifted to higher temperatures. By measuring the transitions by differential calorimetry, dielectric depolarization spectroscopy or dilatometry, the peaks appear at lower temperatures. Examples for these will be given somewhat later in the present chapter.

The transitions in PMMA are assigned to the following molecular motions:

$T_{ll}(α')$ A transition where the viscoelastic fluid becomes viscous. It involves large-scale mobility of large parts of the polymer chains. Probably there is a structural change at T_{ll}, but it has not been verified yet.

$T_g(α)$ The glass–rubber transition, which is assigned to microbrownian motion of main-chain segments, submolecules containing about 50—100 C−C bonds. In the high-temperature region of this transition the Williams–Landel–Ferry (WLF) equation (1.166) is valid. At lower temperatures the deviation from the WLF equation becomes large; a modified equation is fulfilled (see section 2.2). The Arrhenius equation for the temperature dependence of the relaxation time is not valid. The glass transition temperature is strongly dependent on the thermal history of the sample.

$T_{gg}^{(1)}(β)$ This transition is assigned to the rotation of the ester side-group about the C−C bond which links it to the main-chain. The corresponding

76

experimental activation energy is 17 kcal/mole (Mikhailov, 1955). According to recent studies by Koppelman (1972), there is an interaction between the α-process main-chain submolecule motion and the β-process (ester-group rotation), so this transition is also somewhat dependent on the thermal history of the sample.

$T_{gg}^{(2)}(\gamma)$ This transition is assigned to the rotation of the methyl group attached to the main chain (α-methyl group). It produces a weak dielectric spectrum band as the α-methyl group is slightly polar. The γ-transition is usually detected by mechanical relaxation and by NMR; the corresponding dielectric band is very weak. The measured activation energy is 14 kcal/mole.

$T_{gg}^{(3)}(\delta)$ This transition has been detected by the NMR spin-echo method (Filipovich et al., 1965; Powles and Hunt, 1965).

According to these measurements this transition is assigned to the rotation of the methyl group attached to the ester side group ($\beta - CH_3$ group). It appears that this methyl group exhibits hindered rotation even at the temperature of liquid helium 4 K.

Two additional transitions are shown in Fig. 2.3 in the temperature range of about $-90\,^\circ C$, $-50\,^\circ C$ and below $-196\,^\circ C$ respectively. The higher temperature one is due to the small amount of water which is always present in the conventional polymer unless it is carefully vacuum dried. Water also has a plasticizing effect in PMMA which appears as a shift of T_g to lower temperature by increasing water content. The peak detected at temperatures below $-196\,^\circ C$ (the boiling point of nitrogen) is due to impurities and is absent in a carefully purified polymer.

In Fig. 2.4 the dielectric absorption spectrum of conventional amorphous PMMA is shown as a function of the temperature at a constant frequency of 20 Hz in comparison with the mechanical relaxation spectrum, measured at 1 Hz. It is seen that the α-peak (T_g) appears smaller in the dielectric spectrum than the β-peak, while in the mechanical spectrum the situation is reversed. This is due to the fact that the polar group in this polymer is located in the ester side-group, not in the main chain, correspondingly the liberation of the rotation of this group involves higher effective dipole moment than that of the large-scale motion of the main chain to which the rotating side groups are attached. The γ-relaxation appears in the mechanical spectrum but is weak in the dielectric spectrum as the effective dipole moment of the α-methyl group is small.

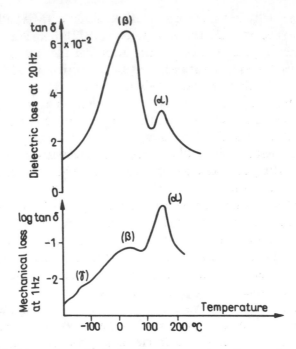

Fig. 2.4 Dielectric and mechanical relaxation spectra of amorphous polymethyl methacrylate, recorded as a function of the temperature

Transitions associated with the crystalline phase

Crystallinity means long-range symmetry, i.e. repetition of a unit of specific symmetry in a macroscopic range. Such a repetition produces sharp X-ray diffraction patterns superimposed on the broad amorphous background. A typical example of this is shown in Fig. 2.5, where the Debye–Scherrer type of X-ray diffractograms of high-density and low-density polyethylene are represented in comparison with that of amorphous polystyrene (Martynov and Bylegzhanina, 1972). The crystallinity is defined as the relative area of the sharp maxima with respect to the broad amorphous band. No information can be derived from the X-ray diffraction measurements about how the amorphous phase is distributed in a highly crystalline polymer.

According to the two-phase model introduced by Gerngross et al. (1930), the crystalline and amorphous phases are separated in space. In a polymer of low and intermediate crystallinity the crystallites would form separate regions in the disordered amorphous system. This view

78

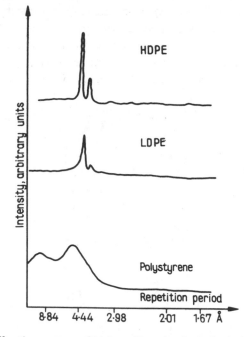

Fig. 2.5 X-ray diffraction pattern of high and low density polyethylene and polystyrene

has been modified by Hosemann (1950), at least for polymers of high crystallinity. According to the paracrystalline model of Hosemann the amorphous band observed by X-ray diffraction in highly crystalline polymers is due to the defects, especially at the boundaries of the crystallites. This means that in such systems the amorphous phase is not separated from the crystalline one in space; it is scattered throughout the system. The problem has been discussed in detail by Stuart (1959). Recent model experiments of Bodor (1972) show that even in polymers exhibiting relatively low crystallinity $\sim 20\%$ the X-ray diffraction patterns can be simulated by introducing defects in a crystalline structure rather than by separating the amorphous disordered phases from the crystalline ordered ones in space. On the other hand Yeh (1972) showed by the electron diffraction method that in such a classically amorphous polymer as atactic polystyrene ordered regions of 20–40 Å were present.

As the amorphous and crystalline phases are not well defined in polymers it is difficult to decide which transitions belong to which phase. We shall consider as belonging to the crystalline phase those transitions

79

which are appreciably increased by increasing crystallinity, crystal form or size. This does not necessarily mean that the units, the motion of which is reflected by the particular transition, are actually arranged in a crystal lattice. For example, in Fig. 2.6 the lamellar crystal structure of polyethylene is shown schematically (Keller, 1957). The chains of the

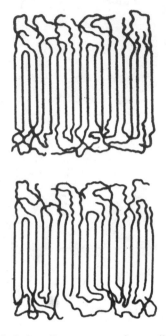

Fig. 2.6 Folded-chain lamellar structure of crystalline polyethylene

polyethylene molecule are folded to form a lamellar configuration, the interlamellar spacings being in the order of 100 Å. At the surface of the lamellae the mobility of the chain segments is different from that inside the lamellae (Peterlin, 1965). The transition corresponding to the motion of the surface groups of the lamellae (usually referred to as α-transition) will be considered here as a crystalline transition as it is highly increased by increasing crystallinity. The chain segments the motion of which produces the transition, are evidently not arranged periodically; in a strict sense the corresponding thermodynamical subsystem should be considered as amorphous.

The lamellar configuration tends to arrange in a spherically symmetric form shown in Fig. 2.7. This formation is referred to as spherulite and can be easily observed under a light microscope.

Transitions attributed to the crystalline phase are:

T_m, (α_m) Crystalline melting temperature. This is an order-disorder transition involving a large enthalpy and entropy change, where the long-range order is destroyed. At this transition the sharp X-ray diffraction peaks vanish; an abrupt change in the thermal dilatation curve and a large DSC peak are observed.

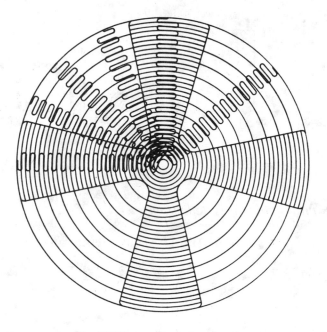

Fig. 2.7 The structure of a spherulite

T_c, (α_c) Transitions which are appreciably increased by increasing crystallinity and crystal size but are not due to the mobility of groups inside the crystal. These transitions usually correspond to the mobility of the groups at the surfaces or at the lattice defects.

T_{cc}^1, $T_{cc}^2 \ldots$, (α_{cc}) Transition of one crystal form into another. It is evidently a structural transition involving long-range rearrangement of the system. A typical crystal–crystal transition is observed in poly-tetrafluoroethylene at 19 °C, where the triclinic crystal form rearranges to the hexagonal form.

γ_c Transitions involving local motion (vibration or rotation) of groups of the main chain arranged in the crystal lattice. These are not structural transitions as the equilibrium position of the vibrating or rotating units is unchanged; the long-range symmetry is not affected by the transition. The local rotations and vibrations are in crystals'collective phonon states; the spectrum of such motions determines the specific heat of crystalline polymers at low temperatures.

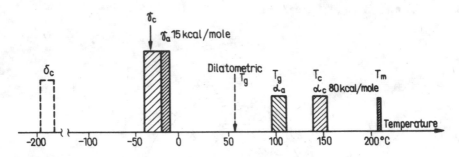

Fig. 2.8 Transitions in polychloro trifluoro ethylene

Example of transitions in crystalline polymers: PCTFE

Polychloro trifluoro ethylene (PCTFE) has a repeat unit

$$\begin{array}{cc} F & F \\ | & | \\ \sim C - C \sim \\ | & | \\ F & Cl \end{array}$$

It is a partially crystalline polymer having a degree of crystallinity varying from 0.1 up to 0.80 depending on thermal pretreatment. By undercooling from the melt it can be prepared in an amorphous form. PCTFE has a hexagonal crystal form of unit cell dimensions 6.5 Å and 35 Å. The chains in the crystal are helical. The crystalline melting temperature is 220 °C. The main transitions in PCTFE at 1 Hz are schematically illustrated in Fig. 2.8, for an 80 % crystalline polymer. There are two transitions at about 120–150 °C (α_a, α_c) which are resolved only for polymers of high crystallinity. By decreasing the crystallinity the α_c-peak decreases and the α_a-peak increases. According to Scott *et al.* (1962) the α_c-peak corresponds to motion at the surface of the chain-folded lamellar spherulites, which are formed only in samples of high crystallinity. The temperature dependence ot the α_c relaxation time can

82

be described by the Arrhenius equation; the corresponding activation energy is 80 kcal/mole. No break in the thermal dilatation curve is observed at this temperature.

The α_a-transition appearing at about $100\,^{\circ}\text{C}$ is attributed to the glass–rubber transition of the amorphous phase. This transition has been formerly denoted by β (see, for example, McCrum et al. 1967). We would rather use here the notation suggested by Ishida (1969), according to which the transitions corresponding to the amorphous and crystalline phases are distinguished.

In the α_a-relaxation region of PCTFE the dilatometric curve shows a sharp change; instead of the Arrhenius equation (1.165) the WLF equation (1.166) is to be used in order to describe the temperature dependence of the relaxation time.

The γ-relaxation region is also divided into crystalline γ_c and amorphous γ_a parts which are overlapping. The assumption that this relaxation is due to two overlapping processes has been proposed by McCrum (1962) on the basis of analyzing the shape of the mechanical relaxation curves measured at different crystallinity. The dielectric studies of Scott et al. (1962) support this view. The molecular motion attributed to the γ_a- and γ_c-processes in PCTFE are evidently local mode vibration or crankshaft-type local motion in the main chain, as this polymer does not contain side groups. The fact that the crystalline and amorphous local mode transitions are overlapping shows that for this type of motion energetically it makes little or no difference whether the groups involved are arranged in a crystal lattice or not.

A very low temperature (high frequency) transition δ_c in PCTFE has been discovered by Scott et al. (1962). This transition was found to be affected greatly by changing the crystallinity. From the temperature dependence of the oscillator strength of the dielectric transition $\varepsilon_0 - \varepsilon_\infty$ Scott et al. (1962) suggested that this process involves local uncoiling of the helix configurations in the crystalline phase.

2.2 The glass–rubber transition

The most prominent change in the macroscopic behaviour of amorphous polymers occurs at the glass transition where the rigid glassy solid material becomes a viscoelastic fluid. At this transition the mechanical strength of the material decreases rapidly; there is an abrupt change in the thermal dilatation versus temperature curve. The thermal conductivity, mechanical loss at a periodic stress, dielectric loss and static dielectric constant also change appreciably by passing through this transition. This is illustrated in Figs. 2.9, 2.10 and 2.11 where some mechanical, thermal and electric properties of unplasticized polyvinyl-

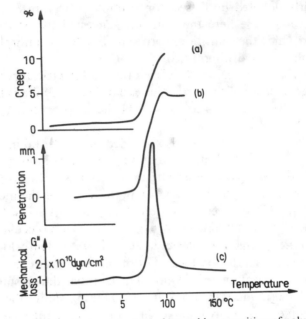

Fig. 2.9 Thermomechanical curves at the glass–rubber transition of polyvinylchloride

chloride are shown as a function of the temperature by passing through the glass–rubber transition at constant speed, starting from the low temperature side. Curve (a) in Fig. 2.9 represents the expansion of the sample at a constant load recorded at a constant rate of heating. Curve (b) represents the penetration of a cylindrical profile into the polymer at constant load. Curve (c) represents the mechanical loss at a torsional periodic stress of constant frequency (10 Hz), i.e. the temperature dependence of the loss modulus G'' (T). It is seen that the mechanical parameters show very drastic change at about 80 °C where the glass–rubber transition of this polymer is. The shifts of the position of the abrupt changes are due to the difference in the effective frequency of the transition. This is why the mechanical loss peak (curve c) appears at higher temperature than that corresponding to the abrupt change in thermal expansion (dilatometric) transition. The abrupt changes

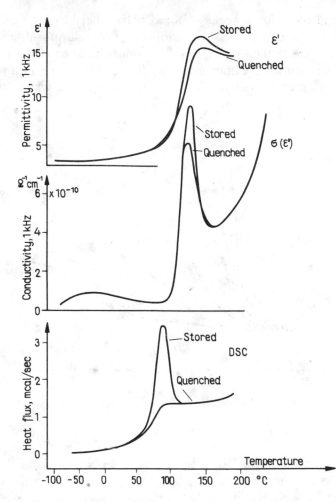

Fig. 2.10 Dielectric spectra and differential scanning calorimetric curve of polyvinyl-chloride in the glass-transition region. The effects of storage at 65 C for 2 weeks are also shown

observed in the thermo-mechanical curves (a) and (b) are dependent on the load; by increasing the load the transition temperature is shifted to lower temperatures.

In Fig. 2.10 the dielectric spectrum of unplasticized PVC is shown in comparsion with the differential scanning calorimetric (DSC) curve. Results of two representative samples are illustrated. One has been cooled down from 150 °C at a low speed of 0.1 °C/min and annealed at 65°C for 2 weeks—'stored'; the other has been quenched from 150°C to 0°C at a rate of 10°C/min before measuring—'quenched'. The change in the thermal expansion is much higher for the annealed sample than that for the quenched one. In the DSC curve a rather high endothermic peak is observed for the annealed material which is absent

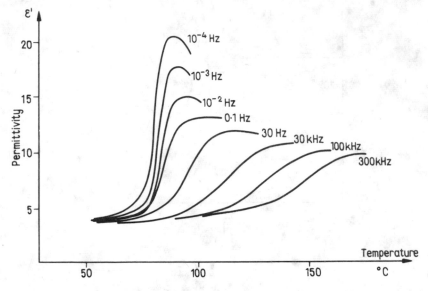

Fig. 2.11 Dielectric $\varepsilon'(T)$ spectra of unplasticized polyvinylchloride measured at different frequencies (after Reddish, 1965)

in the quenched sample. An arbitrary sample selected from the stock shows effects in between these two extreme cases. It has been shown by Illers (1969a) and Retting (1969) that the room-temperature tensile strength is highly influenced by storage of unplasticized PVC below the glass–rubber transition temperature.

The dielectric spectra $\varepsilon'(T)$ of unplasticized PVC are shown in Fig. 2.11, measured at different frequencies. The effect of annealing is observed only at low frequencies; the oscillator strength of the dielectric spectrum band measured as a function of the temperature $(\varepsilon_0 - \varepsilon_\infty)_T$ increases when the sample is annealed (Hedvig, 1969b). It is seen that $(\varepsilon_0 - \varepsilon_\infty)_T$ increases by decreasing the frequency. This effect becomes even more prominent at low frequencies below 1 Hz. A dielectric absorption curve measured by the time dependent polarization method introduced by Hamon (1952)—see Chapter 3— for a quenched sample has a much lower maximum at T_g than an annealed sample. As $(\varepsilon_0 - \varepsilon_\infty)_T$ is proportional to the area under the $\varepsilon''(T)$ curves it is concluded that by annealing $(\varepsilon_0 - \varepsilon_\infty)_T$ is considerably increased. A similar effect is observed by measuring dielectric depolarization currents (see Chapter 3). By this technique the sample is polarized above the transition temperature by a dc electric field, cooled down under field,

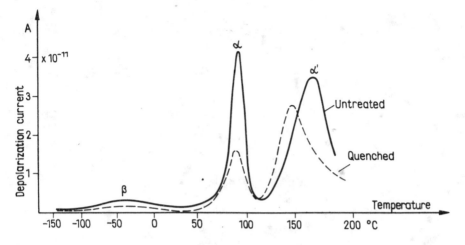

Fig. 2.12 Dielectric depolarization spectrum of polyvinylchloride. Polarization field 10 kV/cm, temperature 150°C, rates of heating and cooling 1°C/min; for the dashed curve rate of cooling 10°C/min

and subsequently heated up without external field to record the depolarization current. The curves shown in Fig. 2.12 have been recorded this way by using slow 1°C/min and fast 10°C/min cooling rates and identical heating rates. It is seen that the area under the depolarization current peak is highly increased for the sample cooled down slowly, which roughly corresponds to the 'annealed' case of the previous measurements.

The change in the mechanical, thermal and electrical parameters at the glass–rubber transition illustrated for unplasticized PVC is generally characteristic of amorphous polymers. From the experimental facts it is concluded that at T_g large parts of the polymer chain become mobile and, besides change in mobility, a structural change also occurs. According to the concept discussed in the previous section we shall regard T_g as a structural transition.

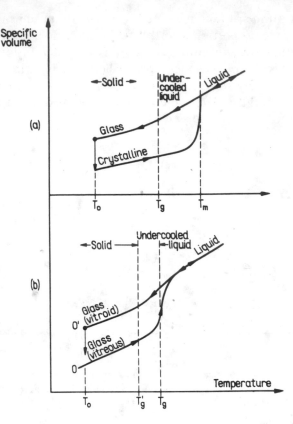

Fig. 2.13 Change of specific volume of a crystalline (a) and an amorphous (b) material on passing through melting, i.e. glass–rubber transition, up and down

The non-equilibrium character of the glassy state

Formation of a glassy state is not a unique property of polymers. It is very often encountered in organic and inorganic low-molecular-weight systems too: Generally, if a crystallizable material is cooled down from the melt, first an undercooled liquid state is formed below the melting temperature (T_m) and then a glassy solid is generally formed which can crystallize if the thermodynamical conditions are favourable. A complete heating–cooling cycle for a crystallizable material is shown schematically in Fig. 2.13a where the specific volume is plotted against temperature. Such a cycle is quite commonly found for various inorganic and organic systems. The time scale of the crystallization process is different for different systems, which means that the temperature range between T_0 and T_m may be small, i.e. the sample would crystallize rapidly very near to the melting temperature. It is evident that the undercooled liquid state and the glass state are thermodynamically non-equilibrium states. When the glass transition (T_g) is passed through from above, no abrupt changes in volume and in other thermodynamical variables are expected or found. From this point of view the glassy state may be considered as being a very-high-viscosity undercooled liquid state.

In Fig. 2.13b a heating–cooling cycle is shown for materials which would not crystallize under the usual conditions. In this case by passing through T_g from the low-temperature side an abrupt change in volume is observed, while by passing through T_g from above there is practically no such change. Evidently the points marked by 0 and 0' at T_0 correspond to two different glassy states. There is a time-dependent process for the system to pass from 0' to 0 when it is kept at the temperature T_0. This process is the analogue of crystallization. The system being in state 0' or in-between 0' and 0 evidently cannot be considered as an undercooled liquid. The state 0' is referred to as vitroid and that which is near to thermodynamical equilibrium 0 as vitreous (Eckstein, 1966, 1967, 1969). There are very few studies on the kinetic process by which a vitroid system approaches equilibrium. It is known, however, that this process is generally very slow, especially in polymeric systems. (Kovács, 1958a; Rusch, 1968.) Therefore it is concluded that an amorphous polymer, unless specially treated, is far from being in an equilibrium state; a slow time dependent structural change should always be considered to proceed. As this process involves drastic changes in the macroscopic properties it may be referred to as physical (structural) ageing (Hedvig, 1973).

As the temperature of a system is generally defined as the derivative of the informational entropy with respect to the total energy

$$\frac{1}{kT} = \frac{\partial S}{\partial \langle H \rangle_{av}}$$

where H is the total hamiltonian of the system, it is evident that the temperature of the thermostate (T_0 in Fig. 2.13b) is not representative of the system in the glassy (vitroid) state. Partial temperatures should be introduced which represent the configuration of the system being in a state 0′ or somewhere between 0′ and 0. The partial temperature depends on the thermal history of the sample. The general equation of the thermodynamical state of a vitroid is therefore generally expressed as (Eckstein, 1969)

$$f(p, v, T_{ph}, T_c^1 \ldots T_c^n) = 0$$

where p is the pressure, v is the volume, T_{ph} is the phonon-temperature arising from the collective vibrational modes, $T_c^1 \ldots T_c^n$ are the configurational temperatures.

Theories of the glass–rubber transition

There are several approaches for a molecular interpretation of the glass–rubber transition in polymers. A detailed critical survey of these is beyond the scope of this book. For a detailed discussion and historical survey see McCrum et al. (1967), Volkenstein (1963), Bartenev et al. (1969) and Goldstein (1969).

One of the early approaches introduced by Kargin and Slonimsky (1948, 1949) is based on the statistical theory of the microbrownian motion of polymer chains in dilute solution. In this approach the polymer molecules are divided into submolecules of lengths varying according to the gaussian probability distribution. The motions of these 'gaussian' submolecules are kinetically treated; the submolecules themselves are thought to be unchanged during thermal motion. This approach is referred to as 'normal mode theory based on gaussian submolecular model' (McCrum et al. 1967; see also in Bueche 1961; Tobolsky 1960; Ferry 1961).

The interpretation of the dielectric glass–rubber transition in terms of the normal mode theory has been discussed by Zimm et al. (1956 and Zimm 1960), Van Beek and Hermans (1957), Kästner (1961, 1962), Stockmayer and Baur (1964). The early theory of Kirkwood and Fuoss (1941) is also based mainly on the normal-mode aspect. Yamafuji and Ishida (1962) extended these theories by accounting for local motions.

The normal-mode theories are of little practical use. Because of the enormous mathematical difficulties, calculations cannot be performed exactly, so several semi-empirical parameters have been introduced. Moreover the normal-mode theories could not describe the non-equilibrium behaviour of the glassy state, which is its most important property. One thing, however, can be deduced from these calculations

90

which is of some practical importance: at the glass transition parts of the polymer molecule containing about 50—100 $C-C$ bonds become mobile.

Another approach originated by Debye (1945) and developed further by Fröhlich (1949) and Hoffman (1952, 1955, 1965) is referred to as the barrier-theory of the glass–rubber transition. In this approach the system is represented by a series of potential valleys. In order to change configuration the system must overcome a certain potential barrier. The probability for this can be described by simple kinetic equations used in general theory of rate processes.

The theory was first developed for describing rotational motions. Later Goldstein (1969) proposed a generalized barrier theory to describe configurational changes at the glass–rubber transition. The barrier picture has the advantage over the normal-mode theories that the non-equilibrium behaviour of the glassy state can be accounted for. It can also be readily connected with non-equilibrium statistical thermodynamics. Its quantitative application is hindered by our lack of knowledge on the forms of the intermolecular interaction potentials by which the potential barriers are formed.

Another difficulty is that at T_g the structure changes so drastically that the potential barriers themselves are strongly temperature dependent. This is why the barrier model is most frequently applied to describe local-mode transitions rather than glass transition.

A quite different approach has been introduced by Williams *et al.* (1955) based on the concepts of Doolittle (1951) about the free-volume theory of the viscosity of liquids. This approach is based on the properties of the liquid, so the glass–rubber transition is approached from the high-temperature side. According to this view the viscosity of liquids is determined mainly by their structure, which is characterized by the free volume

$$v_f = v - v_0$$

where v is the actual specific volume, v_0 is termed as the 'occupied volume' which would correspond to closest packing; v_0 has been approximated in the original (Doolittle) concept of free volume as the extrapolated volume to the absolute zero temperature. This definition can be extended by requiring also infinite time of storage at 0 K, i.e. equilibrium structure, and considering only that part of the volume as free by which the molecules can redistribute without additional energy (Cohen, Turnbull, 1959).

By the free-volume theory the high-temperature part of the glass–rubber transition could be satisfactorily explained. Below the dilatometric T_g in the glassy state the original theory fails, because it is based on the assumption that by passing through T_g from above the free volume is frozen-in and is unchanged in the glassy state. By accounting for the time and temperature dependence of the free volume in the glass the non-equilibrium behaviour can be qualitatively interpreted (Kovács, 1958a). The free-volume theory has been further developed by Rusch (1968) to be able to describe the temperature dependence of the relaxation time in the region below dilatometric T_g by introducing non-equilibrium free volume.

This theory will be discussed in some detail later in the present section in connection with the interpretation of the Williams–Landel–Ferry equation in the glassy state.

The problem of the glass–rubber transition has also been approached from the point of view of thermodynamics. Gibbs and di Marzio (1958) treated the glass–transition as a second-order thermodynamic transition. The way of regarding the problem in this approach is, in principle, similar to the barrier theory. The glass transition is approached from the high-temperature side using equilibrium thermodynamical partition function. At T_g the system is thought to be completely frozen; the entropy is considered to be zero. This is an oversimplification which makes the theory of no use in the glassy state, when it is in a non-equilibrium thermodynamical state.

Several attempts have been made to connect the thermodynamical aspects with some molecular theories and with the free-volume theory. A kinetic theory based on statistical considerations has been developed by Volkenstein and Ptytsin (1956). This theory is computationally rather complicated and does not seem to be practically useful.

The correlation between the kinetic and free-volume theories has been investigated by Bartenev in a series of papers (1951, 1955, 1956, 1969, 1970). He found that by introducing a temperature-dependent activation energy for the kinetic process the same results are obtained as by the free-volume theory.

Nose (1971, 1972) developed further the hole theory of liquids in order to account for the non-equilibrium behaviour of the glassy state. In his treatment a configurational entropy S^c is defined in the glassy state which is different from zero.

Instead of discussing the various T_g theories in detail, we select an illustrative problem: the application of the free-volume theory to interpret the empirical Williams–Landel–Ferry equation, above and below the dilatometric T_g. To the author it seems that at the present state of our knowledge about intra- and intermolecular interactions in a polymeric system one should be statisfied with such semi-phenomenological interpretations as the hole theory or the free-volume theory, which can be easily connected with the basic aspects of non-equilibrium statistical thermodynamics.

Interpretation of the Williams–Landel–Ferry equation in terms of free-volume theory

The empirical WLF equation is

$$\ln \frac{\tau(T)}{\tau(T_0)} = -\frac{A(T-T_0)}{B+T-T_0} \tag{2.1}$$

where τ is the relaxation time, T_0 is an arbitrary reference temperature, which can be chosen as the dilatometric T_g, A and B are constants having approximate values of $A=17$, $B=51$ respectively more or less independently of the polymer structure. The WLF equation has been found valid in many polymer systems above dilatometric T_g up to about $T_g+100°C$. Below dilatometric T_g the WLF equation in the original form is not fulfilled. This is illustrated in Fig. 2.14 for conventional amorphous polymethyl methacrylate after Rusch and Beck (1970). It is seen that above 105°C the experimental curve fits excellently to the WLF curve (solid line); below 105°C the deviation is appreciable. The dashed line in Fig. 2.14 represents a generalized WLF equation derived by Rusch and Beck (1970). It will be discussed somewhat later in the present section.

The viscosity of liquids is expressed in terms of the fractional free volume $\bar{v}_f = \frac{v_f}{v}$ according to Doolittle (1951) as

$$\eta = a \exp\left(\frac{b}{\bar{v}_f}\right) \tag{2.2}$$

where a and b are constants.

It is assumed that the fractional free volume depends linearly on temperature

$$\bar{v}_f(T) = \bar{v}_f(T_g) + \Delta\alpha(T-T_g) \tag{2.3}$$

93

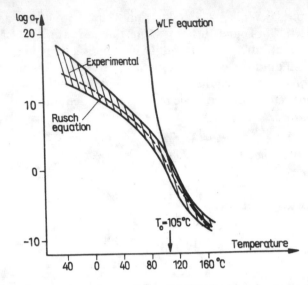

Fig. 2.14 Temperature dependence of the dielectric and mechanical relaxation time shift factor of amorphous polymethyl methacrylate in the glass–rubber transition region. The experimental (shaded) curve is compared with that obtained by the Williams–Landel–Ferry (WLF) equation and by the Rusch-equation (1968)

where $\Delta\alpha$ is the difference in the thermal expansion coefficients below and above T_g (Saito, 1963). Considering

$$\ln \frac{\tau(T)}{\tau(T_g)} \approx \ln \frac{\eta(T)}{\eta(T_g)} \tag{2.4}$$

from Eqns (2.2) and (2.3) the WLF-equation (2.1) is obtained with the following constants

$$A = \frac{b}{2.303 \, \bar{v}_f(T_g)}$$

$$\tag{2.5}$$

$$B = \frac{\bar{v}_f(T_g)}{\Delta\alpha}$$

By considering the universal approximate values of constants A and B it is obtained that, according to the original free-volume theory, the fractional free volume at T_g and the thermal expansion coefficient

94

difference $\Delta\alpha$ would be universal constants for all kinds of polymers. The values are

$$\bar{v}_f(T_g) \approx 2.5 \ 10^{-2}$$

$$\Delta\alpha \approx 4.8 \ 10^{-4} \ \text{deg}^{-1}$$

The considerations outlined above are evidently valid only if T_g is approached from the high-temperature side where the thermal expansion is linear. By approaching T_g from the low-temperature side after a certain storage time, starting from a quasi-equilibrium state (cf. Fig. 2.13) the thermal expansion coefficient rises suddenly when T_g is approached; thus Eqn. (2.3) is not valid.

The dependence of the viscosity on the free volume Eqn. (2.2) is based on the molecular picture of Cohen and Turnbull (1959), which, in turn, is based on the assumption that molecular transport in liquids is due to movement of molecules into voids (holes) of the structure which are greater than a certain critical volume v^*. The probability of formation of a hole exceeding this critical volume is

$$P(v^*) = A_V \exp\left(-\frac{\gamma v^*}{v_f}\right) \tag{2.6}$$

where γ is a numerical constant between 0.5 and 1 accounting for the overlap of free volume, A_v is a temperature-independent or slightly temperature-dependent factor.

The shift factor in the WLF equation is expressed as

$$a_T = \frac{\tau(T)}{\tau(T_g)} = \frac{P(T_g)}{P(T)} \tag{2.7}$$

as the relaxation time is defined as the reciprocal of the probability of formation of a hole.

From the Cohen–Turnbull theory of free volume the following general relations for the viscosity or for the relaxation time follow

$$\tau(T) = \tau_0 \exp\left(\frac{C}{T-T_\infty}\right) \tag{2.8}$$

$$\eta(T) = \eta_0 \exp\left(\frac{C}{T-T_\infty}\right) \tag{2.9}$$

where C is a constant and T_∞ is a reference temperature below which the free volume is considered to be zero. T_∞ is thus, roughly speaking, a hypothetical transition temperature at which the close-packed system gets loose. It is definitely below the dilatometric T_g.

When $T_\infty = 0$ the Arrhenius-like dependence of the viscosity or relaxation time is obtained. When $T_\infty > 0$ the relaxation time predicted by Eqn. (2.9) becomes infinite when the transition temperature T_∞ is approached. This means that in order to observe T_∞ one should cool down the sample at an infinitely low rate.

Extension of the free-volume theory to temperatures below dilatometric T_g

As has been shown earlier, the original free-volume theory is not valid in the temperature region below dilatometric T_g as the non-equilibrium state of the glass is not accounted for. Rusch (1968) extended the validity of this concept to this temperature region and derived a modified WLF equation.

The basic assumption of Rusch is that in the glassy state the free volume can be considered to consist of two parts, the equilibrium one v_f and the non-equilibrium one w_f. It is also assumed that the relaxation process is mainly controlled by the free volume. By cooling down the substance infinitely slowly a break of the volume–temperature curve should be obtained at a temperature T_∞. At this transition temperature the equilibrium free volume is considered to become zero. When the sample is cooled down at a finite rate the corresponding break occurs at higher temperature T_0 and the corresponding total free volume will be higher than the equilibrium one as a result of the contribution of the non-equilibrium part. The change of the equilibrium and non-equilibrium free volumes on approaching T_g are schematically illustrated in Fig. 2.15.

The free volume is defined as a part of the total volume change in thermal expansion. The total specific volume at equilibrium is

$$v(T) = v_e(T) + v_f(T) + w_f(T) \tag{2.10}$$

where v_e arises from the anharmonicity of the lattice vibrations, v_f is that part which corresponds to redistribution of the system with no accompanying energy change, at equilibrium, w_f is that at non-equilibrium condition.

96

For a molecular rearrangement it is necessary that the random redistribution of free volume results in a hole of volume exceeding a certain value v^*, and that the submolecule has enough kinetic energy to overcome the potential barrier set by the restraining forces. The proba-

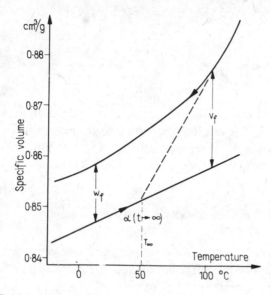

Fig. 2.15 Definition of equilibrium and non-equilibrium free volumes (after Rusch, 1968)

bility for this is expressed by the Cohen–Turnbull equation considering the additional, non-equilibrium part of the free volume. Assuming $\gamma = 1$ and

$$v_f = \Delta\alpha(T - T_\infty) \tag{2.11}$$

the probability of formation of a hole exceeding a critical volume v^* according to Rusch and Beck (1970) is the following

$$P(v^*) \propto \exp\left\{\frac{-v^*}{\Delta\alpha(T - T_\infty) + w_f(T)} - \frac{E}{kT}\right\} \tag{2.12}$$

Here E represents the activation energy for the molecular rearrangement, i.e. the potential barrier to overcome the attractive forces.

By using these assumptions the WLF equation can be derived. For the constants the following expressions are obtained

$$A = \frac{v^*}{2.303 \ \Delta\alpha(T_g - T_\infty) + w_f(T_g)}$$

$$B = T_g - T_\infty + \frac{w_f T_g}{\Delta\alpha}$$

(2.13)

By comparison of this result with Eqns (2.5) it is seen that as a result of the contribution of the non-equilibrium free volume the WLF constants are not universal any more; through the non-equilibrium part of the free volume a time dependence is introduced.

Effective temperature. The generalized WLF equation

The modified WLF equation containing the non-equilibrium part of the free volume can be formulated also by introducing an effective temperature for the non-equilibrium glass as

$$T_{eff} = T \qquad\qquad \text{if } T > T_g$$

$$T_{eff} = T + \frac{w_f(T)}{\Delta\alpha} \quad \text{if } T_g > T > T_\infty$$

$$T_{eff} = T_\infty + \frac{w_f(T)}{\Delta\alpha} \quad \text{if } T \leq T_\infty$$

(2.14)

The equation is

$$\ln\frac{\tau(T)}{\tau(T_g)} = \frac{-A(T_{eff} - T_g)}{B + T_{eff} - T_g} + \frac{E}{kT}$$

(2.15)

where

$$A = \frac{v^*}{2.303 \ \Delta\alpha(T_g - T_\infty)}$$

$$B = T_g - T_\infty$$

(2.16)

98

By definition the effective temperature represents that temperature at which the glass would have an equilibrium free volume v_f equal to the total free volume $v_f + w_f$

$$w_f(T) + v_f(T) = v_f(T_{eff}) \qquad (2.17)$$

By introducing the effective temperature into the WLF equation, the time dependence arising from the non-equilibrium state of the glass is taken into account. Equation (2.15) has been found to fit the experimental data in the case of polymethyl methacrylate by setting $E = 0$ in the temperature region from $T_g - 100°C$ up to $T_g + 100°C$. This indicates that the relaxation process is mainly governed by the free volume. Once the free volume greater than a critical one (v^*) is assured the submolecules can rearrange without additional energy.

The fit of the generalized WLF equation to the experimental data is shown in Fig. 2.14—solid line— for $E = 0$ and $T_g = 105°C$. The original WLF equation is shown in the figure by a dashed line.

Rusch and Beck (1969, 1970) described the yielding of glassy substances by Eqn. (2.15). By this view yielding is initiated at a point when the dilational component of the applied stress is sufficient to create a free volume which would be in the unstressed polymer at T_g. Again by comparing Eqn. (2.15) with the data it was found that yielding could be explained by setting $E = 0$, so this process is also mainly governed by free volume.

Volume relaxation

From the expression for the effective temperature (Eqn. 2.14) in the glassy state at constant external temperature ($T = $ const) the following kinetic equation can be derived (Davies and Jones, 1953)

$$\frac{1}{v} \frac{dv(T, t)}{dt} = \frac{\Delta\alpha(T - T_{eff})}{\tau(T)} \qquad (2.18)$$

$$\frac{dv(T, t)}{dt} = -\frac{w_f(T)}{\tau(T)} \qquad (2.19)$$

where $v(T, t)$ is the total specific volume of the system. Equation (2.18) describes the temperature dependence of the volume change in the non-equilibrium glass at constant temperature T. This equation was first derived by Kovács (1961) on the basis of a model network; the phenomenon is referred to as volume relaxation.

According to Eqn. (2.19) the time dependence of the specific volume in the non-equilibrium glass is expressed as

$$v(t) = v_0 \left[1 - \exp \left(-\frac{t}{t_0} \right) \right] \tag{2.20}$$

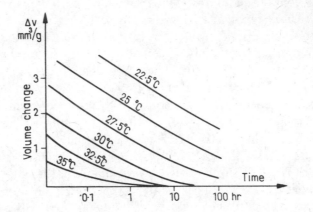

Fig. 2.16 Volume relaxation in polyvinyl acetate (after Kovács, 1958)

where v_0 is the equilibrium specific volume at temperature T,

$$t_0 = \frac{w_f(T)}{\tau(T)} \tag{2.21}$$

It is evident that the rate of change of volume as a function of time will be higher the more the system is deviated from equilibrium. (w_f high) and the closer it is to the glass transition ($\tau(T)$ small).

Volume relaxation in polyvinyl acetate is illustrated in Fig. 2.16 after Kovács (1958 b). The change of the specific volume with respect to the equilibrium volume Δv is shown as a function of time at different temperatures. The dilatometric glass-transition temperature of PVAC is 35° C. It is seen from Fig. 2.16 that as the temperature of storage approaches T_g the volume-relaxation process gets faster according to Eqns (2.20) and (2.21).

100

2.3 Transitions in the glassy state

Transitions in the glassy state, involving local molecular motion (T_{gg}-type transitions) are more easy to handle theoretically than T_g transitions, as the motion of a relatively small group of atoms is to be considered in the framework of a frozen-in structure. A side group such as, for example, the ester group in polymethyl methacrylate discussed in section 2.1, cannot rotate freely in the solid about the $C-C$ bond which links it to the main chain because of the large inter- and intramolecular forces. As a result of this the side group has different energy in different rotational positions. The problem is very similar to that of rotational isomerism of small molecules. Below T_g the main chain is approximately rigid; its structure determines a potential field in which the side group moves. This potential may exhibit minima at certain positions. Rotation of the side group from one minimum to another requires an energy of activation equivalent to the height of the potential barrier to overcome. Rotation of a side group is thus a rate process; the probability of transition from one equilibrium (minimum energy) position to another is expressed in terms of rate constants similar to the case of chemical reactions.

Another typical possibility of local motion in polymers is conformational transition in the main chain. Any change in the stereochemical configuration of short-chain segments would result in such a transition. Conformational transitions of the main chain involve rotation and/or translation. In this case again the potential barrier picture is very useful for interpretation of the corresponding dielectric transition. The general problem is to calculate the potential barrier between two configurations and to determine the rate constants for the transition by statistical theory.

The transitions involving rotation of side groups (β-transitions) usually appear at low frequencies in the temperature range from about $-50\,^\circ$C up to $+50\,^\circ$C. The transitions involving conformational isomerization of the main chain appear at lower temperatures, typically near -150°C. The transitions observed below the temperature of boiling nitrogen down to near absolute zero are referred to as cyrogenic transitions. They are also interpreted as being due to rotation and vibration of short-chain segments and to hindered rotation of small side groups, such as the methyl group, for example. At very low temperatures the collective vibrational phonon states of the solid become more and more important.

A general formulation of conformational transitions

For discussing conformational transitions a sequence of bonds containing n-units is selected which is between the rest of the chain. A conformational transition of such a group is generally represented as (Helfand, 1971)

$$P(s_1, s_2 \ldots s_n) Q \rightleftarrows P'(s_1, s_2, \ldots s_n) Q' \qquad (2.22)$$

where $s_1 \ldots s_n$, $s_1 \ldots s_n$ represent rotational states, P, Q, P' O' are the rest of the polymer molecule, the 'tails'. It is possible to fix the coordinate system to one of the tails, thus $P = P'$. In general $s_1 \ldots s_n$ represent any rotational state, i.e. the corresponding bond angle can have any value. In saturated hydrocarbon polymers as a result of the sp^3 hybridization of the carbon orbitals all bond angles can be approximately taken as $109.5°$ the tetrahedral angle, and the rotational isomers might be described in *trans* (t) *gauche* (g^+) rotated from *trans* by an angle of $120°$ in a positive direction, *gauche* (g^-) rotated from *trans* by $120°$ in a negative direction.

As a specific example, in Fig. 2.17a the crankshaft-type transition is illustrated. This transition is symbolized as

$$P(t\, tg^+\, tg^+\, tt)\, Q \rightleftarrows P(g^+\, tg^+\, tg^+\, tg^-)\, Q \qquad (2.23)$$

It is seen that this is really a local motion, the coordinates of the tails are unchanged. Most of the transitions observed in hydrocarbon polymers near $-150°C$ are interpreted as being due to crankshaft transition.

A similar transition is:

$$P(t\, tg^+\, tt)\, Q \rightleftarrows P(g^+\, tg^-\, tg^+)\, Q \qquad (2.24)$$

Such conformational transitions by which the coordinates of the tails are unchanged are referred to as Type 1 transitions (Helfand, 1971).

Figure 2.17b shows a local motion that involves a shift of the tail Q, the *gauche*-migration:

$$P(g\, t\, t)\, Q \rightleftarrows P'(t\, t\, g)\, Q' \qquad (2.25)$$

By this transition the total lenght of the unit changes, which makes the tail shift. Such conformational transitions which involve translation of the tail are referred to as Type II transitions. Another Type II conformational transition is shown in Fig. 2.17c:

$$P(tt\, t)\, Q \rightleftarrows \dot{P}(g^+\, tg^-)\, Q' \qquad (2.26)$$

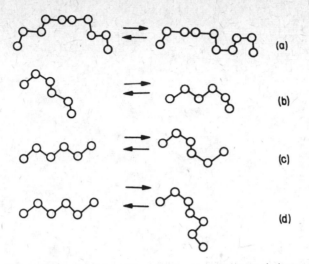

Fig. 2.17 Typical conformational transitions of the main polymer chain: (a) crankshaft rotation: (b) gauche migration: (c) zigzag-gauche pair formation: (d) bond rotation (after Helfand, 1971)

In this transition the zigzag *trans-trans* chain transforms into a *gauche* pair.

Another class of transition (Type III) involves rotation of the tail. Such is a simple bond rotation shown in Fig. 2.17d

$$P(t) Q \rightleftarrows P(g) Q' \tag{2.27}$$

and

$$P(g^+ t) Q \rightleftarrows P(tg^-) Q' \tag{2.28}$$

Such a transition evidently cannot be local, as the rotation of the tail involves large-scale rearrangement of the molecular configuration which is highly hindered in the glassy state.

The intensity of the dielectric absorption depends on the position of the most polar bond in the molecule with respect to the specific units involved in the thermal motion. Some typical cases are represented in Fig. 2.18 for the case of a simple *trans-trans* chain; (a) represents the case when the polar bond is built in the chain. A typical example for this is polyethylene oxide

$$
\begin{array}{c c c}
\text{H} & \text{H} & \text{H} \\
| & | & | \\
\sim\text{C} - \text{C} - \text{O} - \text{C}\sim \\
| & | & | \\
\text{H} & \text{H} & \text{H}
\end{array}
$$

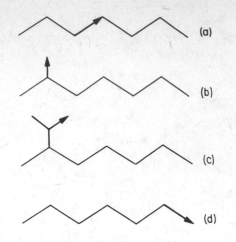

Fig. 2.18 Typical position of the polar groups (indicated by arrows) in polymers: (a) polar main-chain; (b) apolar main chain with polar groups rigidly attached; (c) polar main chain flexible polar side group (d) apolar main chain with polar end

where the polar $C-O$ group is built in the helical main chain, the sequence of the polar $C-O-C-C-O$ groups is *ttg* (Tadokoro, 1963). By studying transitions in the glassy state, systems which exhibit polar groups at the chain ends (Fig. 2.18d) are of specific interest. In the solid state chain ends are usually more mobile as the other part of the chain. Thus a structural defect is formed around the ends where the packing in the amorphous as well as in the crystalline phase is loose. A specific example of this, polyoxymethylene, will be discussed in some detail in section 2.4.

Another important class of polar polymers is represented in Fig. 2.18b where the polar bond is rigidly attached to the main chain. A typical example for this is polyvinylchloride

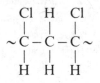

In this case the dielectric spectrum should reflect the large-scale micro-brownian motion as well as the local conformational transitions of the polymer.

104

Figure 2.18c shows the case when the polar group is attached flexibly to the main chain as a side group. The motion of the side group is approximately independent of that of the main chain; polar side groups can be treated as an independent thermodynamical system. A classical example for this is poly-methyl methacrylate discussed already in some detail in section 2.1. In this polymer the polar ester side group is attached through a $C-C$ bond to the main chain. The configuration of the main chain is helical in the isotactic polymer and zigzag-*trans* in the syndiotactic one.

Mixing of the α-, β- and γ-processes

The picture that the dielectric transitions are uniquely governed by specific molecular motions is evidently oversimplified. In reality all the possible conformational transitions contribute to the relaxation of the dipole moment system. This can be easily formulated in terms of the correlation theory, which has been discussed in Chapter 1. The decay function of a dipole system can be written down as

$$\Phi(t) = \frac{\langle \mu(0)\,\mu(t)\rangle}{\mu^2} = \sum_i P_i q_i^\alpha \phi_i^\alpha + \sum_i P_i q_i^\beta \phi_i^\beta + \sum_i P_i q_i^\gamma \phi_i^\gamma + \ldots \quad (2.29)$$

where μ is the dipole moment, $q_i = \left(\dfrac{\langle \mu_i \rangle^2}{\mu^2}\right)_{\alpha,\,\beta,\,\gamma}$ where superscripts α, β, γ, ... mean that the averaging of the corresponding dipole moment located at site i should be taken over the time scale of the α, β, γ... processes respectively, P_i is the probability for obtaining site i, ϕ_i^α, ϕ_i^β, ϕ_i^γ are the dipole autocorrelation functions for the subsystems α, β, γ respectively.

The complex dielectric permittivity is correspondingly expressed (cf. Chapter 1) as

$$\frac{\varepsilon^*(i\omega) - \varepsilon_\infty}{\varepsilon_0 - \varepsilon_\infty} = \sum_i P_i q_i^\alpha \mathscr{L}(-\dot{\phi}_i^\alpha) + \sum_i P_i q_i^\beta \mathscr{L}(-\dot{\phi}_i^\beta)$$

$$+ \sum_i P_i q_i^\gamma \mathscr{L}(-\dot{\phi}_i^\gamma) + \ldots$$

$$(2.30)$$

where $\mathscr{L}$ means Laplace transformation.

Equation (2.30) indicates that the total relaxation process of a dipole system is divided into partial relaxations of subsystems α, β, γ ... At a certain temperature–frequency range only a single mechanism dominates, but there are temperature–frequency regions where the processes coalesce. This is experimentally observed. In many amorphous polymers the dielectric transition frequencies plotted against reciprocal temperature for the $\alpha(T_g)$ and β (side group) transitions coalesce at high temperatures into an $(\alpha\beta)$ process. For $T < T_g$ $\phi^\alpha(t) \approx 1$, the time scale of the slow reorientation in the non-equilibrium vitroid state is much longer than the time scale of the measurement. In this case, from Eqn. (2.30) neglecting the higher frequency γ, δ ... relaxation terms, it follows that

$$\frac{\varepsilon^*(i\omega) - \varepsilon_\infty}{\varepsilon_0 - \varepsilon_\infty} = \sum_i P_i q_i \mathscr{L}(-\dot\phi_i^\beta) \tag{2.31}$$

On the other hand if $T > T_g$ $\phi_i^\alpha(t) \approx \phi_i^\beta(t) \approx 1$

$$\frac{\varepsilon^*(i\omega) - \varepsilon_\infty}{\varepsilon_0 - \varepsilon_\infty} = \mathscr{L}[-\dot\phi^\alpha(t)] \tag{2.32}$$

In this case the large-scale microbrownian motion dominates and the processes β, γ, δ ... remain unrelaxed during the time scale of the measurement, i.e. $\phi_i^\beta(t) \approx \phi_i^\gamma(t) \approx 0$.

The correlation theory has been applied for studying the $\alpha - \beta$ interaction in polyvinylchloride by Williams and Watts (1971a) in order to interpret the pressure dependence of the spectra.

The interaction of two relaxation processes can be conveniently treated by using the mechanical and dielectric models discussed in section 1.3. The α- and β-relaxation processes are simulated in Fig. 2.19 by a combination of springs and dashpots. The compliances for the springs corresponding to the α-process are chosen 100 times greater than those of the β-process in this model. At first approximation the spring-constants can be regarded as being independent of the temperature; the temperature dependence of the whole system is thought to be governed by that of the viscosities of the dashpots. Further, it is assumed that the viscosity of dashpot (α) has a strong exponential temperature dependence, while dashpot (β) has none. Using these assumptions, Koppelman (1961) calculated the frequency spectrum of the system at different temperatures. The result is shown in Fig. 2.19. It is seen that at low temperatures the α- and β-peaks are well separated. By increasing the temperature the α-peak is shifted to higher frequencies until it is merged into the β-peak which is unchanged at low temperatures and is shifted to lower frequencies at higher temperatures as a result of the α − β interaction.

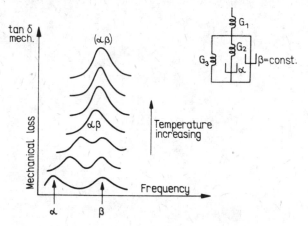

Fig. 2.19 Koppelman s (1972) model for interpretation of the mixing of α- and β-processes

The model, although it is very approximative, agrees rather well with the observed spectra. Figure 2.20 shows a series of mechanical relaxation spectra measured as a function of the temperature at different temperatures for amorphous poly-methyl methacrylate. It is seen that below T_g the β-peak is shifted to high temperatures, as is expected from the usual relaxation behaviour. Above T_g the β-peak is intensified and is shifted to lower temperatures while the α-peak begins to merge with it. This shift is attributed to the α − β interaction and is exhibited also by the simple model of Fig. 2.19. At low temperatures the α-peak is not shown in the experimental curves because the frequency range is not broad enough.

The potential barrier theory

In order to interpret dielectric absorption caused by conformational transitions in the main chain, or those caused by rotation of side groups, one should know exactly the chain conformation in the glassy or crystalline state. In the crystalline state some information about this is

107

gained by X-ray diffraction measurements and by neutron scattering. In the amorphous, glassy, state the chain conformation is generally not known. It is assumed that the conformation in the amorphous solid state would be the same as in solution, and some recent neutron-scattering experiments seem to support this view (Flory, 1972).

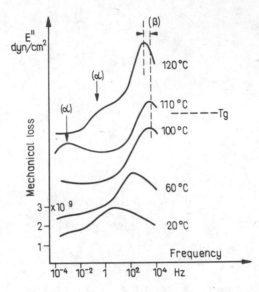

Fig. 2.20 α—β mixing in polymethyl methacrylate (after Koppelman, 1972)

Since the available information about chain conformation in the amorphous phase is vague, it is not possible yet to calculate the potential energy surfaces for a particular motion exactly. The only way to approach the problem presently is to postulate a potential surface from the known chemical structure and packing parameter of the polymer, calculate the complex dielectric permittivity for the particular transition, and compare the results with the experiment. It is hoped that through this semi-empirical procedure information about the chain conformation in the glassy state can be gained.

Once the potential energy surface is known or postulated, the problem can be solved in a relatively simple way by solving a set of kinetical equations. This method has been developed by Hoffman (1952, 1955) and is referred to as the barrier theory. A particular example for a 3-site potential barrier has already been discussed in Chapter 1 in terms of correlation theory. The rotation of the ester side group in polymethyl methacrylate, for example, can be described by the potential picture of Fig. 2.21. In the syndiotactic polymer the energy difference· is

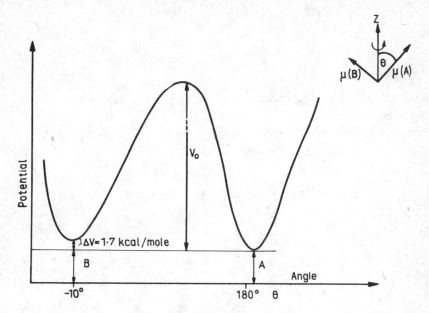

Fig. 2.21 Potential barrier for calculating ester group rotation in polymethyl metha-crylate (after Shindo, 1969)

$V = 0.6$ kcal/mole, in the isotactic polymer $V = 1.7$ kcal/mole (Shindo, 1969). The height of the barrier $V_0 = 10$ kcal/mole. This means that for the ester group there are two stable configurations corresponding to two rotational isomers. The transition rate constants are

$$k_1 = A \exp \left(-\frac{V_0}{kT} \right)$$

$$k_2 = A \exp \left(-\frac{V_0 + V}{kT} \right)$$

The probability for the group to occupy sites 1 and 2 at a time t is

$$P_1(t) = \exp(-k_1 + k_2)t$$

$$P_2(t) = \exp(k_1 - k_2)t$$

(2.34)

109

The dipole correlation function can be expressed in terms of the occupational probabilities as (Cole, 1965)

$$\Phi(t) = \sum_i P_i^0 \phi_i(t) = P_1^0 \phi_1(t) + P_2^0 \phi_2(t)$$

(2.35)

$$\phi_i(t) = \sum_j P_{ji}^0(t) \mu_i \mu_j$$

where $\phi_i(t)$ is the dipole decay function, P_i is the equilibrium occupational probability of site i.

The complex permittivity can thus be evaluated in terms of the transition rate constants by using the general equation

$$\frac{\varepsilon^*(i\omega) - \varepsilon_\infty}{\varepsilon_0 - \varepsilon_\infty} = \int_0^\infty \exp(-i\omega t) \, [-\dot{\phi}(t)] \, dt = \frac{C}{1 + i\omega\tau}$$

(2.36)

where

$$C = \frac{1 - \cos\theta_{12}}{2 \left[1 + \frac{k_1}{k_2}(1 + \cos\theta_{12}) \right]}; \quad \tau = \frac{1}{k_1 + 2k_2}$$

The relaxation time of the transition, correspondingly, can be expressed in terms of the rate constants of transition through the potential barriers hindering the rotation of the ester side group.

In general, where more sites are to be considered Eqn. (2.34) may be written in a matrix form as

$$P(t) = \exp(Tt)$$

(2.37)

where the elements of the P probability matrix P_{ij} represent the probability for the group to occupy site j at time t on the condition that it was in site i at time zero. The elements of the T-matrix (transition matrix) are the combinations of the rate constants.

The transition-rate Eqn. (2.37) can be solved relatively easily by the use of group theory (Williams and Cook 1971) whenever the system of sites shows some symmetry. The problem is essentially the same as in the case of chemical reactions; once the potential energy surface is given, the problem can be solved rather easily by classical non-quantum methods.

110

The local motion of the main-chain segments and that of the side groups may be expected to be completely frozen at very low temperatures near absolute zero. The rotation of a methyl group attached to the main chain, for example, is expected to be frozen-in completely at about 40 K, as its potential barrier is in the order of 2 kcal/mole. As mentioned in section 2.1, this is not observed experimentally. The ester-methyl group of polymethyl methacrylate undergoes hindered rotation even at 4 K, at the temperature of boiling helium, according to mechanical, nuclear magnetic, and thermal measurements. Similarly, in several amorphous and semicrystalline polymers, transitions are found at cryogenic temperatures.

The heat capacity and thermal conductivity of amorphous polymers also show characteristic anomalies which indicate that, besides acoustic vibrational modes, another kind of mobility is present in such polymers at very low temperatures.

At such low temperatures the 3-dimensional vibrations dominate in determining the cubic temperature dependence of the heat capacity.

At temperatures below 10 K, according to Debye (1921), the heat capacity is related to the sound velocity as

$$\frac{c_p}{T^3} = \frac{2\,\pi^2\,k^4}{15\,h^3}\left(\frac{1}{v_l^3}+\frac{2}{v_t^3}\right) \tag{2.38}$$

where v_l, v_t respectively are the longitudinal and transversal sound velocities.

This very simple approximation agrees rather well with the experiment in the case of crystalline polymers, but deviates considerably in amorphous ones (Reese, 1969).

In Fig. 2.22 the temperature dependence of the specific heat of highly crystalline polyethylene, amorphous polymethyl methacrylate and polystyrene are shown in the c_p/T^3 versus T representation, after Tucker and Reese (1967).

According to the Debye theory, c_p/T^3 should be constant. It is seen that for the amorphous polymers the deviation from the $c_p/T^3 = \text{const.}$ line is significant, indicating that besides the 3-dimensional (acoustic) vibrational modes other kinds of motion must be present. Such a deviation is typical in amorphous polymers; it may be attributed to local motion at structural defects. Indeed cryogenic relaxations in amorphous and partially crystalline polymers were found to depend on the thermal and thermomechanical history of the sample (Hiltner and Baer, 1972). Relaxation transitions observed near to 20−40 K in polystyrene, poly p-xylene, polyethylene, polyoxymethylene and in polyethylene

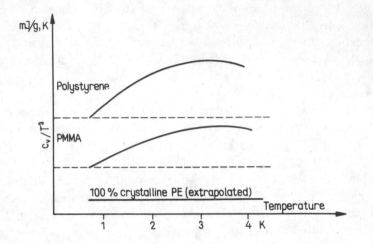

Fig. 2.22 The temperature dependence of the specific heat at very low temperatures for amorphous and crystalline polymers (after Tucker and Reese, 1967)

terephthalate were found to be dependent on annealing and on previous deformation.

Cryogenic relaxations are also found to be extremely sensitive to impurities (Sauer and Saba 1969), which also supports the view that these relaxations correspond to motion at the structural defects.

Figure 2.23 shows some dislocation mechanisms suggested by Hiltner and Baer (1972) for interpreting cryogenic transitions (δ-transitions) observed near 40 K in polyethylene polyoxymethylene and polyethylene terephthalate; (a) represents a screw-dislocation; (b) shows a possible translation of a jog in the linear chain and (c) shows a possibility of redistribution of kinks.

Screw dislocations will be discussed in detail in section 2.4 in connection with transitions in crystalline polymers.

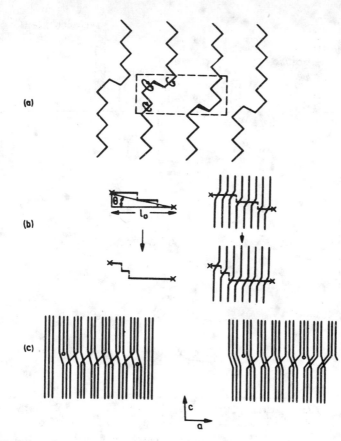

Fig. 2.23 Dislocation types for explaining cyrogenic relaxations in crystalline polymers;
(a).screw dislocation; (b) translation of a jog; (c) redistribution of kinks (from Hiltner
and Baer, 1972)

*Example of cryogenic T_{gg}-type transitions in amorphous polymers: poly-
styrene*

Styrene polymerized by a free radical mechanism produces an atactic,
amorphous polymer

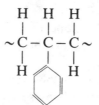

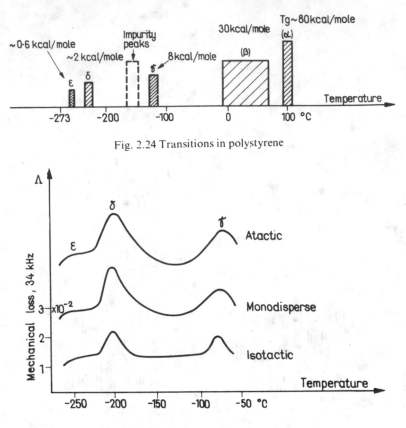

Fig. 2.24 Transitions in polystyrene

Fig. 2.25 Cryogenic transitions in polystyrene

commonly known as polystyrene. By using specific catalysts, crystalline, isotactic polystyrene can also be prepared.

The main transition regions in amorphous polystyrene are illustrated in Fig. 2.24. The apparent activation energies corresponding to Arrhenius-like temperature dependence are also shown.

α-transition is the glass–rubber transition having an apparent energy of activation of 80 kcal/mole; it is observed near 100°C. The β-transition is a broad peak observed between $-10°$C and $+60°$C; it is usually mixed-up with T_g and the corresponding activation energy is 30 kcal/mole.

The γ-transition is observed at about $-120°$C with an activation energy of 8 kcal/mole.

The cyrogenic δ-transition is observed near 40 K; it is shown in Fig. 2.25 for atactic, monodisperse (fraction) and isotactic poly-

114

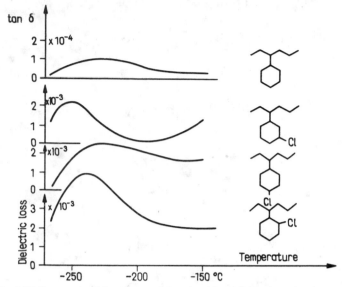

Fig. 2.26 Cryogenic dielectric transitions in chlorine substituted polystyrenes

styrenes. The dielectric and mechanical loss peaks coincide. The activation energy of the transition is in the order of 2 kcal/mole. The corresponding motion has been interpreted as being due to the wagging motion of the phenyl ring (Yano and Wada, 1971). This view is supported by the results obtained for chlorine-substituted polystyrenes, shown in Fig. 2.26. It is seen that by *ortho* and *meta* substitution the cryogenic δ-peak is considerably shifted to lower temperatures while it is practically unaffected by chlorine substitution of the *para*-ring hydrogen. Chlorine substitution evidently increases the dielectric loss peaks as the C−Cl bond is polar (Sauer and Saba, 1969).

Reich and Eisenberg (1972) calculated the hindered rotation of the phenyl ring in polystyrene by assuming 6-fold symmetric potential barrier by using the model of Fig. 2.27. The interaction potential of the shaded H-atoms has been calculated by using the empirical potential of Hirschfeld and Linnett (1950)

$$V(r) = 3.7164 \times 10^3 \exp(-3.0708r) - 82.52r^{-6} \qquad (2.39)$$

where $V(r)$ is the potential in kcal/mole, r is the intermolecular distance, Å.

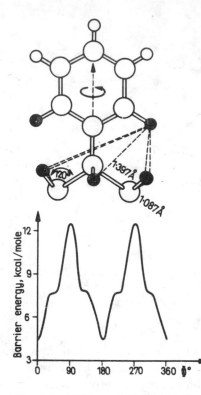

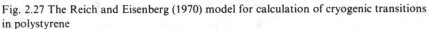

Fig. 2.27 The Reich and Eisenberg (1970) model for calculation of cryogenic transitions in polystyrene

The calculated potential as a function of the rotation angle ϕ is also shown in Fig. 2.27.

The probability for the rotator having an angular velocity between ω and $\omega + d\omega$ is (O'Reilly and Tsang, 1967)

$$P(\omega, \omega + d\omega) = \frac{\exp\left(\dfrac{E_k}{kT}\right)}{\displaystyle\int_0^\infty \exp\left(\dfrac{E_k}{kT}\right)\, d\omega} \qquad .(2.40)$$

where $E_k = \dfrac{\theta \omega^2}{2}$ is the kinetic energy at the top of the barrier, θ is the moment of inertia, T is the temperature.

116

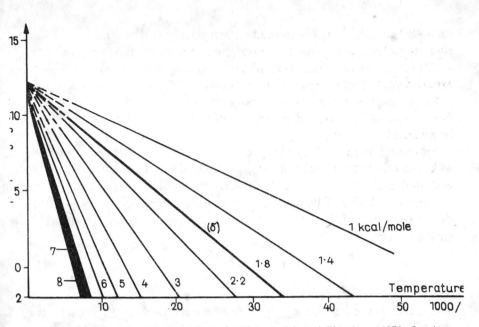

Fig. 2.28 Transition frequency map calculated by Reich and Eisenberg (1970) for ring rotation of polystyrene; numbers indicate barrier energies in kcal/mole

The mean angular velocity of the rotator is

$$\langle \omega \rangle = \frac{\omega \int_0^\infty \exp(-\theta \omega^2 / 2\, kT)\, \omega\, d\omega}{\int_0^\infty \exp(-\theta \omega^2 / 2\, kT)\, d\omega} \tag{2.41}$$

In an N-fold symmetric potential the rate of transition is

$$\nu_t = \frac{N(\omega)}{2} = \frac{B}{2\pi} \left(\frac{2\, kT}{\pi \theta} \right)^{1/2} \exp(-V_0 / kT) \tag{2.42}$$

where V_0 is the height of the barrier.

In the present case $\theta = 14.72\ 10^{-39}\ \mathrm{g/cm^3}$ considering the carbon–hydrogen and carbon–carbon distances shown in the figure.

Figure 2.28 shows $\ln \nu_t$ as a function of the reciprocal temperature according to Eqn. (2.42) for different barrier heights. One group of the experimental results coincides well with the shaded area corresponding

117

to 7—8 kcal/mole: this is the activation energy of the rotation of the ring in polystyrene. The corresponding transition is observed near $-120°C$; it is the γ-transition of polystyrene. Another group of the experimental results coincides with the 1.8 kcal/mole line of Fig. 2.28.

In order to interpret the low value for the activation energy of the δ-transition in polystyrene, a certain defectiveness of the structure is to be assumed.

Yano and Wada (1971) observed a very low temperature ϵ-peak at 34 kHz by the mechanical resonance technique in polystyrene. This peak is shifted from about 10 K up to 60 K by increasing the frequency from 34 kHz to 152 kHz. This peak has been tentatively interpreted as being due to motion of defects from one site to another. The rate constant of such a process is

$$k = \exp\left(-\frac{\Delta E_0}{kT}\right) \tag{2.43}$$

where ΔE_0 is the free-energy difference between two sites. The oscillator strength of the mechanical relaxation in such a case is expressed (Tsuge, 1964) as

$$G_\infty - G_0 = \frac{2E}{\pi k} G''_{max}\left(\frac{1}{T_1} - \frac{1}{T_2}\right) \tag{2.44}$$

where G_∞, G_0 are the unrelaxed and relaxed moduli, E is the activation energy of the process, and T_1, T_2 are the temperatures corresponding to $G_{max}/2$ values.

By considering two defect sites, the oscillator strength is expressed by Lamb (1964), as

$$\frac{G_\infty - G_0}{G_\infty} = \frac{NG_0}{kT}(\Delta v)^2 \frac{E}{(1 + \Delta E)^2} \tag{2.45}$$

where N is the concentration of the defects, Δv is the volume change of the system when the defect moves from site (1) to (2).

The assumption that the cryogenic relaxation of polystyrene is connected with structural defects is supported by the enormous effects of the additives on these peaks. In rigorously purified atactic polystyrene no dielectric transition is observed in the $\epsilon'' \geq 10^{-4}$ level (Nozaki et al., 1970). When the polymer contains traces of water a broad dielectric peak is observed at about 120 K, which is shifted to higher temperatures by increasing the frequency, but the apparent activation energy

118

of this shift is low. By addition of 2 % mineral oil to polystyrene in the mechanical spectrum a broad peak appears at about 120 K, as is shown in Fig. 2.29 (Armeniades *et al.*, 1970).

The heterogeneity of the structure of amorphous polystyrene has been revealed by Yeh (1972) by the electron-diffraction method. Yeh found

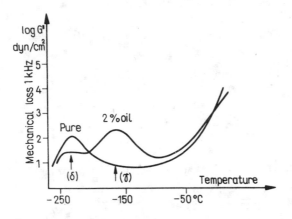

Fig. 2.29 The effect of impurity (mineral oil) on the mechanical γ-relaxation of polystyrene (after Nozaki and Shimada, 1970)

ordered regions ranging from 15 to 45 Å in size in the amorphous polymer. In these regions the chain segments are aligned parallel at a distance of about 9—10 Å. Such a short range order essentially involves high defect concentration.

Quantum tunneling

The apparent activation energy for the methyl group rotation in such polymers as polymethyl methacrylate, polyethyl methacrylate, polyoxy-methylstyrene, polyprophylene and in polyethylene oxide is found to be between 4 and 8 kcal/mole by dynamic mechanical and nuclear magnetic resonance studies (Eisenberg and Reich, 1969). Even by considering the effect of structural defects the potential barrier of the methyl-group rotation is definitely higher than 2 kcal/mole in most polymers. On the other hand, rather intense mechanical losses have been measured at very low temperatures, even below 4 K (see, for example, Sauer and Saba, 1969). An explanation of this has been suggested by Eisenberg and Reich (1969; see also Reich and Eisenberg,

1970). They assumed that the methyl group would undergo the hindering potential by quantum-mechanical tunnel effect.

According to simple quantum-mechanical calculations, a particle having a kinetic energy smaller than the potential barrier has a finite probability of penetrating it. The probability for penetration of a rectangular barrier of height V_0 and width d is

$$P \propto \exp[-2\,(V_0-E)^{1/2}\,d] \tag{2.46}$$

where E is the kinetic energy of the particle. If the potential barrier has an arbitrary shape

$$P \propto \exp\left[-2\int_{x=0}^{d} (V-E)^{1/2}\,dx\right] \tag{2.47}$$

The transition frequency for the methyl group rotation about the 3-fold symmetry axis, as in the case of the α-methyl group in poly methyl-methacrylate is given as (Das, 1957)

$$v = \frac{3}{2\pi}\left(\frac{V_0}{2\theta}\right)^{1/2}\exp\left(-\frac{V_0}{kT}\right) + \frac{2\pi}{9}\frac{\sum\limits_{n}\Delta v_n \exp\left(-\dfrac{hv_n}{kT}\right)}{\sum\limits_{n}\exp\left(-\dfrac{hv_n}{kT}\right)} \tag{2.48}$$

where v_n is the n-th vibrational energy level, $n = 0, 1, 2 \ldots$, Δv_n is the splitting of the n-th pair, it is proportional to the rate of tunneling, θ is the moment of inertia of the methyl group, V_0 is the height of the potential barrier.

The first term in Eqn. (2.48) represents the classical effect: jumping over the barrier. The second term represents tunneling. The tunneling rate Δv_n and the vibrational levels v_n can be expressed in terms of V_0 and θ (Stejskal and Gutowsky, 1958).

The transition frequencies as a function of reciprocal temperature calculated by Tanabe et al. (1970) are shown in Fig. 2.30 for four types of methyl groups: ester-methyl as in poly-methyl methacrylate—$V_0 = 2.4$ kcal/mole; end-methyl group of an alkyl chain—$V_0 = 3.9$ kcal/mole; α-methyl group attached to a tertiary carbon atom as in the case of polypropylene—$V_0 = 4.4$ kcal/mole; and for the α-methyl group attached to a quaternary carbon atom, as the α-methyl group in poly-methyl methacrylate—$V_0 = 6.4$ kcal/mole.

120

It is seen from Fig. 2.30 that tunneling becomes effective at $10^3/T \approx 10$, $T \leq 100$ K and is manifested by the reduction of the activation energy.

The basic problem in quantum tunneling is how the particle (in this case the methyl group) obtains the kinetic energy from the external force field. In the case of an electrical field it is probably a direct coupling

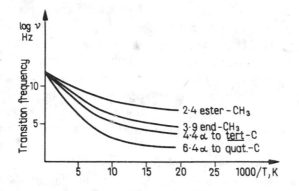

Fig. 2.30 Transition frequency map for methyl group rotation involving quantum tunnelling: numbers indicate barrier energies in kcal/mole (after Tanabe and Hirose. 1970)

between the field and the methyl group dipole. In the case of a mechanical-force field the solid structure, the lattice, is thought to be modulated, which means that the potential barriers are periodically deformed. The tunnel effect in mechanical relaxation is, therefore, mainly determined by intermolecular interactions.

According to this picture a significant difference between dynamic-mechanical, dielectric and NMR transitions should be observed. This problem has, however, so far not been solved because of lack of low-temperature data.

2.4 Transitions in crystalline polymers

As mentioned in section 2.1, there is no 100% crystalline polymer. Even in single polymer crystals a considerable amorphous background is found by the X-ray diffraction method. Correspondingly, by studying transitions involving change in molecular mobilities in crystalline polymers, it is difficult to separate the motion occurring in the amorphous phase from those of the crystalline phase.

From studies on low molecular weight inorganic and organic crystals (see, for example, Fox *et al.* 1964) it can be deduced that in the hypothetical perfectly ordered phase only local vibrations and rotations may occur. The spectrum of such vibrations can be approximately calculated from the temperature dependence of the specific heat. In section 2.3 the specific heat of polymers was discussed in terms of the simplest Debye (1921) model. It has been shown that in crystalline polymers at low temperatures the specific heat is well described by the simple Debye theory which predict T^3-dependence. For deducing information about lattice vibrations from specific heat data, however, more detailed theoretical analysis is needed. This problem has been discussed in detail by Tarasov (1950), Stockmayer and Hecht (1953) and Baur (1970, 1971).

Only the basic approach will be outlined here.

The specific heat of a solid due to harmonic lattice vibrations is generally expressed as

$$c_v(T) = k \int_0^{max} \left(\frac{h\nu}{kT}\right)^2 \frac{\varrho(\nu) \exp\left(\frac{h\nu}{kT}\right)}{\left[\exp\left(\frac{h\nu}{kT}\right) - 1\right]^2} \, d\nu \qquad (2.49)$$

where ν is the frequency of the lattice vibrations $\varrho(\nu)$ is the density of the vibrational states.

Equation (2.49) corresponds to the harmonic lattice-vibration approximation; for a more general treatment anharmonicity should also be taken into account.

By approximating the density of states by a power series of the modes, the heat capacity can be expressed in terms of Debye functions defined as

$$D_n\left(\frac{\theta_n}{T}\right) = n \left(\frac{T}{\theta_n}\right)^n \int_0^{\theta_n/T} \frac{x^{n+1}\exp x}{(\exp x - 1)^2} \, dx \qquad (2.50)$$

where $\theta_n = h\nu_n/k$ is referred to as the characteristic temperature.

The Tarasov (1950) approximation involves that the interaction along the polymer chains through the covalent bonds is much higher than the intramolecular interactions. Correspondingly at elevated temperatures the one-dimensional vibrations would dominate; the $\varrho(\nu)$ spectrum is thus approximated by a one-dimensional continuum.

At lower temperatures, when the intramolecular interactions begin to contribute appreciably, three-dimensional vibrations are considered. In this approximation, correspondingly, two characteristic temperatures are introduced, θ_1 and θ_3 corresponding to the one-dimensional and 3-dimensional vibrations respectively. The corresponding expression is

$$c_v = 3\,R\,D_1\left(\frac{\theta_1}{T}\right) + \frac{\theta_3}{\theta_1}\left[D_3\left(\frac{\theta_3}{T}\right) - D_1\left(\frac{\theta_3}{T}\right)\right] \qquad (2.51)$$

The vibration spectrum corresponding to the Tarasov approximation is shown in Fig. 2.31a; Eqn. (2.51) is referred to as the Tarasov formula. At low temperatures:

$$c_v = \frac{12\,\pi^4\,R}{5}\,\frac{T^3}{\theta_3^2\,\theta_1}, \qquad T \le \theta_3; \qquad (2.52)$$

at higher temperatures:

$$c_v = \frac{3\pi^2\,R}{2}\,\frac{T}{\theta_1}, \qquad \theta_3 \le T \le \theta_1 \qquad (2.53)$$

The Tarasov approximation thus predicts a T^3 dependence of c_v at low temperatures where the 3-dimensional approximation is valid. As shown in Fig. 2.22, this approximation is valid for crystalline polyethylene in the low temperature range below 15 K. At higher temperatures the Tarasov approximation, which predicts linear temperature dependence, fails. According to Baur (1970, 1971a) this disagreement with the experiment is due to the stiffness of the polymer chains, which makes transversal acoustic waves (phonons) effective. For a more detailed analysis the different vibrational modes—bending, stretching—are to be taken into account and also the corresponding acoustic waves phonons which propagate in the polymer anisotropically.

According to the calculations of Baur (1970, 1971a), the heat capacity at low temperatures is expressed as

$$c_v = a_3\,T^3 \qquad (2.54)$$

where $a_3 = 2.64 \times 10^{-5}$ cal/mole $\cdot$ degree^4 for crystalline polyethylene.

123

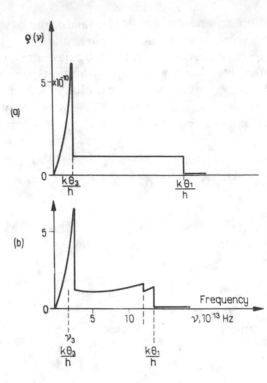

Fig. 2.31 Vibrational energy density spectra of polyethylene: (a) Tarasov theory; (b) experimental (after Baur, 1971)

This T^3-dependence follows also from the simple Tarasov model, and has been experimentally observed, as shown in Fig. 2.22.

At somewhat elevated temperatures, between 10 K and 50 K, the specific heat is

$$c_v = a_3 T^3 + a_n T^n \qquad (2.55)$$

where n is between 3/2 and 3; it decreases with increasing temperature.

The appearance of the T^n term is due to the contribution of the transversal phonons (bending vibrations).

At higher temperatures between 100 K and 200 K,

$$c_v = a_1 T + a_{1/2} T^{1/2} \qquad (2.56)$$

For polyethylene $a_1 = 1.08 \times 10^{-2}$ cal/mole degree^2, $a_{1/2} = 0.1186$ cal/mole degree$^{1/2}$. The additional term $T^{1/2}$ which appears in Eqn.

124

(2.56) with respect to the Tarasov approximation is also attributed to the effect of transverse phonons.

Figure 2.31 b shows the actual vibration density-spectrum of crystalline polyethylene obtained by the best fit with the c_v data, using the equations of Baur (Eqns 2.55 and 2.56). It is seen that the highest contribution to the spectrum is still due to the 3-dimensional vibrational modes which do not depend on the length of the molecule; they are approximately the same for the monomer or hydrogenated monomer and for the polymer. The rest of the spectrum is a continuum.

From the comparison of the heat-capacity data with lattice dynamical calculations it is concluded that mechanical or dielectric transitions are not expected to occur in perfect crystals. The experimental fact that many transitions are strongly dependent on the crystallinity is attributed to local motion at dislocations and defects.

Transitions due to defects in polymer crystals

The types of defects and dislocations in polymer crystals are not well known. The main reason for formation of defects in polymers is that they are never homogeneous in composition. Even in 'monodisperse' pure polymers the molecular weight is distributed along a certain range. A polymer is usually a mixture of molecules of different steric structure, stereo-isomers, so it is chemically inhomogeneous. In some polymers even side chains and branches are found distributed statistically (for example, low density polyethylene). Correspondingly, a polymer is to be regarded as being like a multi-component alloy (Lindenmeyer, 1972) rather than a uniform substance. The other factor which influences defect formation is the very low diffusion of polymers as a result of their high molecular weight. The diffusion constant of macromolecules is about $10^3 - 10^6$ times smaller than that of the small organic molecules. As a result, the size of ordered regions in polymers·formed after crystallization is much smaller, 10—100 Å, than those of the low-molecular-weight compounds. This essentially involves defect structure.

Possible defects and dislocations have been discussed by Predecki and Statton (1967). Although the mechanical and dielectric properties of crystalline polymers are known to be determined mainly by the defects and dislocations, this problem has not yet been investigated in detail. Very little theoretical work has been done in this field and the interpretation of the experimental results is rather vague. Instead of reviewing this field we select an illustrative example based on the pioneering work of Matsui *et al.* (1971): the theoretical interpretation of screw dislocations in polymer crystals, especially in polyethylene.

Screw dislocations in polymer crystals

The main sources of dislocations in polymer crystals are the chain ends. As the polymer molecules have different lengths the chain ends cannot be arranged regularly; they would be scattered randomly inside the crystal. The intrachain energy produced by the interaction among the units of the chain and the interchain energy produced by neighbouring chains are much lower at the chain end than in the perfectly ordered regions. This causes a strain resulting in formation of screw dislocations by deforming the neighbourhood of the chain end. The energy of such dislocations has been analyzed by Matsui *et al.* (1971) by assuming the intermolecular potential as being the Lennard–Jones potential

$$V(r) = -2r_0 \left(\frac{V_0}{r}\right)^6 + r_0 \left(\frac{V_0}{r}\right)^{12} \tag{2.57}$$

where r_0 and V_0 are constants and r is the distance between the interacting units. For a perfectly ordered linear array of polyethylene chains the values of the constants are $r_0 = 4.86$ Å, $V_0 = 6.62 \times 10^{-15}$ erg (Matsui *et al.*, 1971).

The potential energy of a unit in a perfectly ordered array of molecules is

$$V(l) = \int_{-\infty}^{+\infty} V(r)\, \varrho \, dr = -\frac{3\pi}{4}\varrho\, r_0 \left(\frac{V_0^6}{l^5} - \frac{21}{64}\frac{V_0^{12}}{l^{11}}\right) \tag{2.58}$$

where ϱ is the density of the units, l is the interchain distance.

Figure 2.32 shows some simplified cases to illustrate the formation of screw dislocations. In the hypothetical case (a) the structure is not affected by the presence of the chain end. This would result in a much smaller potential energy for the end group, and neighbouring groups, than for the others. As a result, the neighbouring chains are bent to form a dislocation row which travels up to another chain end or to the surface of the crystallite—Fig. 2.32 (b) and (c). The bending of the chains near to the chain end might involve partial rotation, i.e. local isomeric transformation (for example, *trans-gauche* isomerisation in polyethylene). This way a screw-like structure is formed which is known as the screw dislocation.

Matsui *et al.* (1971) calculated the change of the total potential energy due to different types of screw dislocations caused by chain ends. The

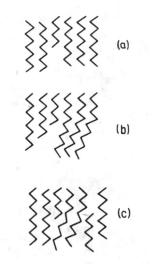

Fig. 2.32 Formation of srew dislocation by the perturbation of a chain-end (after Matsui *et al.*, 1971)

total energy change is generally expressed as

$$\Delta V = \Delta V_{inter} + \Delta V_{intra}$$

where ΔV_{inter} is the change of the intermolecular potential energy due to the bending, ΔV_{intra} is the change in the intramolecular energy.

The change of the intermolecular energy due to a screw dislocation is expressed as

$$\Delta V_{inter} = 1.73 \pi \varrho^2 r_0 l^* X_e + 0.060 \pi m \varrho^2 r_0 l^* X_i \qquad (2.59)$$

where l^* is the Burgers vector, X_e is the curved part of the chain in the vicinity of the chain end; the value of X_e for polyethylene is 14.7 Å corresponding to 11.6 CH_2-units, generally

$$X_e = 1.58 \times 10^2 \left(\frac{l^*}{\varrho} \frac{k_0}{r_0} \right)^{1/4} \frac{1}{\varrho} \qquad (2.60)$$

127

X_i is the curved part of the chains in two adjacent planes along the dislocation

$$X_i \propto \left(\frac{l^* k_0}{\varrho\, r_0}\right)^{1/4} \frac{1}{\varrho} \tag{2.61}$$

The values of X_i for polyethylene for two types of screw dislocations are $X_{i\ jogged} = 24.4\ \text{Å}$, $X_{i\ coupled} = 20.5\ \text{Å}$.

m in Eqn. (2.59) is the number of chain intersects along the screw dislocation, k_0 in Eqns (2.60) and (2.61) is the torsional spring force constant.

The change of the intramolecular energy caused by a screw dislocation is expressed as

$$\Delta V_{intra} = \frac{12\,n\,k_0\,l^{*2}}{\varrho^3\,X_i^3} \tag{2.62}$$

where n is the number of bent chains.

By analysis of the energies of different types of dislocations, Matsui et al. (1971) conclude that the screw dislocations are energetically most stable among the possible kinds of dislocations.

From the lengths of the chain bends X_e and X_i it is possible to estimate the anisotropy of Young's modulus at a dislocation, i.e. the strain introduced to the crystal by the dislocation. The length of the bend is

$$X \propto \left(\frac{l^* k_0}{\varrho\, r_0}\right)^{1/4} \tag{2.63}$$

Young's moduli parallel and perpendicular to the chain axis are respectively

$$E_{\parallel} \propto \frac{k_0}{\varrho\, l^{*2}}$$

$$E_{\perp} \propto \frac{\varrho^2\, r_0}{l^*} \tag{2.64}$$

$$\varrho\, X \propto (\varrho\, l^*)^{1/2} \left(\frac{E_{\parallel}}{E_{\perp}}\right)^{1/4}$$

This means that the strain field introduced by the dislocation is larger in the direction of the chain than in the perpendicular direction.

128

On the basis of this analysis, Matsui *et al.* (1971) attempted to interpret the γ-relaxation peak in polyethylene near $-120°C$ as a transition corresponding to a motion of screw dislocations. In polyethylene single crystals two γ-peaks, $γ_1$ and $γ_2$, are found (Arai *et al.* 1968). On oxidation with fumic nitric acid one of them $(γ_2)$ vanishes; the other $(γ_1)$ does not. On the basis of this it is concluded that the $γ_2$-peak corresponds to the motion of the disordered 'phase' between the crystal lamellae, while $γ_1$

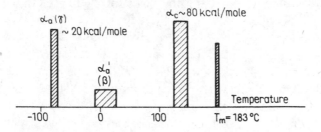

Fig. 2.33 Transitions in polyoxymethylene

is due to dislocations in the ordered phase. Low-temperature transitions due to the motion of dislocations have been detected in simple, non-polymeric, materials, such as aluminium at $-170°C$; it is referred to as the Bordoni peak (see Flügge, 1955). It seems that the various dislocation mechanisms which are rather well understood in metals and alloys can be applied to polymeric systems too, but with suitable reformulation for the specific case of macromolecules.

Example of dielectric absorption due to end-group rotation in crystals:
polyoxymethylene

Polyoxymethylene (polyformaldehyde) has a chemical composition of

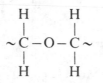

This polymer crystallizes with a hexagonal unit cell. The main transitions are shown in Fig. 2.33. The crystalline melting temperature is 183°C. Here the static dielectric constant $ε_0$, or the oscillator strength, changes abruptly by about one unit (Porter *et al.*, 1970). The α_c-relaxation appearing at about 130°C, well below T_m, is attributed to the crystalline phase. A weak β-transition is exhibited near 0°C and a strong γ-peak is observed near $-80°C$, by the mechanical and dielectric methods. The γ-peak has been found to decrease in intensity when density is increased (McCrum, 1961) and to be affected by plasticizing additives (dioxane; Read and Williams, 1961). The oscillator strength of the dielectric transition increases with increasing temperature. The thermal expansion coefficient shows abrupt change at this transition (Leksina and Novikova, 1959). All these features indicate that γ is a kind of structural transition. Read and Williams (1961) considered it as the glass–transition temperature of the amorphous phase.

Recent studies of Tanaka and Ishida (1972) on single crystals prepared by radiation polymerization of a monomer crystal show that the dielectric absorption peak is anisotropic. In the single crystal prepared in this way, practically no amorphous background could be detected, the γ-transition was, nevertheless, strong enough. The γ-transition was attributed to hindered rotation of the terminal hydroxyl groups near the defects caused by the chain ends:

$$
\begin{array}{ccc}
H & & H \\
| & & | \\
\sim C & - O - & C - OH \\
| & & | \\
H & & H
\end{array}
$$

This view is supported by infrared measurements which indicate dichroism due to the terminal $-OH$ groups.

Figure 2.34 shows two possible conformations of the POM chain ends as calculated by Tanaka and Ishida (1972) by considering Van der Waals potentials and coulombic interactions between the units shown in the figure. By assuming that the surrounding groups exhibit like-conformations in the trigonal cell, the potential energy map and barrier were calculated as shown in Fig. 2.35. The potential energy map shows the interaction energy contours as a function of the angles φ_3 and φ_4 (cf. Fig. 2.34) while the other angles are fixed, their values, determined by X-ray diffraction being $\varphi_1 = \varphi_2 = 78°$; $\varphi_5 = 0°$ by calculation.

The anisotropy of the dielectric loss is, according to this model, due to the different conformations corresponding to sites (A) and (B). The ratio of the dipole polarizabilities for an applied field parallel and

130

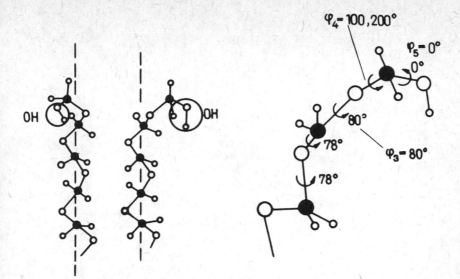

Fig. 2.34 Equilibrium conformations of the −OH chain ends in polyoxymethylene (after Tanaka and Ishida, 1971)

perpendicular to the axis of the helical chains is

$$a = \frac{\mathscr{P}_\perp}{\mathscr{P}_\parallel} = \frac{\mu_{A\perp}^2 + \mu_{B\perp}^2 + 2\,\mu_{A\perp}\mu_{B\perp}\cos\theta_0}{2(\mu_{A\parallel} - \mu_{B\parallel})} \qquad (2.65)$$

where θ_0 is the angle between the projection of the dipole moments for configurations A and B. The individual dipole moment projections can be calculated from the known bond moments and bond-angles. The result is $\mu_{A\perp} = 1.02$ D, $\mu_{B\perp} = 1.20$ D, $\mu_{A\parallel} = 0.85$ D, $\mu_{B\parallel} = 0.73$ D, $\theta_0 = 160°$. By using these values for the anisotropy parameter $a = 3.6$ is obtained.

Equation (2.65) has been calculated on the basis of the barrier theory using the potential shown in Fig. 2.35. The dielectric spectra measured by the external field in the 110 crystal axis and that perpendicular to it are shown in Fig. 2.36.

The interpretation of the γ-relaxation in POM as being due to the motion of terminal −OH groups seems to be supported by the Arrhenius plot shown in Fig. 2.37 in a wide temperature range up to the crystalline melting point. It is seen that at low temperatures a straight line is exhibited, while near to the melting range the ln v versus $1/T$ curve bends.

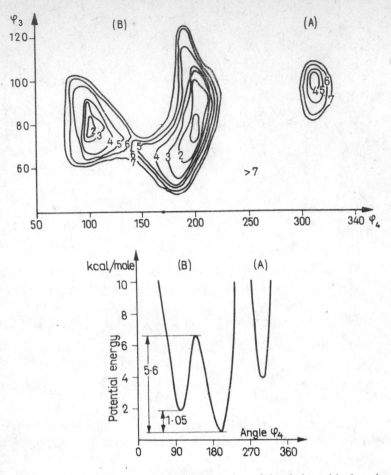

Fig. 2.35 Potential energy maps for hindered rotation of the chain-end hydroxyl group in polyoxymethylene; numbers on contour lines indicate barrier energies, in kcal/mole (after Tanaka and Ishida, 1971)

This can be explained by considering the deformation of the potential barriers for $-OH$ rotation. Porter *et al.* (1970) measured dielectric spectra of POM in the molten state and found transitions in the microwave 0.5—10 Gc/s region. The results could be well described by considering a single relaxation time which indicates the presence of a single dipole in a uniform environment. It is likely that the relaxation observed in the melt is also due to the chain end $-OH$ groups, while the contribution of the chain $-C-O-C$ dipoles is small.

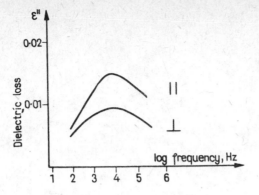

Fig. 2.36 Anisotropy of the γ-relaxation in polyoxymethylene (after Tanaka and Ishida, 1971)

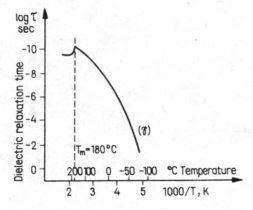

Fig. 2.37 The temperature dependence of the γ-relaxation time of polyoxymethylene in a wide temperature range (after Porter, 1970)

Effect of orientation

The dielectric as well as the mechanical relaxation spectra show considerable anisotropy in stretched polymers. Figure 2.38 shows, for example, the dielectric spectrum of stretched polyethylene terephthalate (PET) measured at different electric field direction with respect to the direction of stretching by Ward (1972). Mechanical relaxation spectra show similar anisotropy. From the figure it seems that the dielectric spectrum is anisotropic in the β-transition. This indicates that in the

133

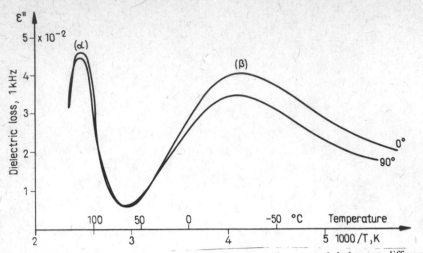

Fig. 2.38 The dielectric spectra of stretched polyethylene terephthalate at different orientations with respect to the draw direction (after Ward, 1973)

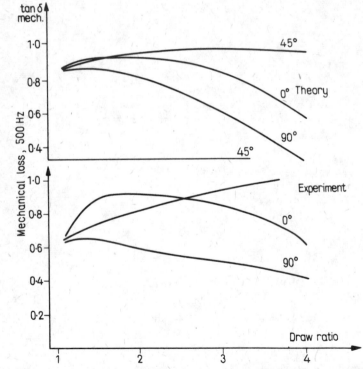

Fig. 2.39 Comparison of the calculated anisotropy of the dielectric loss in stretched polyethylene terephthalate with the experiment (Statchurski and Ward, 1969a, b)

134

amorphous phase to which the α-transition in PET is attributed the dipoles themselves are not oriented by stretching; only the internal field changes. The fact that the dielectric β-transition is anisotropic shows that in the crystalline phase dipole orientation due to stretching occurs.

In order to interpret the effects of stretching on the relaxation spectra it is assumed that the deformation does not change (orient) the molecules themselves, but the supermolecular structures, aggregates, formed by them. In the crystalline phase of polyethylene terephthalate or in polyethylene the lammellar structure is considered to be deformed. In the amorphous phase the aggregates might deform without the change of the molecular conformation. Correspondingly, in the amorphous phase no effect is to be observed from the chain dipoles. A considerable effect in the dielectric depolarization spectra is to be expected, however, but this has not been studied yet.

In the crystalline phase transitions connected with the crystallite surfaces or with defects are expected to become dielectrically anisotropic by stretching.

The anisotropy of the mechanical relaxation has been treated by Ward (1962) and Stachurski and Ward (1969) by assuming that there are units in the system which are unchanged upon stretching; only the aggregates formed by these units are deformed.

By considering a unit of an aggregate as a visco-elastic system the stress—strain relationship is expressed in a tensor form as

$$\gamma_i = s_{ij}\sigma_i$$

(2.66)

$$\sigma_k = c_{kl}\gamma_l$$

where s_{ij} are the components of the complex compliances, c_{kl} those of the moduli, σ_k is the stress component and γ_i is the strain component.

The complex compliance matrix is

$$s_{ij}^* = \begin{pmatrix} s_{11}^* & \cdots\cdots & s_{16}^* \\ \vdots & & \vdots \\ s_{61}^* & \cdots\cdots & s_{66}^* \end{pmatrix}$$

(2.67)

135

The complex modulus matrix has the same form. For an isotropic system

$$
s_{ij} = \begin{pmatrix}
s_{33}^* & s_{12}^* & s_{12}^* & 0 & 0 & 0 \\
s_{12}^* & s_{12}^* & s_{33}^* & & & 0 \\
s_{12}^* & s_{12}^* & s_{33}^* & 0 & 0 & 0 \\
0 & 0 & 0 & 2(s_{33}^*-s_{12}^*) & 0 & 0 \\
0 & 0 & 0 & 0 & 2(s_{33}^*-s_{12}^*) & 0 \\
0 & 0 & 0 & 0 & 0 & 2(s_{33}^*-s_{12}^*)
\end{pmatrix}
\tag{2.68}
$$

The macroscopic moduli are

$$
E^* = 1/s_{33}^*
\tag{2.69}
$$

$$
G^* = 1/2\,(s_{33}^* - s_{12}^*) = 2\,(c_{33}^* - c_{12}^*)
\tag{2.70}
$$

For a transversally isotropic solid, i.e. a polymer stretched in one direction

$$
s_{ik} = \begin{pmatrix}
s_{11}^* & s_{12}^* & s_{13}^* & 0 & 0 & 0 \\
s_{12}^* & s_{11}^* & s_{13}^* & 0 & 0 & 0 \\
s_{13}^* & s_{13}^* & s_{13}^* & 0 & 0 & 0 \\
0 & 0 & 0 & s_{44}^* & 0 & 0 \\
0 & 0 & 0 & 0 & s_{44}^* & 0 \\
0 & 0 & 0 & 0 & 0 & 2(s_{11}^*-s_{12}^*)
\end{pmatrix}
\tag{2.71}
$$

The corresponding macroscopic moduli for orientations parallel $0°$ and perpendicular $90°$ to the draw axis are

$$
E^*(0°) = 1/s_{33}^*
\tag{2.72}
$$

$$
E^*(90°) = 1/s_{11}^*
\tag{2.73}
$$

A further simplification is used by Statchurski and Ward (1969a,b) by regarding only the s_{44}^* component as complex (lossy), as this is the main factor determining the response of the transversally isotropic aggregate.

136

Thus the compliance matrix is

$$
s_{ik} = \begin{pmatrix}
s_{11}^{\circ} & s_{12}^{\circ} & s_{13}^{\circ} & 0 & 0 & 0 \\
s_{12}^{\circ} & s_{11}^{\circ} & s_{13}^{\circ} & 0 & 0 & 0 \\
s_{13}^{\circ} & s_{13}^{\circ} & s_{33}^{\circ} & 0 & 0 & 0 \\
0 & 0 & 0 & s_{44}^{*} & 0 & 0 \\
0 & 0 & 0 & 0 & s_{44}^{*} & 0 \\
0 & 0 & 0 & 0 & 0 & 2(s_{11}^{\circ} - s_{12}^{\circ})
\end{pmatrix}
\tag{2.74}
$$

Here the superscripts $^{\circ}$ indicate that the coordinate system is fixed within the aggregate. The draw axis is (3), the (3°) axis is the axis of the transverse isotropy of the unit, the angle between axes (3°) and (3) is θ.

By averaging with respect to θ, Ward (1962) expressed the averaged components of the aggregate in terms of the orientation functions

$I_1, I_2 \ldots I_5$ defined as

$$
I(\theta) = \int_0^{\pi/2} f(\theta') \sin \theta' \, d\theta'
\tag{2.75}
$$

where $f(\theta)$ is the orientation distribution function which describes deformation of an aggregate. The averaged components of the compliance tensor are

$$
\langle s_{33}^{*} \rangle_{av} = I_1 s_{11}^{\circ} + I_2 s_{33}^{\circ} + I_3 (2 s_{13}^{\circ} - s_{44}^{*})
\tag{2.76}
$$

$$
\langle s_{44}^{*} \rangle_{av} = (2 I_3 + I_4) s_{11}^{\circ} - I_4 s_{12}^{\circ} - 4 I_3 s_{13}^{\circ} + 2 I_3 s_{33}^{\circ}
$$
$$
+ \frac{1}{2} (I_1 + I_2 - 2 I_3 + I_5) s_{44}^{*}
\tag{2.77}
$$

$$
\langle s_{11}^{*} \rangle_{av} = \frac{1}{8} (3 I_2 + 2 I_5 + 3) s_{11}^{\circ} + \frac{1}{4} (3 I_3 + I_4) s_{13}^{\circ}
$$
$$
+ \frac{3}{8} I_1 s_{33}^{\circ} + \frac{1}{8} (3 I_3 + I_4) s_{44}^{*}
\tag{2.78}
$$

$$
\langle s_{12}^{*} \rangle_{av} = \frac{1}{8} (I_2 - 2 I_5 + 1) s_{11}^{\circ} + I_5 s_{12}^{\circ} + \frac{1}{4} (I_3 + 3 I_4) s_{13}^{\circ}
$$
$$
+ \frac{1}{8} (I_3 - I_4) s_{44}^{*} + \frac{1}{8} I_1 s_{33}^{\circ}
\tag{2.79}
$$

137

$$\langle s_{13}^* \rangle_{av} = \frac{1}{2} I_3 s_{11}^\circ + \frac{1}{2} I_4 s_{12}^\circ + \frac{1}{2} (I_1 + I_2 + I_5)\, s_{13}^\circ$$

$$+ \frac{1}{2} I_3 (s_{33}^\circ - s_{44}^*) \tag{2.80}$$

The orientation functions I_i depend on the draw ratio, so it is possible to express the dependence of the mechanical loss tangent as a function of draw ratio. The loss tangent in the direction of the stretching is

$$(\tan \delta)_{0^\circ} = (\tan \delta)_{33} = \frac{\langle s_{33}'' \rangle_{av}}{\langle s_{33}' \rangle_{av}} = \frac{I_3 s_{44}''}{I_1 s_{11}^\circ + I_2 s_{33}^\circ + I_3 (2 s_{13}^\circ + s_{44}')} \tag{2.81}$$

and that perpendicular to it

$$(\tan \delta)_{90^\circ} = (\tan \delta)_{11} = \frac{\langle s_{11}'' \rangle_{av}}{\langle s_{11}' \rangle_{av}} \tag{2.82}$$

The loss in a direction 45° from the direction of stretching is

$$(\tan \delta)_{45^\circ} = (\tan \delta)_{44} = \frac{\langle s_{44}'' \rangle_{av}}{\langle s_{44}' \rangle_{av}} \tag{2.83}$$

By using the aggregate model outlined above, Statchurski and Ward (1969 a, b) calculated the dependence of the mechanical $\tan \delta$ for different orientations as a function of the draw ratio of high-density polyethylene and compared it with the experimental results. Figure 2.39 shows such a comparison for orientations parallel and perpendicular to the direction of the stretching, and also for the $45^\circ -$ orientation. The $\tan \delta$ values are given in relative units assuming

$$[(\tan \delta)_{44}]_\infty = \frac{\langle s_{44}'' \rangle_{av}}{\langle s_{44}' \rangle_{av}} = \frac{s_{44}^\circ}{s_{44}^\circ} = 1 \tag{2.84}$$

for the most highly oriented specimen, where the subscript ∞ indicates infinite draw ratio. It is seen that by increasing draw ratio $(\tan)_{11}$ and $(\tan \delta)_{90^\circ}$ are predicted to decrease. The experimental curves shown in Fig. 2.39 have been determined by measuring the $(\tan \delta)_{45^\circ}$ value at extremely high draw ratio equal to unity. It is seen that the tendencies

138

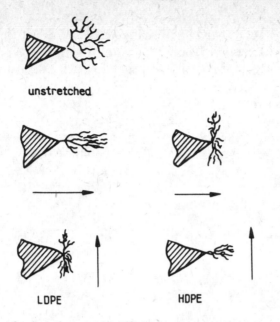

unstretched

LDPE

HDPE

Fig. 2.40 Effect of orientation on the high voltage treeing in high and low density polyethylene; arrows indicate draw direction (after Lobanov *et al.*, 1969)

predicted by the theory are well exhibited, but the absolute values are smaller for the experimental curves. This is due to the incorrectness of the averaging process used for evaluation of the compliance components.

Besides the effect of stretching on the deformation of the aggregates, the defect concentration is also highly influenced by stretching. On the boundaries of the crystallites and aggregates microcracks are formed, which have been observed by low angle X-ray diffractometry and by polarization microscopy as well (Martynov and Bylegzhanina, 1972). This heterogeneity introduced by stretching has a considerable influence on the d.c. properties, but is less effective on the high-frequency a.c. properties. It is therefore expected that from very-low-frequency Fourier transformation spectra important information about microcracks may be obtained.

Experimental evidence of the effect of microcracks has been published by Lobanov *et al.* (1969) in connection with high-field treeing (McMahon and Perkins, 1964) in oriented polyethylene. Treeing is partial deterioration of the polymer under high field strength formed near the point of a needle inserted in the polymer sample. The treeing patterns have been found to be highly anisotropic in stretched samples: the patterns of deterioration would follow the direction of stretching in polyethylene of low crystallinity but are perpendicular to it for the highly crystalline, high-density polymer. This is illustrated in Fig. 2.40, after Lobanov *et al.* (1969).

From X-ray electron-microscope studies it has been concluded that in the highly crystalline polymer microcracks would propagate perpendicularly to the direction of stretching, while for low-crystalline samples parallel to it. Treeing patterns would thus indicate the orientation of microcracks in stretched samples. The influence of the microcracks on the dielectric properties has not yet been studied. To draw possible parallels between the propagating of cracks and crazes in polymers, resulting in mechanical deterioration, and the electrical properties is very tempting.

CHAPTER 3

EXPERIMENTAL TECHNIQUE

Although the technique of measuring dielectric permittivity and loss is very old, only a few sophisticated recording methods have been developed and used for studying polymeric systems. A recording technique is needed not only in order to save time but because the transitions in polymers are dependent on the thermal history of the sample. It would be desirable to record spectra as a function of the frequency at different, fixed, temperatures, i.e. using the frequency as a variable and the temperature as a parameter. The frequency range to be swept through in such an experiment should be from about 10^{-3} Hz up to 10^9 Hz in order to obtain a reasonable spectrum. This cannot be done in one run. Only by the Fourier transformation method which will be discussed in section 3.1 is it possible to cover wide enough frequency ranges. This technique involves the use of a computer directly coupled to the spectrometer. By the conventional direct-current or alternating-current techniques only a small part of the frequency range can be swept through. Because of this technical difficulty, in practice the frequency is very often used as a parameter and the temperature as a variable. This method has the advantage that the dielectric spectra can be easily compared with mechanical relaxation spectra, and with thermomechanical, dilatometric and differential scanning calorimetric curves which are essentially recorded as a function of the temperature.

The use of the temperature as a variable has the disadvantage that the shape of the spectrum bands is difficult to interpret because, especially at structural transitions, the oscillator strength $\varepsilon_0 - \varepsilon_\infty$ changes abruptly at the transition temperature (see Chapter 2).

The classical bridge methods of measuring dielectric permittivity will not be discussed here. These methods may be found in the textbooks of von Hippel (1954), Smyth (1955), McCrum *et al.* (1969), Sazhin (1970), Daniel (1967), Hill *et al.* (1969), and Hedvig (1969c). The present discussion will be concentrated on the recording techniques by which spectra can be measured under well defined conditions in a reasonably short time. In section 3.4 some related techniques will be reviewed briefly, with special emphasis on their correlation with the dielectric spectra and with the possibility of combining them with the dielectric technique.

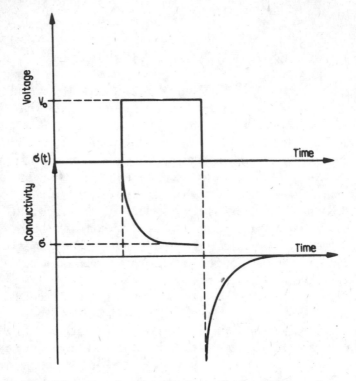

Fig. 3.1 Response of a polymer to the application of a step-voltage

3.1 Direct-current methods

In principle, the simplest way of measuring the dielectric properties of a material is to subject it to an electric field and see how the polarization develops with time. This problem has been theoretically treated in section 1.2, where the correlation between the response of a dielectric material to mechanical and electric force fields was discussed. The experimental procedure is very simple; a step-voltage is applied to the sample, as shown in Fig. 3.1, and the current (conductivity) is recorded as a function of time. A decreasing curve is obtained in this way until a

142

stationary level corresponding to the ohmic conductivity is reached. When the voltage is switched off a similar current response of opposite sign is recorded. A step-function represents a wide Fourier spectrum containing components of very low frequencies (depending on the duration of the recording) up to very high frequencies (depending on the rise time of the step-function). This means that the recorded current response contains all the information needed for constructing the dielectric spectrum in a wide frequency range from about 10^{-4} Hz up to 10^{10} Hz. In order to extract this information from the recorded curve a mathematical transformation should be used; the recorded current response $I(t)$ or transient conductivity $\sigma(t)$ is subjected to a Fourier type transformation to get the dielectric loss ε'' versus frequency curve, i.e. the dielectric spectrum. This method was first introduced by Hamon (1952) and developed further by Williams (1962).

According to the general correlation theory outlined in Chapter 1, the complex permittivity can be expressed in terms of the dipole autocorrelation function ϕ as

$$\frac{\varepsilon^*(i\omega)-\varepsilon_\infty}{\varepsilon_0-\varepsilon_\infty} = \int_0^\infty -\dot\phi \, \exp(-i\omega t) \, dt \qquad (3.1)$$

When the internal field correction is neglected, the dipole auto-correlation function $\phi(t)$ is equal to the normalized decay function $\varphi(t)$ which represents the decay of the polarization of the system as a function of time when the external field is removed. This decay function is related to the recorded current–time function $I(t)$ and transient conductivity $\sigma(t)$ as

$$\varphi(t) = \dot\phi(t) = \frac{I(t)}{C_0 V_0} = 4\pi \, \sigma(t) \qquad (3.2)$$

where C_0 is the capacitance of the sample and V_0 is the amplitude of the applied step voltage.

For the real and unreal parts of the dielectric permittivity

$$\frac{\varepsilon'(\omega)-\varepsilon_\infty}{\varepsilon_0-\varepsilon_\infty} = 4\pi \int_0^\infty \sigma(t) \cos \omega t \, dt \qquad (3.3)$$

$$\frac{\varepsilon''(\omega)}{\varepsilon_0-\varepsilon_\infty} = 4\pi \int_0^\infty \sigma(t) \sin \omega t \, dt \qquad (3.4)$$

143

Thus from the recorded $\sigma(t)$ curve the $\varepsilon'(\omega)$ and $\varepsilon''(\omega)$ functions can be determined by integration. The integrals, in general, cannot be solved analytically; for numerical integration computers should be used.

For evaluating the integrals, different approximation methods have been introduced by postulating a certain form for the current decay function $I(t)$, or transient conductivity $\sigma(t)$.

Hamon (1952) assumed that

$$I(t) = I_0 t^{-n} \tag{3.5}$$

where $0 < n < 2$.

This assumption is fulfilled only in a limited range of the decay. By using this form of $I(t)$

$$\frac{\varepsilon'(\omega) - \varepsilon_\infty}{\varepsilon_0 - \varepsilon_\infty} = \omega^{n-1} I_0 \Gamma(1-n) \sin \frac{n\pi}{2} \qquad 0 < n < 1 \tag{3.6}$$

$$\frac{\varepsilon''(\omega)}{\varepsilon_0 - \varepsilon_\infty} = \omega^{n-1} I_0 \Gamma(1-n) \cos \frac{n\pi}{2} \qquad 0 < n < 2 \tag{3.7}$$

where ω is the angular frequency which is related to time, I_0 is a dimensional constant and $\Gamma(1-n)$ is the Gamma function, which is tabulated (see, for example, Hodgman et al. 1958).
Hamon found that for

$$0.1 \leq n < 1.2 \tag{3.8}$$

Equations (3.6) and (3.7) can be simplified as

$$\varepsilon''(\omega) = \frac{I(t)}{C_0 V_0 \omega} = \frac{4\pi \sigma(t)}{\omega} \tag{3.9}$$

$$\varepsilon'(\omega) - \varepsilon_\infty = \varepsilon''(\omega) \tan \frac{n\pi}{2} \tag{3.10}$$

$$\omega t = \frac{\pi}{5} \tag{3.11}$$

Equations (3.9), (3.10), (3.11) are referred to as the Hamon equations. They were extensively used to convert current-decay functions into dielectric spectra when computer facilities were not so widely available.

144

By measuring the decay of the transient conductivity $\sigma(t)$ in $\Omega^{-1}\,cm^{-1}$ units the simplified Hamon transformation formulae are

$$\varepsilon''(v) = t\,\sigma(t)\,1.8 \times 10^{13} \tag{3.12}$$

$$v = \frac{0.1}{t} \tag{3.13}$$

where t is expressed in seconds, $\sigma(t)$ in $\Omega^{-1}\,cm^{-1}$ units and v in Hz.

Another approach for finding a simplified approximation for the integrals of Equations (3.3) and (3.4) is due to Williams (1962), who expressed the decay function $\varphi(t)$ from the Cole–Cole approximation for characterizing relaxation time distributions – see section 1.3

$$\frac{\varepsilon^*(i\omega) - \varepsilon_\infty}{\varepsilon_0 - \varepsilon_\infty} = (1 + i\omega\tau_0)^{-b} \tag{3.14}$$

The corresponding form of the decay function is

$$\varphi(t) = \begin{cases} \dfrac{\varepsilon_0 - \varepsilon_\infty}{\tau_0} \dfrac{1}{\Gamma(b)} \left(\dfrac{t}{\tau_0}\right)^{-(1-b)} & \text{for } \dfrac{t}{\tau_0} \ll 1 \\[3mm] \dfrac{\varepsilon_0 - \varepsilon_\infty}{\tau_0} \dfrac{b}{\Gamma(1-b)} \left(\dfrac{t}{\tau_0}\right)^{-(1+b)} & \text{for } \dfrac{t}{\tau_0} \gg 1 \end{cases} \tag{3.15}$$

The transformation formulae according to Williams (1962) are

$$\varepsilon''(\omega) = \frac{I(t)}{C_0 V_0 \omega} = \frac{\sigma(t)}{\omega} \tag{3.16}$$

$$\omega t = \left[\Gamma(1-z) \cos\frac{z\pi}{2}\right]^{-1/z} \tag{3.17}$$

where

$$z = \begin{cases} 1-b & \text{for } t/\tau_0 \ll 1, \ \omega\tau_0 \gg 1 \\ 1+b & \text{for } t/\tau_0 \gg 1, \ \omega\tau_0 \ll 1 \end{cases}$$

The Williams equations (3.16) and (3.17) are identical with the Hamon equations (3.9) and (3.11) for $1 < z \le 2$. The Williams approximation is useful when there is a transition in the frequency range of the decay and, correspondingly, the slope of the decay curve changes abruptly. Equations (3.16) and (3.17) are easy to program for an electronic computer.

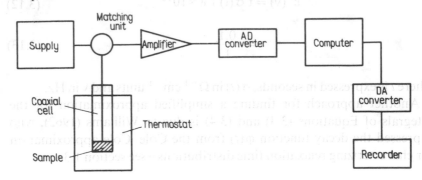

The Fourier-transform dielectric spectrometer (Time-domain-technique)

Although the principle of the Fourier-transformation method of dielectric spectroscopy has long been known, the large-scale application of the method has only recently gained ground (Fig. 3.2). A voltage pulse of short rise time, 12 nsec, is applied to the sample and the current response is measured by a fast-response electrometer. As the very short rise time represents the presence of very-high-frequency components, the sample is placed in a coaxial waveguide and is matched to the generator and to the electrometer. The electrometer, of course, has to have a large frequency band. The signal of the electrometer is digitalized by an analogue-digital (AD) converter and fed into a small computer, programmed to do the transformation according to Eqns (3.16) and (3.17). The transformed digital signal is then converted back to an analog signal by a DA-converter.

In this way, after the application of a step voltage of rise time in the order of 10 nsec, the device records the transformed spectrum, i.e. the frequency dependence of ε'' from 10^{-5} Hz up to the microwave region 10^{10} Hz. The very high frequency, microwave, band of this spectrum is not very reliable because the correct match is difficult to assure in such a wide frequency range. The method clearly has the great advantage that a complete dielectric spectrum can be recorded in a reasonably short time, depending on the lowest frequency components to be measured. The experiment can be repeated at different temperatures and thus a fair picture about the dielectric transitions is obtained.

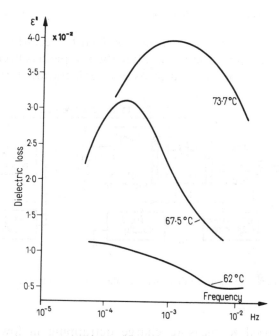

Fig. 3.3 Dielectric spectra obtained by the transformation of discharge current in unplasticized PVC below the glass–rubber transition region

The scheme of Fig. 3.2 can be simplified when the high-frequency components are omitted. In this case only an analog-digital converter and a puncher is to be used, coupled to the electrometer. The frequency limit of such a system depends on the speed of the puncher; it is usually in the order of 10^2—10^3 Hz. The punched tapes are then sent to a computer programmed to perform the transformation. The low-frequency range from about 10^{-5} Hz up to 10^3 Hz which can be covered by this simplified inexpensive method is especially interesting for polymers.

Figure 3.3 shows a typical series of transformed discharge curves for unplasticized polyvinylchloride at different temperatures. The discharge curve after charging in a field of 10 kV/cm for 1 hour has been recorded and transformed to $\varepsilon''(\omega)$ curves by using Eqns (3.12) and

(3.13). The ε'' peaks obtained this way correspond to the β-transition. The shift of the maxima to higher frequencies when the temperature is increased is clearly shown.

Dielectric depolarization spectroscopy

Very special kinds of spectra can be obtained by recording the short-circuit current during warming-up after the sample has been polarized at a constant d.c. field above a transition temperature. Originally this

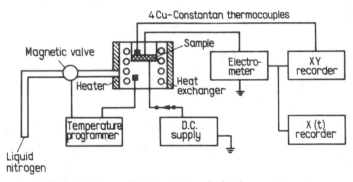

Fig. 3.4 Scheme of dielectric depolarization spectrometer

method was used to measure charge detrapping in low-molecular-weight organic and inorganic compounds. This method has been referred to as thermostimulated current or thermocurrent, or electric-depolarization current method. Only recently has this technique been applied to the study of structural transitions in polymers—see, for example, Van Turnhout (1971) and Hedvig (1973).

The scheme of a dielectric depolarization spectrometer is shown in Fig. 3.4. A high-voltage (1—3 kV) stabilized d.c. supply is used for polarizing the sample above its main transition temperature, usually above T_g. By a temperature program unit the virgin sample is heated up at a constant rate to the polarization temperature under an electric field of about 10 kV/cm. The sample is kept there for a certain time and then cooled down at a constant rate to −150°C. Here the external field is removed and an electrometer is connected to the sample to record the short-circuit current. The internal resistance of the electrometer must be much smaller than that of the sample. The time-dependent discharge current at −150°C levels off in a few minutes, so the warm-up program can be

148

started immediately to record the short-circuit current as a function of the temperature at constant heating rate.

The current peaks recorded this way are found to correlate well with the transition temperatures measured by mechanical relaxation or by conventional (a.c.) dielectric spectroscopy. The depolarization peak

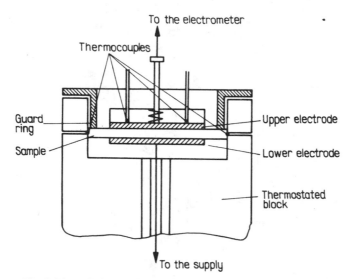

Fig. 3.5 Sample holder for dielectric depolarization spectroscopy

intensities (areas under the peaks) are highly dependent on the thermal history of the sample. Peak intensities are reproducible only if the temperature program is rigorously fixed.

In order to avoid electrode effects, vacuum-evaporated aluminium electrodes are used, deposited on the clean surface of the sample. Figure 3.5 shows a sample holder especially designed for depolarization work. The internal connection is at high voltage when the sample is polarized and is grounded when depolarization current is recorded. The upper electrode contacts the evaporated aluminium layer and is pressed against it lightly by a spring. A guard ring evaporated on to the upper surface of the sample contacts the aluminium ring of the cap, which is grounded. The insulator of the cap is of special quartz-filled epoxy resin. The temperature of the sample is measured by four copper-constantan thermocouples connected in series and positioned as shown in Fig. 3.5 in order to average the temperature. The upper part of the cap is covered with silicon grease in order to avoid moisture precipitation. The sample area be evacuated or subjected to inert-gas atmosphere from pipes running through the wall of the thermostating copper block.

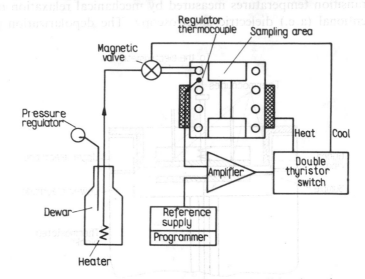

Fig. 3.6 Scheme of temperature programmer for dielectric work

Temperature programming

For recording dielectric depolarization spectra it is essential to have the temperature correctly programmed during the heat-up cycle, polarization, cool-down and second heating up. The correct temperature program is also desirable for the dielectric measurements when the temperature is chosen as a variable. Figure 3.6 shows the block scheme of a universal temperature program unit for the range from $-150°C$ to $500°C$. The sample holder is a copper block equipped with a heat-exchanger and an electric heater as shown in the figure. By use of an alternative thyristor switch, in one period the heater is switched on and the magnetic valve is off; in the other period the heater is off and a magnetic valve is on, which makes liquid nitrogen flow through the heat exchanger, where it is evaporated. The thyristor switch is driven by an integrated differential amplifier. One input of this amplifier is driven by the e. m. f. of a copperconstantan thermocouple, the other by a digital program-unit which supplies a reference signal. By means of this digital reference unit, practically any kind of program can be adjusted—linear heating and cooling, constant temperature, temperature steps, etc. The temperature of the block would follow the reference voltage supplied by the digital programmer with a time constant depending on its total heat capacity. For depolarization and conventional dielectric measurements this time

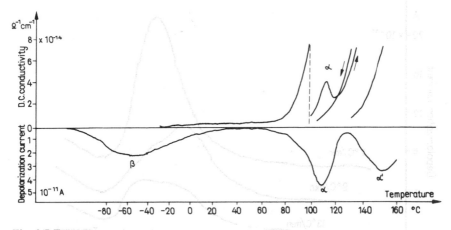

Fig. 3.7 Temperature dependence of the d.c. conductivity and dielectric depolarization spectrum of amorphous polymethyl methacrylate

constant is in the order of 5 min. Typical rates of heating in depolarization spectroscopy are 0.5, 1, 2 and 3°C/min.

Figure 3.7 shows a typical dielectric depolarization spectrum of amorphous polymethyl methacrylate (Solunov and Hedvig, 1972a, b).

First the temperature dependence of the conductivity was recorded when the sample was heated to the polarization temperature at a rate of 2°C/min. As this heating-up period is to be linear it is worth while to record the conductivity which supplies valuable additional information. The ohmic part of the conductivity should rise exponentially with the temperature. It is seen that it does except for the break occurring near T_g, which is due to polarization currents. After polarization at 150°C for 30 minutes in a field of 10 kV/cm the sample was cooled down at a rate of 2°C/min and then reheated at the same rate without external field; the depolarization current was then recorded. It is seen that this current is of opposite sign to that recorded in the heat-up period in field. The highest peak corresponds to the glass-transition temperature which can be measured by mechanical relaxation and by the conventional dielectric method. Another current peak observed near $-20°$ C which corresponds to the β-transition assigned to the rotation of the polar ester side group (see Fig. 2.3) is also shown in Fig. 3.7. The dielectric depolarization α-peak is more intense than the β-peak, while in the conventional (a.c.)

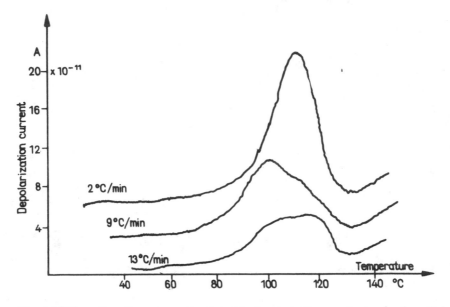

Fig. 3.8 Effect of cooling rate on the dielectric depolarization spectrum of polymethyl methacrylate in the T_g-region; rate of heating 2 C/min; polarization at 100 C with 10 kv/cm

dielectric spectrum the β-peak is more intense (cf. Fig. 2.4). A peculiarity of the depolarization method is shown in Fig. 3.8, namely the very strong dependence of the peak intensity on the rate of cooling. The spectra were recorded under identical conditions; only the rate of cooling from the polarization temperature (150° C) was subsequently increased. By increasing the cooling rate the polarization peak intensities decrease (Solunov and Hedvig, 1972 a, b).

Part of the dielectric depolarization currents may be attributed to thermal depolarization of the dipoles attached to the polymer chain. As all the depolarization peaks exhibit very strong dependence on the thermal history of the sample another mechanism has been suggested (Hedvig, 1972 a, b): depolarization of the dipoles of Maxwell–Wagner–Sillars (MWS) type formed at the structural heterogeneities of the

152

polymer. Another part of the depolarization current may arise from detrapping charge carriers which have been injected to the polymer from the electrodes during polarization and which have been trapped in the defects of the system. The measured depolarization spectrum is a superposition of these three effects. It is believed that at structural transitions involving change in free volume the Maxwell–Wagner–Sillars mechanism is determining.

By considering depolarization as being entirely due to the chain dipoles, the relaxation equation for the instantaneous dielectric polarization $\mathscr{P}(t)$ is

$$\frac{d\mathscr{P}(t)}{dt} = \frac{1}{\tau} (\mathscr{P}_0 - \mathscr{P}_\infty) \tag{3.18}$$

where τ is the relaxation time, $\mathscr{P}_0$ is the Boltzmann-equilibrium polarization and $\mathscr{P}_\infty$ is the instantaneous polarization. The temperature dependence of the relaxation time is assumed to be Arrhenius-like

$$\tau(T) = \frac{\tau_0}{\tau} \exp\left(\frac{E}{kT}\right) \tag{3.19}$$

By the depolarization experiment the temperature is swept linearly in time

$$T(t) = T_0 + qt \tag{3.20}$$

where T_0 is the starting temperature, q is the rate of heating.

The polarization Eqn. (3.18) is thus

$$\frac{d\mathscr{P}}{dT} = \frac{1}{q\tau_0} T(\mathscr{P}_0 - \mathscr{P}_\infty) \exp\left(-\frac{E}{kT}\right) \tag{3.21}$$

When the polarization decays, i. e. there is a change $\Delta\mathscr{P}$, a current flows through the resistance R_0 coupled to the sample. The voltage drop across the resistance R_0 produces a polarization of opposite sign to $\Delta\mathscr{P}$

$$\mathscr{P}_0' = \frac{N_0 \mu_{eff}^2}{3\,kT} \cdot \frac{3\,\varepsilon_0}{2\,\varepsilon_0 + 1}\, \mathscr{E} \tag{3.22}$$

which tends to maintain the original polarization. The depolarization current $I(T)$ is expressed as

$$I(T) = -Aq\,\frac{d\mathscr{P}}{dT} \tag{3.23}$$

Here A is the area of the sample, $\mathscr{E}$ is the field strength.
By using Eqns (3.23) and (3.20), Eqn. (3.18) becomes

$$\frac{d\mathscr{P}}{dT} = \frac{\mathscr{P}}{q\tau(1+c/\tau)} \qquad (3.24)$$

where

$$c = \chi_0 R_0 C_0$$

χ_0 being the macroscopic dielectric susceptibility of the sample, C_0 is the capacitance of the empty cell and R_0 is the input resistance of the electrometer.

Equation (3.24) can be easily solved for the limiting case

$$\frac{c}{\tau} \ll 1, \quad kT \ll E$$

This is fulfilled when after polarization the sample is cooled down to a low temperature T_0, where the depolarization rate is small. In this case

$$\mathscr{P}(T) = \mathscr{P}_0 \exp\left[-\frac{k}{qE\tau_0}\left\{T^3 \exp\left(-\frac{E}{kT}\right) - T_0^3 \exp\left(-\frac{E}{kT_0}\right)\right\}\right]$$

$$(3.25)$$

The depolarization current is

$$I(T) = BT \exp\left[-\frac{E}{kT} - \frac{k}{Eq\tau_0}T^3 \exp\left(-\frac{E}{kT}\right)\right] \qquad (3.26)$$

where the factor B is independent of the temperature.

At low temperatures the first exponential of the equation dominates, so

$$I(T) \approx T \exp\left(-\frac{E}{kT}\right) \qquad \text{for } T \ll T_{max} \qquad (3.27)$$

This makes it possible to estimate the activation energy of the 'depolarization process.

The maximum of the depolarization current is shifted to higher temperatures when the rate of heating q is increased. By recording the depolarization current peak at two different speeds q_1 and q_2 the activation energy of the process can be calculated from

$$\frac{E}{k} = \frac{\ln\left[\dfrac{T_m^3(1)}{q_1}\right] - \ln\left[\dfrac{T_m^3(2)}{q_2}\right]}{1/T_m(1) - 1/T_m(2)} \qquad (3.28)$$

Here $T_m(1)$, $T_m(2)$ are the temperatures corresponding to current maxima for heating rates q_1 and q_2 respectively.

The integral of the depolarization current over the total range of the temperature clearly should be equal to the total charge involved, i.e. the initial equilibrium polarization

$$\int_{T_0}^{\infty} I(T)\, dT = A\mathscr{P}_0 \qquad (3.29)$$

where A is the area of the sample.

The dielectric permittivity

$$\varepsilon_0' = 1 + 4\pi\chi_0 = 1 + \frac{4\pi}{A\mathscr{E}_0} \int_{T_0}^{\infty} I(T)\, dT \qquad (3.30)$$

where $\mathscr{E}_0$ is the polarizing field.

The effective dipole moment involved is

$$\mu_{eff}^2 = \frac{2\,kT}{N_0 \mathscr{E}_0 A} \int_{T_0}^{\infty} I(T)\, dT \qquad (3.31)$$

where N_0 is the concentration of the polar units.

Equation (3.31) is valid only if the effective dipole moment is unchanged at the transition, and if the transition is mainly determined by orientational polarization of the dipoles attached to the polymer chains. The very high depolarization currents found in apolar polymers and the strong dependence of the depolarization peaks on the thermal history suggest, however, that the main factor determining the depolarization peaks is the the Maxwell–Wagner–Sillars (MWS) polarization arising from the structural inhomogeneities. This problem has not yet been studied in detail. To the author it seems that in both polar and in nonpolar polymers the basic factor determining dielectric depolarization peaks is the Maxwell–Wagner–Sillars effect. As this effect is sensitive to structural inhomogeneities of the system the depolarization method offers a new possibility for studying defect structures.

3.2 Recording spectrometers in the low and intermediate frequency range (20 Hz to 2 MHz)

For studying transitions in polymers by dielectric spectroscopy the low frequency range is most useful. One reason for this is that by measuring transitions at high frequencies as a function of the temperature the dielectric loss peaks are shifted to high temperatures at which

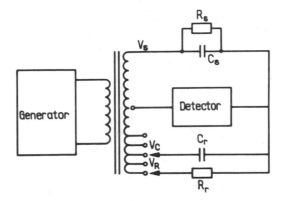

Fig. 3.9 Scheme of a transformer ratio arm bridge

chemical decomposition processes begin to be effective. On the other hand, for studying cryogenic transitions and problems connected with lattice vibrations the high-frequency methods are more useful. To the author it seems that for polymer work the very low frequency range and the very high frequency range are most important.

Automatically balanced bridges

The most straightforward way of constructing a recording dielectric spectrometer is to make a bridge automatically balanced. The most suitable type of bridge for that is the transformer ratio arm bridge (Cole and Gross, 1949) the scheme of which is shown in Fig. 3.9. The resistance and capacitance of the sample R_s and C_s are balanced by the reference elements R_r and C_r which are fed by different voltages V_R and V_c respectively. The conditions for the capacitive (90° out of phase) and resistive (in-phase) balance are

$$V_s C_s = V_C C_r$$
$$V_s R_s = V_R R_r$$

(3.32)

156

It is, correspondingly, possible to balance the bridge by varying the reference resistor R_r and capacitor C_r or by varying the voltages V_R and V_C.

The usual procedure is to balance the bridge first roughly by switching over the reference elements and then smoothly by changing the voltages V_R and V_C until complete balance is achieved.

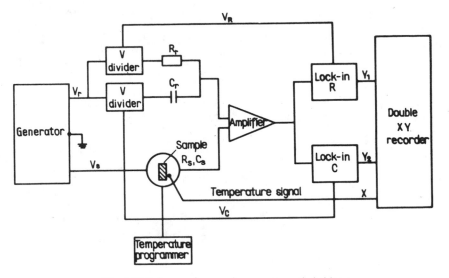

Fig. 3.10 Scheme of an analogue automatic bridge

An automatically balanced bridge based on this principle is shown in Fig. 3.10.

Automatic balance of the resistive component is done by regulating the voltage V_r through a series resistance (voltage divider) by means of a phase-sensitive lock-in detector, the phase of which is adjusted to currents arising from resistive unbalance (lock-in R in Fig. 3.10). The automatic balance of the capacitive current is achieved by a similar phase detector adjusted to the capacitive unbalance signal. The signal of lock-in R can be calibrated in conductivity units, that of lock-in C in ε'-units, as ε' is directly proportional to the sample capacitance. The changes in V_R and V_C are simultaneously recorded by a double XY recorder while the temperature is swept linearly. The X axis of the double recorder is driven by the emf of a thermocouple placed near the sample.

The variation of the voltages V_R and V_C can be done electro-mechanically simply by driving a potentiometer, and also electronically by varying the resistance of a vacuum tube or a transistor. The bridge is balanced manually by adjusting suitable series resistors and parallel capacitors to R_r and C_r respectively with the empty sample holder or with the sample at such a low temperature where it is practically lossless. The sample is then heated up linearly while the bridge is automatically

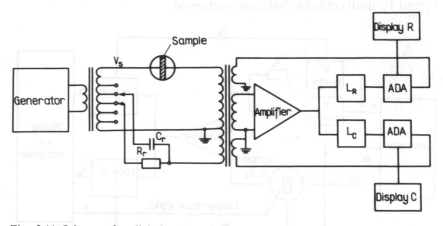

Fig. 3.11 Scheme of a digital automatic bridge; L_R. L_C: lock-in detectors; ADA: analogue-digital-analogue converter

kept in balance. The recorded signals in channels C and R then will be related to the change in ε' and in $\sigma = \sigma_0 + \dfrac{\omega}{4\pi}\varepsilon''$ respectively, where σ_0 is the ohmic conductivity, ω is the angular frequency and ε'' is the dielectric loss factor. The voltage in the sample arm V_s is kept constant.

The bridge shown in Fig. 3.10 is essentially a transformer ratio arm bridge of type shown in Fig. 3.9, but instead of a transformer two variable generator output voltages 90° out of phase are used.

The performance of the automatic balance can be highly improved by using digital circuits instead of the analog ones of Fig. 3.10.

In Fig. 3.11 an automatically balanced, digital, transformer ratio arm bridge is shown schematically. In this case toroid transformers are used for connecting the sample and reference resistors to the generator and also for detection of the residual current. Balancing is by a digital comparator which digitalizes the capacitive and resistive unbalance signals, separates them, and converts them back to a.c. and d.c. analog signals.

The a.c. analog signals are proportional to the unbalance signals; these are fed back to the comparator transformer in order to restore balance.

The d.c. analog signals are proportional to ε' and σ respectively; they are used to drive the XY recorder. The digital unit makes it possible to couple the spectrometer to a computer or to a puncher in order to analyze the spectra. The digital comparator also has the advantage that the bridge can be kept always exactly on balance, while in the analog device of Fig. 3.10 a considerable unbalance is always needed to produce signals large enough to record.

The capacitive unbalance signal V_C, which is recorded by the transformer ratio arm bridges, is related to the permittivity as

$$\varepsilon' = \frac{4\pi d}{A} \frac{C_r}{V_s} V_c \qquad (3.33)$$

where d is the sample thickness, A is the electrode area, C_r is the reference capacitance, V_s is the constant sample voltage.

The total conductivity is

$$\sigma = \sigma_0 + \frac{\omega}{4\pi} \varepsilon'' = \frac{d}{A} \frac{V_s}{R_r} \frac{1}{V_R} \qquad (3.34)$$

where R_0 is the reference resistance, V_R is the recorded signal.

From Eqns (3.33) and (3.34) it is seen that the ε' scale for the automatic bridge is linear while the σ-scale is hyperbolic. This also indicates the advantage of the digital bridge as the scaler in the R-channel of the bridge of Fig. 3.11 can be adjusted so the d.c. analog signal is proportional to σ or $\tan\delta$.

Yalof and Wrasidlo (1972) developed a recording dielectric spectrometer capable of recording the change in the sample capacity as percentage and the loss factor as a function of the temperature at two frequencies which are periodically switched over a certain range. A typical spectrum recorded this way is shown in Fig. 3.12 for amorphous poly 4—4' oxydiphenilene pyromellitimide. This is a special heat-resistant polymer developed by du Pont de Nemours (du Pont H-film). Its chemical structure is

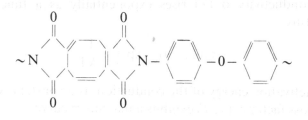

The recorded spectrum of Fig. 3.12 corresponds to frequencies of 110 Hz and 1 kHz, switched over while the temperature is scanned at a rate of 5° C/min. The change in capacity in percentage and the tan δ values are simultaneously recorded. Thus in one run four curves are obtained, ε' (110 Hz, T); ε' (1 kHz, T), tan δ (110 Hz, T) and tan δ (1 kHz, T). The measurement was made under nitrogen atmosphere by Wrasidlo (1972).

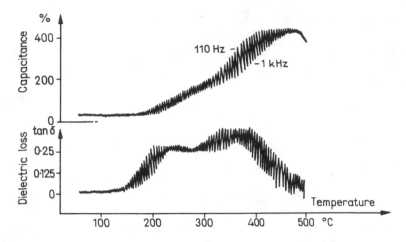

Fig. 3.12 Dielectric spectrum of du Pont's H-film measured by the automatic technique of Yalof and Wrasidlo (1972); frequency is swept from 100 Hz to 1 kHz

A direct recording dielectric spectrometer

When studying dielectric transitions in polar polymers, it is usually not necessary to use the automatic-bridge method because changes in ε' and tan δ or ε'' are large. Sometimes it is even inconvenient to use an automatic bridge because its range of balance is limited and at intense peaks it may run out of balance. Figure 3.13 shows a simple spectrometer for recording the permittivity ε' and the total conductivity $\sigma(\omega, T)$ as a function of the temperature at constant frequency ω (Hedvig, 1971b). The ohmic conductivity $\sigma_0(T)$ rises exponentially as a function of the temperature

$$\sigma_0(T) = \sigma_0^0 \exp\left(-\frac{E_c}{kT}\right) \tag{3.35}$$

E_c, the activation energy of the conduction, is a constant, while the dielectric loss factor $\varepsilon''(\omega, T)$ exhibits a maximum curve.

160

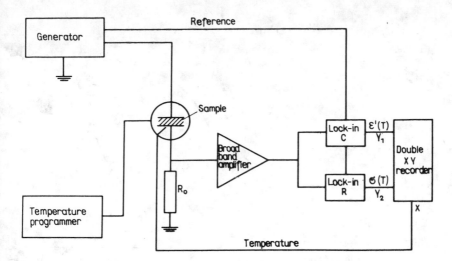

Fig. 3.13 Scheme of a direct recording dielectric spectrometer

The method is based on the phase separation of the current flowing through the sample. This current produces a voltage drop on a small (1 kΩ) series resistor. This voltage drop is amplified and phase detected. The in-phase component of this voltage is

$$V_R = R_r \, V_0 \, \frac{A}{d} \, \sigma \qquad (3.36)$$

where R_r is the reference resistor, A is the sample area, d is the sample diameter, σ is the total conductivity given by Eqn. (3.34), V_0 is the amplitude of the a.c. voltage.

The out-of-phase component is

$$V_c = R_r \, V_0 \, \frac{\omega}{4\pi} \, \frac{A}{d} \, \varepsilon' \qquad (3.37)$$

where ε' is the real part of the permittivity. According to Eqns (3.36) and (3.37) the output signals of the phase-sensitive lock-in detectors shown in Fig. 3.13 can be calibrated in terms of the permittivity ε' and total conductivity σ. The correct phase shift of the lock-in detectors can be obtained simply by replacing the sample alternatively by a capacitor and by a resistor and adjusting the gain of the corresponding channel to fit a prefixed scale.

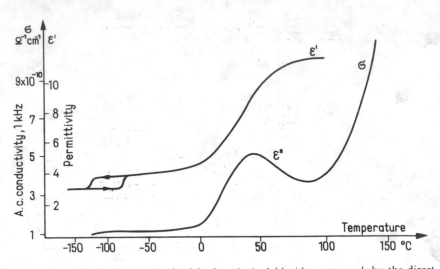

Fig. 3.14 Dielectric spectra of plasticized polyvinylchloride measured by the direct recording technique at 1 kHz

Figure 3.14 shows a typical pair of spectra recorded by using the device of Fig. 3.13 as a function of the temperature for a typical plasticized polyvinylchloride compound. The $\varepsilon'(T)$ curve is a typical dielectric dispersion curve.

The $\sigma(T)$ curve is a superposition of the dielectric loss maximum $\varepsilon''(T)$ and the exponentially increasing ohmic conductivity. From the recorded $\sigma(T)$ curve, correspondingly the T_g-transition of plasticized PVC is shown at the σ-maximum and the activation energy of the ohmic conduction according to Eqn. (3.35) can also be determined. The loss factor is expressed as

$$\varepsilon'' = \frac{2.26}{\nu}\,(\sigma - \sigma_0) \times 10^{13} \tag{3.38}$$

where σ and σ_0 are measured in $\Omega^{-1}\,cm^{-1}$ units, ν in Hz.

As the dielectric loss maximum increases linearly with increasing frequency the loss peak is more separated from the conductivity.

As discussed in Chapter 1, one component of the permittivity is usually enough for characterizing the transition. From the $\varepsilon'(T)$ curves measured at different frequencies the oscillator strength $\varepsilon_0 - \varepsilon_\infty$, the transition temperature and activation energy can be determined. The additional

162

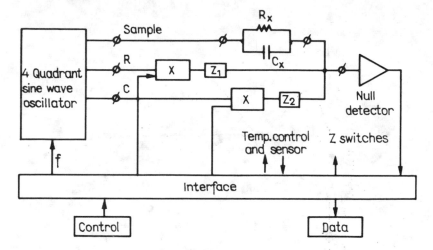

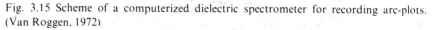

Fig. 3.15 Scheme of a computerized dielectric spectrometer for recording arc-plots. (Van Roggen, 1972)

information gained by recording the temperature dependence of the conductivity is in many cases very valuable. This will be discussed in some detail in Chapter 7 in connection with degradation of PVC.

A computerized set-up for recording arc-diagrams

This device was constructed by van Roggen (1972). It is essentially an automatic bridge fed by an oscillator, the frequency of which can be varied according to a pre-set programme. The block scheme of the apparatus is shown in Fig. 3.15. The generator has 3 outputs; one is for the sample circuit, the amplitude of which is constant (stabilized).

The other two outputs, marked by R and C, are shifted 90° out of phase with respect to each other. The amplitude of outputs R and C are individually controlled by the computer. The currents flow through the sample and through the reference impedances Z_1 and Z_2 which are also controlled, i.e. switched over, by the computer. The mixed signal is amplified and fed into the computer where it is phase-separated and digitalized.

The device can be programmed so that it scans through a frequency range in pre-set steps to record the arc diagram at a given temperature and repeat the measurement at different temperatures. It is, of course, useful

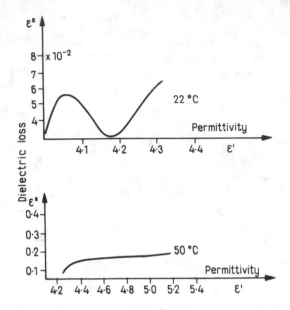

Fig. 3.16 Arc-diagrams of 66-Nylon recorded by van Roggen's computerized spectrometer

to record the ε' and ε'' components separately as a function of the temperature too. When the measurement has been done in a wide frequency and temperature range by using the data stored in the computer the activation energy of the transition can be evaluated by plotting the transition frequencies against $1/T$.

The computerized method opens up a possibility for quantitative study of relaxation-time distributions and changes in these distributions when passing through a transition temperature.

Figure 3.16 shows arc diagrams for 66-nylon measured by the computer technique (van Roggen, 1972) at different temperatures.

3.3 Dielectric work at high frequencies

The bridge method of measuring dielectric permittivities is useful in the frequency range from 0.1 Hz up to about 10 MHz. The direct phase-separation technique can be used also approximately in this frequency range, mainly from 20 Hz up to 1 MHz. At higher frequencies up to the sub-millimeter wave region (10^{12} Hz) the resonance method can be used. The principle of the resonance method is shown in Fig. 3.17. The signal generator feeds a series LC resonant circuit having an inductance L_0, resistance R_0, and variable capacitance C_0. The sample represented by parallel capacitance C_s and resistance R_s is coupled parallel to the variable capacitor of the device. The input signal amplitude and that appearing at the sample is measured by meters M_1 and M_2.

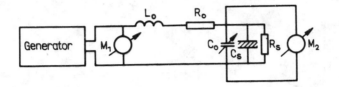

Fig. 3.17 Principle of the resonance method for measuring dielectric permittivity and loss at high frequencies

The series RLC resonant circuit is set to resonance without the sample. The loss factor of the circuit with the empty sample holder is

$$Q_0 = \frac{\omega L}{R_0} \tag{3.39}$$

and that with the sample is

$$Q_s = \frac{\omega L}{R_0 + R_s} \tag{3.40}$$

From Eqns (3.39) and (3.40) it follows that

$$\frac{1}{Q_s} - \frac{1}{Q_0} = \frac{\Delta Q}{Q^2} = \frac{R_s}{\omega L} = \frac{4\pi}{3} \frac{C_s^0}{C_s + C_0} \sigma \tag{3.41}$$

where $Q = Q_0 - Q_s$, C_s^0 is the capacitance of the empty sample holder, σ is the total conductivity of the sample expressed by Eqn (3.34), L is the inductance of the resonant circuit.

From Eqn. (3.41) it is seen that if $\Delta Q \ll Q^2$ the conductivity of the sample is proportional to the change of the Q-value of the circuit. The permittivity is simply expressed as

$$\varepsilon' = \frac{C_s}{C_s^0} \tag{3.42}$$

165

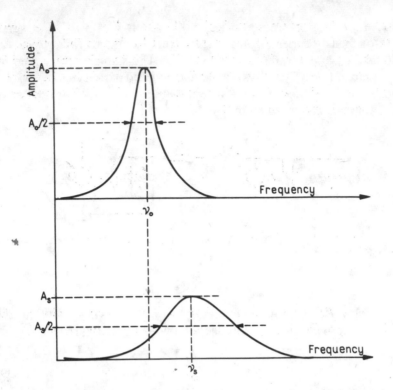

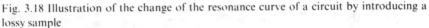

Fig. 3.18 Illustration of the change of the resonance curve of a circuit by introducing a lossy sample

The Q-value of the circuit can be measured accurately by plotting the resonance curve of the circuit with and without the sample. The resonance curves in the absence and in the presence of the sample are shown in Fig. 3.18. The Q-value of the circuit is defined as

$$Q = \frac{\nu_0}{\Delta \nu} \qquad (3.43)$$

where $\Delta \nu$ is the half-width of the resonance curve as indicated in Fig. 3.18, ν_0 is the resonance frequency. By inserting the sample ν_0 and $\Delta \nu$ are increased and the maximum amplitude is reduced, as shown in Fig. 3.18.

By measuring dielectric permittivities and losses after insertion of the sample the circuit is set to resonance again by changing the variable capacitor C_0, leaving the resonance frequency ν_0 unchanged. It is possible to calibrate the reading of meter M_2 in terms of Q-values, as the maximum amplitude is related to the half width $\Delta\nu$ of the resonance curve. This makes it possible to construct an automatic device for measuring ε' and σ as a function of the temperature in the frequency range between 10 MHz and 100 MHz by the resonance method. The block scheme of such a spectrometer is shown in Fig. 3.19.

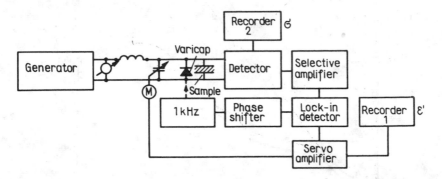

Fig. 3.19 Scheme of an automatic resonance spectrometer for high frequency dielectric work

The resonant circuit is essentially the same as that of Fig. 3.18 with the exception that an additional small variable capacitor (varicap) or a capacitor controlled by an electromechanical mechanism is inserted in order to frequency modulate the signal at low frequencies (1 kHz). In this way a 1-kHz signal is produced at the detector, the phase and amplitude of which is dependent on the tune of the resonance circuit. If it is on resonance the signal is very small and it is increased if the circuit is shifted off resonance. The error signal obtained this way is fed into a phase-sensitive lock-in detector and through a servomechanism it adjusts the capacitor C_0 so that the system is always kept on resonance.

The d.c. signal appearing on the detector is separately amplified and recorded. This d.c. signal can be calibrated in terms of Q-values, i.e. in terms of conductivities σ. The capacitance change ΔC_0 which is needed to keep the circuit on-resonance is proportional to the change of $\Delta\varepsilon'$

$$\frac{\Delta C_0}{C_s^0} = \Delta\varepsilon' \tag{3.44}$$

and can be recorded separately.

167

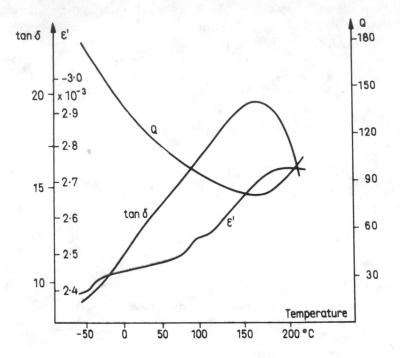

Fig. 3.20 Dielectric spectrum of polyethylene terephthalate measured by the automatic resonance technique at 30 MHz as a function of the temperature

Figure 3.20 shows a typical pair of curves measured by the automatized resonance method for polyethylene-terephthalate at 30 MHz, as a function of the temperature. According to Eqn. (3.44) the $\varepsilon'(T)$ scale is linear; it is directly proportional to the change of the capacitor C_0 which is directly recorded. The calibration of the $\sigma(T)$ scale is somewhat complicated as the recorded change in the output voltage contains the variation of ε' through the C_s term appearing in Eqn. (3.41).

Measurement at microwave frequencies

In the frequency range between 10^8 Hz and 10^{11} Hz the conventional *RLC* circuits are no longer useful; waveguides and resonators are used. The electric $\mathcal{E}$ and magnetic $\mathcal{H}$ field vectors in a coaxial waveguide or in free-space TEM mode are expressed as

$$\mathcal{E} = \mathcal{E}_0 \exp(i\omega t - \gamma^* z)$$
$$\mathcal{H} = \mathcal{H}_0 \exp(i\omega t - \gamma^* z)$$

(3.45)

where ω is the angular frequency, z is the direction of the propagation, γ^* is the complex propagation constant

$$\gamma^* = \gamma' + i\gamma''$$

where γ' describes the attenuation of the wave, $\gamma'' = \dfrac{2\pi}{\lambda_g}$ is the propagation constant, λ_g is the wavelength inside the waveguide. The TEM mode in a coaxial waveguide is shown in Fig. 3.21a. It is seen that

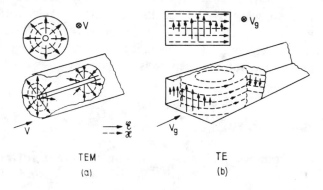

TEM
(a)

TE
(b)

Fig. 3.21 Typical microwave waveguides

the $\mathscr{E}$ and $\mathscr{H}$ vectors are both perpendicular to the direction of propagation: the TEM mode is transversal (Transversal Electric and Magnetic mode).

The propagation constant is directly related to the permittivity of the material with which the waveguide is filled:

$$\varepsilon' = \left(\frac{\gamma''}{\gamma_e''}\right)^2 \left(1 - \frac{\gamma'^2}{\gamma''^2}\right) \tag{3.46}$$

$$\varepsilon'' = 2\left(\frac{\gamma''}{\gamma_e''}\right)^2 \frac{\gamma'^2}{\gamma''^2} \tag{3.47}$$

Here γ_e'' is the unreal propagation constant of the empty waveguide $\gamma_e'' = 2\pi/\lambda_g$. To the TEM mode the free-space wavelength is equal to that in the waveguide $\lambda_0 = \lambda_g$. Figure 3.21 b shows another propagation mode in a rectangular waveguide. In this case one of the field vectors $\mathscr{H}$ is parallel with the direction of the propagation, the other $\mathscr{E}$ is perpendicular. It is usual to denote the propagation modes by specifying which

169

vector is perpendicular to the direction of propagation. Thus TE means Transversal Electric mode, i.e. the electrical field is perpendicular; TM means Transversal Magnetic mode when the magnetic field vector is perpendicular to the direction of propagation.

The wavelength inside a waveguide (λ_g) is longer than that in the free space (λ_0)

$$\frac{1}{\lambda_g^2} = \frac{1}{\lambda_0^2} - \frac{1}{\lambda_c^2} \tag{3.48}$$

Here λ_c is referred to as the cutoff wavelength; the given mode cannot propagate in the given waveguide if its free space wavelength is larger than the cutoff wavelength, $\lambda_0 > \lambda_c$. The value of the cutoff wavelength depends on the dimensions of the waveguide. For a rectangular waveguide in the TE mode the cutoff wavelength is

$$\lambda_c = 2 \left[\left(\frac{m}{a} \right)^2 + \left(\frac{n}{b} \right)^2 \right]^{-1/2} \tag{3.49}$$

where m and n are integer numbers. According to the values of n and m the corresponding propagation mode is denoted as TE_{nm}, a and b being the cross-sectional dimensions of the waveguide. In Fig. 3.21 b, for example, the TE_{01} mode is shown.

In waveguides the complex dielectric permittivity of the material by which the waveguide is completely filled is

$$\varepsilon^* = \left[\frac{1}{\lambda_c^2} - \left(\frac{\gamma^*}{2\pi} \right)^2 \right] \left[\frac{1}{\lambda_c^2} + \frac{1}{\lambda_g^2} \right]^{-1} \tag{3.50}$$

Measurement of the complex permittivity therefore involves measurement of the complex propagation constant of the dielectric-filled waveguide.

The propagation constant is not difficult to measure; see von Hippel (1954) and Hill *et al.* (1969). Considerable computational difficulties arise when only part of the waveguide is filled with the material to be measured. The propagation constant then depends not only on the permittivity but on the shape and dimensions of the sample as well. The propagation constant within the sample in such cases can be determined by measuring the standing waves formed by reflexion of the wave from a short-circuit plunger with the sample present and then absent. The method is schematically shown in Fig. 3.22. The standing waves are measured by a movable sonde which produces at the crystal detector a signal which is proportional to the square of the intensity of the electric field. By moving the sonde along the slotted line the detector shows minima and maxima. The ratio of the minimum to maximum voltages at the detector is referred to as the standing wave ratio r. The complex propagation

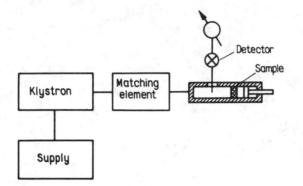

Fig. 3.22 Scheme of the standing wave ratio method for measuring dielectric permittivity and loss at microwaves (von Hippel, 1954)

constant of the sample is expressed in terms of the measurable standing wave ratio r, minimum position z_0, and sample thickness d by the following transcendental equation

$$\frac{\tanh(\gamma^* d)}{\gamma^* d} = -\frac{i\lambda_g}{2\pi d} \frac{r - i \tan\left(\frac{2\pi z_0}{\lambda_g}\right)}{1 - ir \tan\left(\frac{2\pi z_0}{\lambda_g}\right)} \tag{3.51}$$

Once z_0, r and λ_g are directly measured by the slotted line d, is known from Eqn. (3.51); the complex propagation constant γ^* can be calculated and then the complex permittivity can be determined from Eqn. (3.50).

The solution of the transcendental Eqn. (3.51) for different parameters are given by von Hippel (1954) and Dakin and Works (1947) in nomograms.

Resonance method at microwave frequencies

Very high sensitivity can be achieved in the microwave region by using cavity resonators. A typical TE_{012} mode rectangular cavity is shown in Fig. 3.23a. The microwave power is fed into the resonator through a cou-

171

pling hole. The length of the resonator can be changed by moving a short-circuit plunger. The sample can be placed right at the plunger. This resonator is simply made of a piece of waveguide. Its quality factor Q is 2—3000. Very-high-quality cavity resonators can be constructed by utilizing the TE_{011} or TE_{012} cylindrical mode. Such a resonator is shown

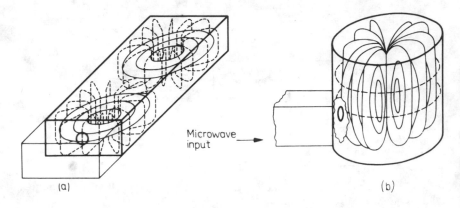

Microwave input →

(a) (b)

Fig. 3.23 Rectangular and cylindrical microwave cavity resonators

in Fig. 3.23b. The TE_{01n}-mode cylindrical resonators have the highest Q-values in the order of 10.000 of all the modes.

For a cylindrical TE_{01n}-type resonator partially filled with a dielectric sample so that all the cross section is filled-up the propagation constant —imaginary part—can be calculated from the following equation

$$\frac{\tan{(\gamma'' d)}}{\gamma''} + \frac{\tan{\gamma''_e (1-d)}}{\gamma''_e} = 0 \qquad (3.52)$$

Here d is the sample thickness, l is the length of the resonator at resonance, γ''_e is the propagation constant in the empty (air-filled) resonator.

The real part of the complex permittivity is

$$\varepsilon' = \frac{(\gamma'')^2 + a^2}{(\gamma''_e)^2 + a^2} \qquad (3.53)$$

where r_0 is the radius of the cylindrical resonator, $a = 3.832/r_0 \cdot \gamma''_e$ is the propagation constant of the empty resonator $\gamma''_e = 2/\lambda^0_g$ where $\lambda^{\tilde{e}}_g$ is the

172

wavelength inside the cavity without the sample; it can be calculated from the dimensions of the resonator or can be measured by moving the plunger to the next resonance.

For determining the loss factor, the Q-values of the resonator when empty and with the sample in position are measured. The Q-value can be measured by changing the resonator length until half value of the output signal is reached

$$Q = \frac{\Delta l}{l}$$

$$Q_e = \frac{\Delta l_e}{l_e}$$

(3.54)

where Δl is the variation of the resonator length for obtaining half output signal. The loss tangent is expressed as (Horner et $al.$ 1946)

$$\tan \delta = (\gamma_e'')^2 \left[2\Delta l - \frac{s' \Delta l_e}{2r_0 (\gamma_e'')^2 + l_e a^2} \right] [p(2d-s)(\gamma''^2 + a^2)]^{-1} \quad (3.55)$$

where
$p = \sin^2 \gamma_e''(1-d)/\sin^2 \gamma'' d; s = \sin 2\gamma'' d/\gamma''$,
$s' = a^2 [p(2d-s) + 2(1-d) - q] + 2r_0 [p(\gamma'')^2 (\gamma_e'')^2]$
$q = \sin 2\gamma_e''(1-d)/\gamma_e''$,
l, l_e are the length of the resonator with and without the sample respectively; d is the sample thickness.

From Eqns (3.54) and (3.55) it is seen that, by measuring the resonance lengths l, l_e and half-power lengths Δl, Δl_e with and without sample, ε' and $\tan \delta$ can be calculated. The calculation is computationally complicated but can be performed without using approximations, unlike the standing-wave method where a transcendental equation is to be solved.

Very accurate Q-measurements can be performed by displaying the resonance curve on the cathode ray oscilloscope or by recording it by an XY recorder. The block scheme of such a device is shown in Fig. 3.24. A reflex klystron oscillator (K_1) is frequency modulated by a sawtooth voltage in the frequency band of about 10 MHz, centred to the resonance frequency of the TE_{011} cylindrical cavity resonator. The resonance signal is recorded by an XY recorder the X axis of which is driven by the frequency sweep voltage. The frequency differences are accurately measured by using an additional klystron oscillator (K_2) operated at constant frequency near the frequency range of Klystron 1. The beat

173

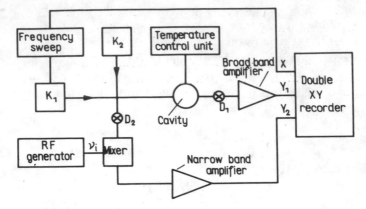

Fig. 3.24 Scheme of a device for displaying resonance curves at microwaves: K_1, K_2 klystron generators, D_1, D_2 crystal detectors

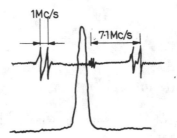

Fig. 3.25 Resonance curve of a cylindrical TE_{011}-mode cavity at 14 GHz

frequency of the two klystrons as detected by the crystal diode (D_2) mixer is thus varied in the range of the frequency modulation. By using another, intermediate, frequency ν_i mixed with this beat signal a series of pulses separated by $2\nu_i$ are obtained from a narrow-band amplifier, resulting from beats with the harmonics of ν_i. In this way a set of frequency markers can be recorded along with the cavity resonance curve. The distances between successive markers are varied by changing the frequency of the intermediate frequency generator. As at 9 Gc/s the empty Q-values are in the order of 10,000 the half width of the empty resonator is in the order of 1 MHz, so a generator of about 1—10 MHz is needed for measurement of frequency difference.

Figure 3.25 shows the resonance curve of a cavity with the frequency markers. The zero-beats between klystrons 1 and 2 are adjusted to be out of scale. From the measured value of the resonance frequency shift Δv_0 and change in the half widths δv, the permittivity and loss can be calculated.

If the volume of the sample V_s is small in relation to that of the cavity V_c according to Labuda and Craw (1961)

$$\varepsilon' \approx \frac{1}{4\pi c} \frac{V_c}{V_s} \frac{\Delta v_0}{v_0} \quad \text{if} \quad V_s \ll V_c \qquad (3.56)$$

where c is a constant. The dielectric loss is calculated from the following expression

$$\frac{1}{Q_s} - \frac{1}{Q_0} \approx 8\pi c \frac{V_s}{V_c} \sigma \quad \text{if} \quad V_s \ll V_c \qquad (3.57)$$

where Q_0, Q_s are the quality factors of the empty cavity and that containing the sample respectively; they can be calculated by the measured values of $\delta v/v_0$ with and without sample; σ is the total conductivity expressed by Eqn 3.38.

The condition $V_s \ll V_c$ is fulfilled even for low-loss materials if cylindrical TE_{01n}-mode resonator is used. The resonator volume is in the order of 20 cc, that of the sample is 0.1—0.5 cc in order to get measurable resonance-frequency shifts and Q-changes in low-loss polymers.

The method of displaying resonance curves as a function of frequency has the advantage over that involving change of resonator length in that fixed resonators can be used; these are easier to construct and have higher empty Q-values than those with a movable plunger.

In order to measure dielectric spectra as a function of the temperature in a wide range the cavity resonator should be thermostated. The scheme for a simple thermostating device for a cylindrical TE_{011} resonator is shown in Fig. 3.26. Temperature control is achieved by flowing nitrogen gas. At low temperatures from about $-150°C$ up to $+10°C$ the gas flows through a heat exchanger cooled with liquid nitrogen. The gas temperature is controlled by an electric heater, regulated by a digital program unit. At the high temperature range the nitrogen flow is switched over a bypass to go directly to the cavity. This way the cavity is always under dry nitrogen atmosphere and this avoids moisture precipitation inside. For insulating the cavity from the rest of the microwave system, stainless-steel waveguide sections are used. The coupling holes are equipped with thin mica windows to seal nitrogen gas from the waveguides.

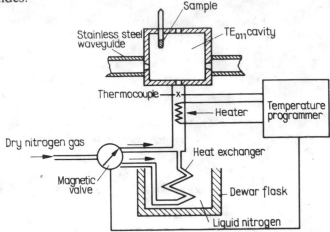

Fig. 3.26 Thermostate for cylindrical microwave cavity resonators programmed for a wide temperature range

3.4 Related techniques

As shown in Chapter 2, dielectric spectroscopy is extremely useful in studying structural transitions and molecular mobilities in polymers. To derive useful information from the measured spectrum, however, dielectric spectroscopy needs to be combined with other methods, especially with dilatometry, mechanical relaxation spectroscopy, differential scanning calorimetry and nuclear magnetic resonance. It would be highly desirable to measure transitions by various methods under identical experimental conditions. Only a few attempts to do this have so far been made.

In this section a very brief survey will be given on the non-dielectric methods of studying transitions in polymers, with special emphasis on the possibility of combining them with dielectric spectroscopy.

Thermomechanical measurements and dilatometry

A simple way of determining transitions in polymers is to record the response of the sample to a constant load as a function of the temperature at a constant rate of heating. Such 'thermomechanical' curves can be

176

measured by applying compressional, extensional or torsional loads. In practice very widely used is the measurement of the penetration of various profiles into the polymeric sample.

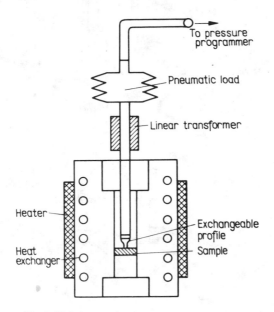

Fig. 3.27 Scheme of a thermomechanical device

In Fig. 3.27 the scheme of a universal thermomechanical set-up is shown. The load is applied by a pneumatic device using compressed nitrogen gas and so the load can be easily programmed; it can be switched on and off during heating or can even be changed stepwise or continuously. The sample is placed in a copper block which is thermostated, i.e. temperature programmed from the temperature of liquid nitrogen up to about 500°C, or higher if it is desirable. The load is applied to the sample through exchangeable heads exhibiting various profiles, as shown in the figure. Penetration of the profile is transformed into an electric signal by a linear transformer which indicates the movement of the rod connecting the pneumatic device with the profile. The same arrangement can be used for recording the thermal expansion of the sample.

Thermomechanical curves can be easily recorded simultaneously with the dielectric spectra or depolarization spectra. In the upper part of the copper block shown in Fig. 3.27 the thermomechanical measurement is

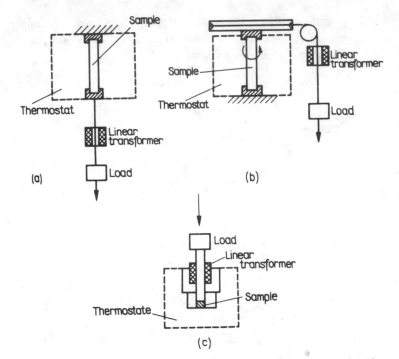

Fig. 3.28 Possibilities for recording thermomechanical curves (a) extension (b) torsion (c) penetration or compression

made at the lower end, the dielectric measurement is made by using an automatic bridge or the direct-recording technique discussed in section 2.2.

Figure 3.28 shows schematically the possibility of recording thermomechanical curves in extension (a), in torsion (b) and in compression (c); (a) and (b) are essentially creep measurements (cf. section 1.3) under conditions when time and temperature are simultaneously changed. The curves obtained exhibit bends at transitions. The position of these bends is dependent on the rate of heating and on the load.

The device shown in Fig. 3.27 is also useful for measuring thermal expansion of polymers in a wide temperature range. For this a fused silica rod with flat end contacts the sample; its extension is directly measured by recording the displacement of the rod. In this way, dilatometric transitions can be detected. The possibility of combining thermal analysis with the thermomechanical method has been reviewed by Miller (1969).

178

Mechanical relaxation spectroscopy

This method is widely used to study transitions in polymers. It involves the application of a periodic stress (tensional or torsional) to the polymer. The response of the polymer to periodic stresses has been discussed in detail in Chapter 1; it is similar to the response to periodic electric fields.

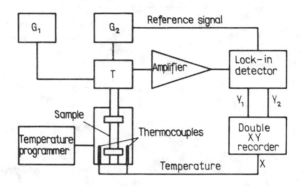

Fig. 3.29 Scheme of a recording mechanical relaxation spectrometer: G_1, G_2: generators. T: electromechanical transducer

Three main types of mechanical relaxation spectrometers are used. In one type the attenuation of a free vibrating pendulum connected mechanically with the sample is measured (torsional pendulum method), in another the sample is forced to oscillate in compression, extension, or torsion (forced oscillation method); in the third method the velocity of the sound travelling through the sample is measured (ultrasound method). The torsional pendulum method is very accurate, but can be used only at very low frequencies (usually 1—2 Hz) and recording is not easy. The forced-vibration method can be used in a wider frequency range from about 10^{-3} Hz up to 10^3 Hz and is not difficult to automatize. At higher frequencies the ultrasound methods can be used. For detailed discussion see McCrum et al. (1967) and Ferry (1961).

Figure 3.29 shows the block scheme of a forced-oscillation device for the low frequency range which is capable of recording the storage torsional modulus G' and the loss modulus G'' in a wide temperature range (Hedvig, 1970a). The method involves the application of a transducer (T), containing two motors, coupled in a bridge, which exhibit oscillations as they are fed by two different frequencies, from generators

G_1 and G_2. The thermostated sample is mechanically connected to one of the motors while the other is vibrating freely. The second balance motor can also be biased by attaching a sample of known modulus to it.

The difference signal of the motors is amplified and phase-separated into two components 90° out of phase with each other. The output signal of the two phase detectors (lock-in) can be calibrated in G' and G'' units.

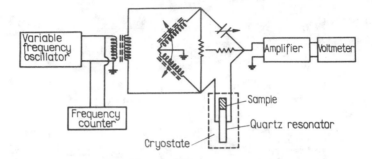

Fig. 3.30 Scheme of a composite oscillator for measuring mechanical losses at ultrasonic frequencies (after Yamamoto and Wada, 1957)

The temperature program of this spectrometer is essentially the same as that of the dielectric spectrometer of Fig. 3.13, the thermo-mechanical device of Fig. 3.28 and the dielectric depolarization spectrometer of Fig. 3.4. It is, correspondingly, possible to perform mechanical relaxation measurements and dielectric measurements by using the same temperature program, even by using the same thermostating block.

The block scheme of a high-frequency piezoelectric mechanical relaxometer is shown in Fig. 3.30 after Yamamoto and Wada (1957). In this device torsional vibrations are excited by a resonant quartz crystal to which the polymer sample is glued. The quartz resonator is coupled in a bridge, as shown in Fig. 3.30. The dimensions of the sample are adjusted so that the resonator can be tuned off only by a maximum of 5% with respect to its resonant frequency. The method is referred to as the 'composite oscillator' technique.

The storage modulus is expressed in terms of the change in the resonance frequency $\nu_0 - \nu_s$ as follows:

$$G' = \left(\frac{2\,l}{n}\right)^2 \left[\nu_s - (\nu_0 - \nu_s)\,\frac{m_0}{m_s}\right]^2 \varrho \qquad (3.58)$$

180

where l is the length of the sample, v_0, v_s are the resonance frequencies without and with the sample, m_0, m_s are the corresponding masses and ϱ is the density of the sample.

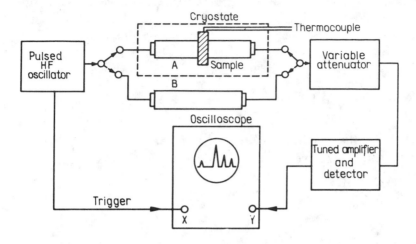

Fig. 3.31 Scheme of a mechanical relaxometer based on sound velocity measurement (after Tanabe et al., 1970)

The loss modulus is

$$G'' = \frac{1+(m_0/m_s)\Delta v_s}{G} \tag{3.59}$$

where Δv_s is the half-width of the resonance curve in the presence of the sample.

v_0, v_s and Δv_s can be measured directly by displaying the resonance curve of the quartz oscillator on the screen of a cathode-ray oscilloscope. The density of the sample should be independently measured, or known, over the temperature range of the measurement.

The composite oscillator method has been used by Yano and Wada (1971) at 34 kHz and 152 kHz. In principle it can be used up to very high frequencies. The spectra of polystyrene shown in Fig. 2.25 have been measured by this method.

Figure 3.31 shows the block scheme of a mechanical relaxometer based on measurement of the propagation of the longitudinal elastic waves through the sample. Longitudinal elastic waves are excited in a quartz rod by a piezoelectric quartz crystal. The sample is placed between two

181

quartz rods, as shown in Fig. 3.31. The acoustic wave travelling through the sample is detected by a receiver quartz crystal. The excitation is pulsed to avoid unwanted reflections from the sample surfaces. The pulse picked up by the receiver is displayed by a cathode-ray oscilloscope. The relative pulse height and the pulse delay are measured. For reference, a similar system without sample is used.

The velocity of the acoustic wave in the sample is

$$V_a = \frac{l}{\Delta t} \tag{3.60}$$

where l is the length of the sample, Δt is the time delay of the received signal pulse observed in the sample line with respect to that of the reference line.

The absorption coefficient of the sample is

$$\alpha = \frac{\Delta A - A_0}{l} \tag{3.61}$$

where ΔA is the difference in the received pulse amplitudes measured at the sample and reference lines, A_0 is a factor arising from the reflexion loss occurring from the quartz—sample interface. It is independent of the temperature.

Differential scanning calorimetry

Differential Scanning Calorimetry (DSC) is a method of measuring differences of heat fluxes between sample and thermostat. The schematic arrangement of a DSC device is shown in Fig. 3.32. The sample is placed on a metallic disc connected to a metallic block (thermostat) by a heat-insulator layer of thickness l. The temperature difference between disc and block is measured by a system of thermocouples. A similar arrangement is used for the reference arm, sample and reference thermocouples being coupled in opposite polarity. A regulating device keeps sample and reference arms in balance while the temperature is swept linearly; this can be assured by regulating the heating power introduced to the reference arm. If the enthalpy or specific heat of the sample changes, the heat flux through the insulating layer of the sample arm will also change, resulting in a change in the temperature between the calorimeter body (disc) and the thermostating block. The e.m.f.

182

of the sample thermocouples is correspondingly changed, upsetting the balance. Balance is restored by increasing or decreasing the heating power of the reference arm. This heating power is recorded as a function of the temperature.

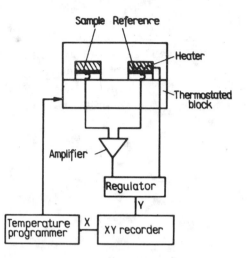

Fig. 3.32 Scheme of a differential scanning calorimeter

The condition of balance between the sample and the block is (see, for example, Calvet and Prat 1956)

$$[c_0 + c_s(t)] \frac{dT_s}{dt} + \lambda_0 A_0 \frac{T_s - T_0}{l_0} = q_s(t) v_s \qquad (3.62)$$

where T_s is the sample temperature, T_0 is the temperature of the thermostated block, t is the time, λ_0 is the thermal conductivity of the heat-insulating layer, l_0 is the length of this layer, v_s is the volume of the sample, $q_s(t)$ is the heat flux produced by the sample, c_s is the specific heat of the sample, c_0 is that of the calorimeter body (disc), and A_0 is the area of the calorimeter body.

If the rate of heating is constant the temperature difference between the calorimeter and block is

$$T_s - T_0 = \frac{q_s(t) v_s l_0}{A_0} \left\{ 1 - \exp\left[\frac{\lambda_0 A_0}{l_0 [c_0 + c_s(t)]} t \right] \right\} \qquad (3.63)$$

183

It is seen that the temperature difference $T_s - T_0$ decays exponentially in time with a time constant

$$\tau_s = \frac{l_0}{\lambda_0 A_0} (c_0 + c_s) \qquad (3.64)$$

if c_s and q_s are constant. τ_s is referred to as the time constant of the calorimeter sample arm. It can be experimentally measured.

It is possible to choose the rate of heating so that in the time scale of the observation $t \ll \tau$. The calorimeter is then said to work in the adiabatic regime and Eqn. (3.63) is simplified to

$$T_s - T_0 = \frac{v_s \lambda_0}{c_0 + c_s} q_s(t) t \quad (t \ll \tau_s) \qquad (3.65)$$

On the other hand, if $t \gg \tau_s$ the calorimeter is operated in the quasi-isothermal regime

$$T_s - T_0 = \frac{v_s l_0}{A_0} q_s(t) \quad (t \gg \tau_s) \qquad (3.66)$$

For the reference arm similar equations are true

$$T_r - T_0 = \frac{v_r \lambda_0}{c_0 + c_r} q_r(t) t \quad (t \ll \tau_r)$$

$$\qquad (3.67)$$

$$T_r - T_0 = \frac{v_r l_0}{A_0} q_r(t) \quad (t \gg \tau_r)$$

The signal of the thermocouples is proportional to $T_r - T_s$. If no heat is introduced to the reference calorimeter, i.e. $q_r = $ const, and the calorimeters are equal ($v_s = v_r = v_0$)

$$\Delta T = T_r - T_s = -\frac{v_0 \lambda_0}{c_0} \left[\frac{q_s(t)}{1 + c_s(t)/c_0} - \frac{q_r(t)}{1 + c_r/c_0} t \right] \begin{array}{c} (t \ll \tau_s, \tau_r) \\ \text{adiabatic} \end{array} \qquad (3.68)$$

$$\Delta T = T_r - T_s = \frac{v_0 l_0}{A_0} [q_s(t) - q_{(t)}] \begin{array}{c} (t \gg \tau_s, \tau_r) \\ \text{isothermal} \end{array} \qquad (3.69)$$

184

The temperature difference ΔT between the sample and reference arm in the isothermal regime is proportional to the heat flux produced by the sample $q_s(t)$. This regime is therefore useful for studying reaction kinetics.

In the adiabatic regime ΔT is also dependent on $q_s(t)$ but depends on the heat capacity $c_s(t)$ of the sample as well. This is correspondingly useful for measuring structural transitions.

By using the device shown in Fig. 3.32, i.e. by keeping ΔT zero,

$$q_s(t) = \frac{c_0 + c_s(t)}{c_0} \, q_r(t) \qquad \begin{array}{c} (t \ll \tau_s, \tau_r) \\ \text{adiabatic} \end{array} \qquad (3.70)$$

and

$$q_s(t) = q_r(t) \qquad \begin{array}{c} (t \gg \tau_s, \tau_r) \\ \text{isothermal} \end{array} \qquad (3.71)$$

As $q_r(t)$ is the heat flux introduced to the reference arm, which is recorded, from Equation (3.70). It is seen that by DSC in the adiabatic regime changes in the specific heat of the sample are detected along with change in heat quantity $q_s(t)$, while in the isothermal regime only $q_s(t)$ is detected.

Adiabatic regime is achieved when the rates of heating or cooling are high and the heat capacity of the calorimeter body c_0 and that of the sample c_s are small. Correspondingly, for measurement of transitions in polymers small quantities of samples (≤ 10 mg) are used with specially designed low-inertia calorimeters. For measuring reaction kinetics, such as polymerization and degradation or curing of resins, calorimeters of high inertia are used and the quantity of the sample may be in the order of 0.1—1 g. The sample and reference cell of an isothermal differential scanning calorimeter developed for reaction kinetic studies by Barkalov *et al.* (1972) is shown in Fig. 3.33. The body of this calorimeter is a copper cylinder. The heat-insulating material is a well cured, quartz-filled epoxy resin; the temperature-difference sensor is a thermopile of 120 copper—constantan thermocouples arranged symmetrically as in Fig. 3.33. In this calorimeter it is possible to use Pyrex glass or fused silica sample holders which can be easily evacuated or filled with gases. The calorimeter is useful for operating in the temperature range between $-150°C$ and $+300°C$; the upper limit can be relatively easily increased to about $600°C$.

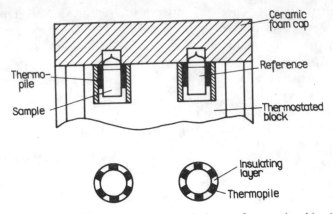

Fig. 3.33 Scheme of a differential scanning calorimeter for reaction kinetical studies (after Barkalov *et al.*, 1972)

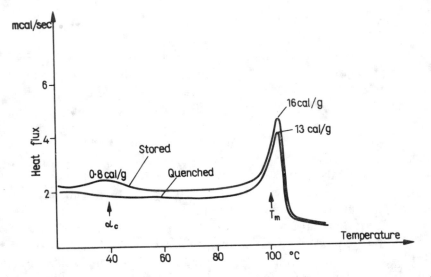

Fig. 3.34 Differential scanning calorimetric curve of low density polyethylene; (a) stored at room temperature for 6 months; (b) quenched from the melt

Figure 3.34 shows typical DSC curves for low density polyethylene measured after quenching from the melt and after prolonged storage at room temperature. The α-transition appears only for the stored sample.

Very fruitful information can be gained by combining the methods of thermal analysis with dielectric spectroscopy. This can be simply done by recording the change in the sample temperature in relation to a reference

186

cell (DTA measurement). Such a measurement can be performed simultaneously with the recording of dielectric spectra, using the same sample. Examples will be given in Chapter 6 in connection with curing of resins.

Yalof and Wrasidlo (1972) combined differential scanning calorimetry with dielectric spectroscopy. The DSC curves can be recorded simultaneously with $\varepsilon'(T)$ and $\tan\delta(T)$ curves in a specially designed combined cell using different but identical samples for the dielectric and DSC measurements.

It is also relatively easy to perform DSC measurement simultaneously with dilatometry or with the recording of thermomechanical curves. An example of this was in Fig. 2.2 in connection with the glass–rubber transition of polyvinylchloride.

Nuclear magnetic resonance

Nuclear magnetic resonance (NMR) is a method involving measurement of magnetic dipole transitions between magnetic sublevels of nuclei which have magnetic moment. In diluted solutions this method is widely used for analysis (High Resolution NMR). In the solid state the method is very useful for studying molecular mobilities, especially rotation of groups, and diffusion. The method is based on the fact that the motion of the nuclear spins and magnetic moments built in the polymer chains (H-atoms, for example) influence the NMR transitions by dipole—dipole interactions among the nuclear spins and by the energy exchange between the nuclear spin system and the surrounding non-paramagnetic lattice. These interactions are expressed by the nuclear spin—spin relaxation time T_2 and by the nuclear spin—lattice relaxation time T_1. These relaxation times are defined by the decay of the transversal and longitudinal net magnetization respectively of the sample caused by the nuclear spins exhibited in response to a strong external magnetic field. The equations of motion for the average transversal xy magnetization in an external field applied to the z-direction are

$$\frac{dM_x}{dt} = -\frac{M_x}{T_2}$$

$$\frac{dM_y}{dt} = -\frac{M_y}{T_2}$$

(3.72)

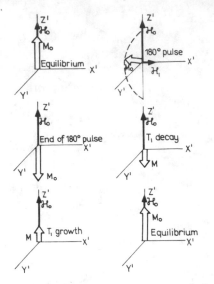

Fig. 3.35 Vector model for explaining the measurement of the nuclear spin lattice relaxation time T_1: restoration of equilibrium magnetization after a 180 H_1-pulse

The longitudinal magnetization is

$$\frac{dM_z}{dt} = M_0 - \frac{M_z}{T_1} \tag{3.73}$$

The solutions of Eqns (3.72) and (3.73) are

$$M_x(t) = M_0 \exp(-t/T_2)$$
$$M_y(t) = M_0 \exp(-t/T_2) \tag{3.74}$$
$$M_z(t) = M_0[1 - \exp(-t/T_1)]$$

The relaxation times T_1 and T_2 thus describe the rate of the exponential decay of the magnetization. T_1 and T_2 can be measured independently by the spin-echo method introduced by Hahn (1950) and developed further by Carr and Purcell (1954).

The motion of the average magnetization of the nuclei in an external, homogeneous magnetic field is shown in Fig. 3.35. The precessional motion of the magnetization is perturbed by the application of an alternating field $\mathcal{H}_1$ perpendicularly to $\mathcal{H}_0$, if the frequency ν_0 of $\mathcal{H}_1$ is in

188

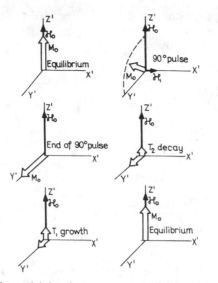

Fig. 3.36 Vector model for explaining the measurement of the nuclear spin-spin relaxation time T_2: restoration of equilibrium magnetization after a 90° H_1-pulse

resonance with the precessional frequency, i.e. $\omega_0 = \gamma_1 \mathscr{H}_0$, where γ_1 is a constant depending on the spin and nuclear moment of the nucleus in question. In fact that component of $\mathscr{H}_1$ which rotates along with the precessing moment has such an effect on the precessing magnetization.

By applying the resonance field (r.f.) $\mathscr{H}_1$ in pulses of different duration, it is possible to turn the magnetization M_0 by 90° or by 180°. In the spin-echo method a 90° pulse and a 180° pulse are applied after each other, turning M to the xy plane and reversing its phase there. An echo pulse is then detected at the receiver after a certain time when the moments scattered by the dipole–dipole interaction get in phase again. By applying a series of 90° and 180° pulses with increasing time lag between them, the echo pulse will decay exponentially; the decay envelope is proportional to $\exp(-t/T_2)$. In this way the spin–spin relaxation time can be measured (see Fig. 3.36).

For measuring the spin–lattice relaxation time, two subsequent 90° pulses are applied with increasing time lag between them. The two 90° rotations will turn the magnetization to the zx plane, so the decay of the longitudinal component can be detected. The echo envelope in this case is related to T_1.

In a more recent technique introduced by Slichter and Ailion (1964) a 90° pulse is followed by a 90° phase shift (see Fig. 3.37). In this way the magnetization in the xy plane is oriented along the direction of $\mathscr{H}_1$. After time t the $\mathscr{H}_1$ field is switched off and the signal at the receiver coil is recorded. The signal amplitude a decays as

$$a(t) \propto \exp(-t/T_{1\varrho}) \tag{3.75}$$

where $T_{1\varrho}$ is referred to as the rotating frame spin–lattice relaxation time.

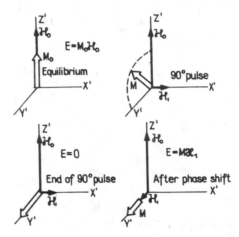

Fig. 3.37 Vector model for explaining the measurement of the rotating frame nuclear spin lattice relaxation time $T_{1\varrho}$

The spin–lattice relaxation time of a nuclear spin system is the measure of the interaction of the spin system with the surroundings, generally referred to as the 'lattice'. The spin–spin relaxation time is a measure of the dipole–dipole interaction between the individual spins which are in phase just after the application of the resonance field $\mathscr{H}_1$ but become out of phase as a result of the dipole–dipole interaction. The rotating frame spin–lattice relaxation time is characteristic for the time needed by the nuclear spin system to warm up to the temperature of the lattice after having been cooled down by the $\mathscr{H}_1$-field. Thus $T_{1\varrho}$ is also a measure of the energy exchange between the nuclear spin system with the lattice, as T_1, but the effective frequency is much smaller.

For details of the application of NMR to polymers see Bovey (1972), Slonim and Ljubimov (1970), Slichter (1963), Connor (1970) and Jones (1976).

The spin–spin relaxation process T_2 connected with the dipole–dipole interaction of the paramagnetic nuclei can be described in terms of correlation functions attributed to stochastic fluctuations of the internal magnetic field caused by the nuclear dipoles which exhibit thermal motion along with the molecules or groups. This correlation function component along the direction of the external field (z-direction) is expressed in terms of spherical functions $Y_0(t)$ as

$$\varphi_z(t) = \frac{9}{16} \gamma_I^2 h^2 \langle Y_0(t) \, Y_0(t+\tau) \rangle_{av} \qquad (3.76)$$

where γ_I is the nuclear magnetogyric ratio $Y_0 = r^{-3}(1 - 3\cos^2\theta)$.

In the simplest case when the molecule containing the paramagnetic nucleus is regarded as being a rigid sphere exhibiting random fluctuations,

$$\varphi_z(t) = \frac{9}{16} \gamma_I^2 h^2 |Y(0)|^2 \exp(-t/\tau_c) \qquad (3.77)$$

where τ_c is referred to as the NMR correlation time. In the rigid sphere approximation

$$\tau_c = \frac{4\pi\eta r^3}{3kT} \qquad (3.78)$$

where η is the viscosity of the surrounding medium. The connection of the nuclear spin–spin relaxation time and the correlation time if only spin–spin interaction is effective as

$$\frac{1}{T_2} = \frac{3}{20} \frac{h^2 \gamma_I^4}{r^6} \left(3\tau_c + \frac{5\tau_c}{1 + \omega_0^2 \tau_c^2} + \frac{2\tau_c}{1 + 4\omega_0^2 \tau_c^2} \right) \qquad (3.79)$$

The dependence of the nuclear spin–spin relaxation time on the correlation time is shown in Fig. 3.38. It is seen that at short relaxation times $1/T_2 \propto \tau_c$. At longer relaxation times in the solid state the T_2 versus τ_c dependence levels off, the nuclear spin–spin relaxation time becomes practically independent of the correlation time.

The spin–lattice relaxation process (T_1) is connected with the energy exchange between the nuclear spin system and the surrounding, non-paramagnetic, lattice. The spin–lattice relaxation is also connected with the random motion of the molecules characterized by the correlation time τ_c as

$$\frac{1}{T_1} = \frac{3\,\gamma_1^4\,h^2}{10\,r^6}\left[\frac{\tau_c}{1+(\omega_0\,\tau_c)^2} + \frac{4\,\tau_c}{1+(2\,\omega_0\tau_c)^2}\right] \qquad (3.80)$$

where ω_0 is the NMR resonance angular frequency.

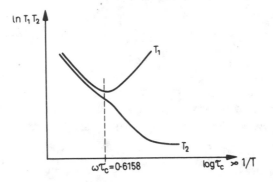

Fig. 3.38 Schematic representation of the dependence of the nuclear spin-lattice (T_1) and spin spin (T_2) relaxation times on the correlation time τ_c

Figure 3.38 also shows the dependence of the NMR spin–lattice relaxation time T_1 on the correlation time τ_c in a logarithmic scale. It is seen that the $\ln T_1$ against $\ln \tau_c$ curve has a minimum. The corresponding correlation time is approximately

$$\tau_c \approx 0.616/\omega_0 \quad \text{(at } T_1 \text{ minimum)} \qquad (3.81)$$

According to Eqn. (3.78) the correlation time is directly related to the temperature.

By measuring T_1 as a function of the temperature in a wide enough temperature range, from the $\ln T_1$ versus $1/T$ plot the correlation time associated with frequency ω_0 can be determined.

The rotating-frame spin–lattice relaxation time T_1 defined by Eqn. (3.75) is also connected with the correlation time:

$$\frac{1}{T_{1\varrho}} = K\left[\frac{3}{2}\frac{\tau_c}{1+4\,\omega_e^2\,\tau_c^2} + \frac{5}{2}\frac{\tau_c}{1+\omega_0^2\,\tau_c^2} + \frac{\tau_c}{1+4\,\omega_0^2\,\tau_c^2}\right] \qquad (3.82)$$

where K is a constant, ω_e is the frequency in the rotating frame and is much smaller than the resonance frequency $\omega_e \ll \omega_0$. This effective frequency is

192

the difference between the resonance frequency ω_0 and the actual precessional frequency of the magnetic moment depends on the magnitude of the $\mathscr{H}_1$ field, $\omega_e \approx \gamma_I \mathscr{H}_1$.

The $T_{1\varrho}$ versus temperature curve exhibits a minimum at

$$\tau_c \approx 0.5/\omega_0 \quad \text{(at } T_{1\varrho} \text{ minimum)} \qquad (3.83)$$

Thus, by measuring $T_{1\varrho}$ against temperature, the correlation time corresponding to the frequency ω_e and temperature T ($T_{1\varrho}$ minimum) can be calculated.

The nuclear spin–lattice (T_1) and spin–spin (T_2) relaxation times are also related to the NMR line widths. It is therefore possible to detect transitions by observing changes in the NMR line widths as a function of the temperature. If spin–spin relaxation processes dominate in determining the line width, which is often the case,

$$T_2 \approx \frac{2}{\Delta\omega} \approx \frac{\tau_c}{2\pi} \qquad (3.84)$$

where $\Delta\omega$ is the NMR line width.

It is also usual to measure the second moment of the NMR line defined as

$$\langle \Delta\mathscr{H}^2 \rangle = -\frac{\frac{1}{3}\int (\mathscr{H}_0 - \mathscr{H})^3 \, G'(\mathscr{H}) \, d\mathscr{H}}{\int (\mathscr{H}_0 - \mathscr{H}) \, G'(\mathscr{H}) \, d\mathscr{H}} \qquad (3.85)$$

where $\mathscr{H}_0$ is the resonance external field, $\mathscr{H}$ is the swept external field, and $G'(\mathscr{H})$ is the derivative of the NMR line, which is usually recorded in the broad line technique.

In terms of the second moment

$$T_2 \approx \frac{2}{\gamma_I} \langle \Delta\mathscr{H}^2 \rangle^{-1/2} \qquad (3.86)$$

By measuring NMR line widths or second moments as a function of the temperature, transitions are indicated by abrupt changes. The effective frequency corresponding to these transitions is the NMR resonance frequency ω_0.

As an illustration, Fig. 3.39 shows the nuclear magnetic relaxation times of unplasticized polyvinylchloride, T_1, $T_{1\varrho}$ and T_2, as a function of the

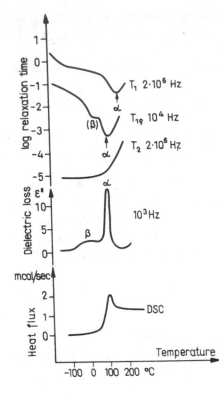

Fig. 3.39 The nuclear relaxation times of PVC compared with the dielectric spectrum and with the differential scanning calorimetric curve

temperature in comparison with the dielectric spectrum measured at 1 kHz and with the differential scanning calorimetric curve. The NMR resonance frequency in this experiment was 30 MHz, the effective rotating-frame frequency for the $T_{1\varrho}$ measurement was 10 kHz. It is seen that, allowing for the different effective frequencies, the T_1 and $T_{1\varrho}$ minima correspond to the α- and β-transitions measured by the dielectric method. By differential scanning calorimetry only the α-transition (T_g) can be detected. The sharp increase exhibited by the spin–spin relaxation time at the glass–rubber transition is an indication of the microbrownian motion, especially rotations in the main chain which averages out the spin–spin interactions resulting in an increase of T_2. According to Eqn. (3.86) this involves line sharpening, which is experimentally observed.

194

Besides the thermomechanical, thermal (DSC) dilatometric, dielectric dynamic mechanical methods and NMR some other methods are also useful for detecting transitions in polymeric solids and visco-elastic fluids. These are the electron spin resonance (ESR) spin labelling technique (see, for example, Hedvig, 1975), infrared spectroscopy (Ogura et al., 1971), radiothermoluminescence (Nikolsky and Buben, 1960; Charlesby and Partridge, 1963, 1965), fluorescence (North and Soutar, 1972), neutron scattering (Berghmans et al., 1971; Safford and Naumann, 1967).

In some special cases when suitable nuclei are present in the polymers, as Fe, Co, Ni . . . the Mössbauer spectroscopic method nuclear gamma resonance is also useful for studying molecular mobilities; especially lattice vibrations. For the principle and applications of Mössbauer spectroscopy, see Goldansky and Herber (1968), and Hedvig (1975).

Recently from laser-Raman spectroscopy (Melveger, 1972; McGraw, 1971) and from laser diffraction (Hashimoto et al., 1973) valuable information has been derived about transitions and molecular mobilities in polymers.

It is apparent that in order to study polymeric structures a fairly wide scale of technique is needed.

CHAPTER 4

DIELECTRIC SPECTRA OF PURE POLYMERS

During the past few decades of dielectric work quite a large body of experimental data on polymers has been collected, and has been comprehensively reviewed by McCrum *et al.* (1967). Additional data may be found in Sazhin (1970), Ishida (1969), Hedvig (1969c).

In the present chapter, instead of reviewing the data on individual polymers, the common features of the spectra exhibited by polymers having polar groups in their main chain, attached rigidly to the chain and to their flexible side chain will be discussed. The dielectric properties of non-polar polymers are also discussed in some detail as they are the most important practical insulating materials. A separate section is devoted to polymers having hydrogen bonds. Thermosetting systems and rubbers are discussed in Chapter 6 with respect to the effect of crosslinking to the dielectric properties.

4.1 Non-polar polymers

Hydrocarbon polymers such as polyethylene, polypropylene, polystyrene and polyisobutylene contain only $C-H$ apolar bonds. The dipole moments of the groups contained in these polymers are in the order of 0.1 debye, which corresponds to oscillator strengths $\varepsilon_0 - \varepsilon_\infty$ of 10^{-2} and maximum loss factors $\varepsilon''(\text{max}) \approx 10^{-6}$ which are hardly detectable by the usual dielectric technique.

Despite this, apolar polymers exhibit measurable dielectric spectra corresponding to the transitions measured by the mechanical-relaxation technique. Especially by the dielectric depolarization method and by the very-low-frequency Fourier transformation method, quite high dielectric loss peaks can be detected in apolar polymers. The reason is probably that these polymers are always slightly oxidized and so contain polar carbonyl, peroxy or hydroperoxy groups. Oxidation proceeds at

196

the structural defects, so oxidized layers are formed in the apolar polymer, resulting in very high Maxwell–Wagner–Sillars polarization. This problem will be discussed in Chapter 7 in connection with the oxidative degradation of polyethylene.

The dielectric properties of polystyrene have already been discussed in Chapter 2 in connection with cryogenic relaxations. In this section two representative polymers will be discussed—one is a pure hydrocarbon polymer, polyethylene (PE), the other is a fluorocarbon polymer polytetrafluoroethylene (PTFE). Although in PTFE the $C-F$ bonds are very polar (1.39 debye) the polymer is to be regarded as apolar because the individual dipole moments are cancelled.

The dielectric spectrum of polyethylene

Polyethylene is available in three basic forms. One is polymerized by a free radical mechanism with peroxide initiators under high pressure; it is referred to as high-pressure or low-density polyethylene (LDPE). Its chemical formula is

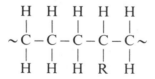

The side branches R are mainly methyl or ethyl groups; they are statistically distributed along the main chain.

In LDPE, the average number of methyl groups per 1000 carbon atoms is 21.5, that of ethyl groups is 14.4; its density is typically $0.91-0.92$ g/cm^3, and its crystallinity is about 60%. The crystalline melting temperature of LDPE is 105—120°C.

Polyethylene is also produced at low pressure by Ziegler–Natta catalysts. The corresponding polymer has high density ($0.93-0.94$ g/cm^3) and high crystallinity (87%); it is referred to as low-pressure or high-density polyethylene (HDPE). The number of methyl groups per 1000 carbon atoms in HDPE is 3, that of ethyl groups is 1. The crystalline melting temperature of HDPE is between 120 and 140°C.

The third typical polyethylene is prepared by as specific catalyst (Phillips) which produces a linear polyethylene (LPE) of high density (0.96 g/cm^3) and high crystallinity (93%) exhibiting few side branches

< 1 per 1000 carbon atoms. The crystalline melting temperature of LPE is 140°C.

Figure 4.1 shows the transitions measured by mechanical relaxation spectroscopy for all the three polymer types. The highest temperature

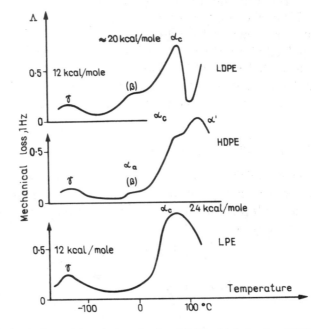

Fig. 4.1 Mechanical transitions in low density (LDPE), high density (HDPE) and linear (LPE) polyethylene

α_c-peak is found in all the three types; it is thought to be due to motion at the crystallite surfaces. The position and intensity of the α_c-peak is dependent on the crystallinity and crystallite size, and so this peak is sensitive to heat treatments. The lower temperature α_a-peak, which is often referred to as the β-peak, is found in LDPE, is less intense in HDPE, and is absent in LPE. At low frequencies (1 Hz) the α_c-peak is found near +50°C. Two subsequent peaks, α_a and α_a', are found at about −30°C and −75°C respectively, usually poorly resolved.

The α_a-peak has been attributed to the motion of the side branches in the amorphous phase but this interpretation has been recently questioned. It is doubtlessly connected with the amorphous phase. In the literature there has been a controversy about the α_a or β_a peaks being the glass transition in polyethylene. This problem has been

198

discussed in detail by Boyer (1973). His conclusion is that both transitions can be regarded as glass–rubber transitions. According to NMR studies by Bergmann and Nawotki (1972) and X-ray studies by Bezjak and Veksli (1970), in low-density polyethylene there are two kinds of amorphous phases. This is also indicated by dielectric depolarization measurements (Hedvig, 1972).

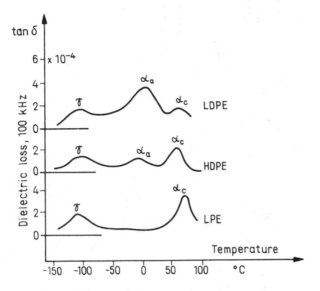

Fig. 4.2 Dielectric transitions in low density (LDPE), high density (HDPE) and linear (LPE) polyethylene

The γ-relaxation at about $-120°C$ is found in all three polymers; it is attributed to a crankshaft-type local motion in the main chain (see Fig. 2.2) but a dislocation–migration mechanism has also been proposed (Matsui et al., 1971; see also section 2.4).

In Fig. 4.1 the activation energies determined by measuring the transitions at different frequencies are also indicated.

Linear polyethylene exhibits a cryogenic transition at 40 K (γ_c-relaxation). It is assigned to the motion of dislocations in the crystalline phase (Hiltner and Baer, 1972) as it vanishes by quenching and is intensified in stretched polymers.

Figure 4.2 shows the dielectric spectra of all three types of polyethylene at a frequency of 100 kHz. It is seen that the α_c, $\alpha_a(\beta)$ and γ-transitions appear in the dielectric spectra and are similar to the mechanical-relaxation spectra. The dielectric loss peaks are found to be intensified by oxidizing the polymer thermally or by ultraviolet illumination or high energy irradiation. This problem will be discussed in Chapter 7 in connection with the change of the dielectric properties by oxidative degradation.

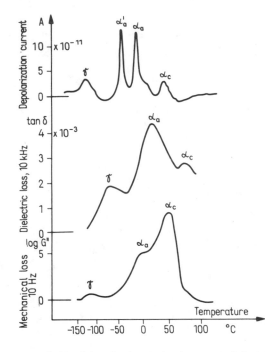

Fig. 4.3 Comparison of the dielectric depolarization, a.c. dielectric and mechanical relaxation spectra of low density polyethylene

Figure 4.3 compares the dielectric depolarization spectrum of slightly oxidized low-density polyethylene with the corresponding high-frequency (10 kHz) dielectric and low frequency (10 Hz) mechanical spectra. The depolarization spectrum has been measured after polarization of the sample at 100°C for 30 minutes. The rates of heating and cooling were 2°C/min. It is seen that in the depolarization spectrum the α_a and α_a' peaks are resolved much better than in the mechanical and in the dielectric spectra.

The depolarization peak intensities α_a and α_a' are found to be very sensitive to previous heat treatments. By quenching, these peaks are strongly reduced similarly to the depolarization peaks of the amorphous polymers at the glass–rubber transition. This supports the hypothesis

200

that this transition corresponds to the glass transition of the 'amorphous phase' in this polymer.

Sasabe and Saito (1972) investigated the dependence of the permittivity of non-polar polymers on pressure and on temperature at a frequency of 1 Hz. If the polymers were really non-polar, only the induced polarizability should be effective in determining the dielectric permittivity, which should correspond to the unrelaxed permittivity of the polar polymers. For such a polarization the Clausius–Mosotti equation is valid

$$\frac{\varepsilon_\infty - 1}{\varepsilon_\infty + 2} \frac{M}{\varrho} = \frac{4\pi}{3} N_A \alpha \tag{4.1}$$

where M is the molecular weight, ϱ is the density, N_A is Avogadro's number, α is the polarizability of the molecule.

For slightly polar polymers the unrelaxed permittivity ε_∞ is approximated as

$$\frac{\varepsilon_\infty - 1}{\varepsilon_\infty + 2} \frac{M}{\varrho} = \frac{4\pi N_A}{3} \left(\alpha + \frac{\mu^2}{3kT} \right) \tag{4.2}$$

where μ is the permanent dipole moment, for example, that of the carbonyl groups formed by oxidation, and T is the temperature.

Figure 4.4 shows the temperature dependence of ε' measured at 1 kHz for high-density and low-density polyethylenes (HDPE, LDPE), polypropylene (PP), polytetrafluoroethylene (PTFE) and polystyrene (PST). Figure 4.5 shows the corresponding pressure dependences. It is seen that the $1/T$ dependence predicted by Eqn. (4.2) is exhibited, but in LDPE and in PST breaks are found in the $\varepsilon'(T)$ and $\varepsilon'(P)$ curves as well. These breaks are due to the change of density at the transitions. By representing the data as $y = \dfrac{\varepsilon_\infty - 1}{\varepsilon_\infty + 2} \dfrac{M}{\varrho}$ against pressure or against $1/T$, fairly constant values of the corresponding low molecular weight compounds are obtained. This means that y is uniquely determined by the polarizability of the repeat unit which is not influenced by changes in molecular weight, conformation and packing.

On the basis of comparison of dielectric data with infrared spectroscopic measurements (Reddish and Barrie, 1959; Sasakura, 1963) it is concluded that the dielectric properties of polyethylene are mainly

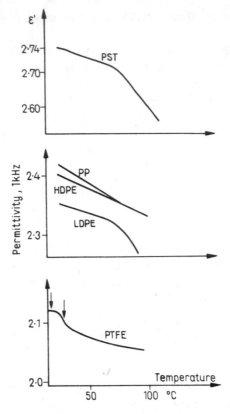

Fig. 4.4 Temperature dependence of the dielectric permittivity at 1 kHz of some apolar polymers: PST: polystyrene, PP: polypropylene, HDPE: high-density polyethylene, LDPE: low-density polyethylene, PTFE: polytetrafluoroethylene (after Sasabe and Saito, 1972)

determined by the carbonyl groups formed by oxidation

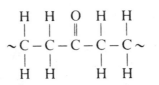

As these groups are statistically distributed in the polymer chain, all the three main transitions are exhibited in the dielectric spectra. This opens up a possibility of studying oxidative degradation processes. This will be discussed in Chapter 7.

202

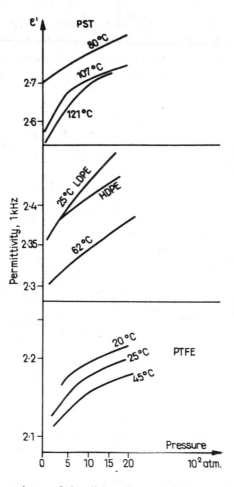

Fig. 4.5 Pressure dependence of the dielectric permittivity at 1 kHz of some apolar polymers (for details see caption to Fig. 4.4)

Polytetrafluoroethylene

This polymer is apolar because the polar $C-F$ groups are symmetrically arranged so the dipole moments would cancel each other. Despite this, a well defined dielectric spectrum is exhibited which correlates well with the mechanical relaxation spectrum. This is shown in Fig. 4.6. The dielectric spectrum was measured at 1 kHz. The mechanical relaxation

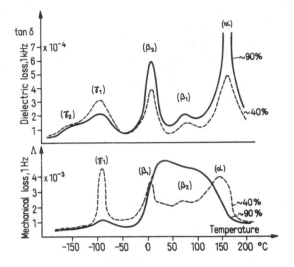

Fig. 4.6 Dielectric and mechanical relaxation spectra of 90% and 40% crystalline polytetrafluoroethylene

spectrum was measured at 1 Hz. Apart from the shift due to the difference in the frequencies, the two spectra are similar. The dielectric loss maximum in the α-transition is rather high for an apolar polymer.

The origin of the dielectric spectra in PTFE is also attributed to the effects of oxidation. In this polymer the peroxy radicals are known to be extremely stable

The electric dipole moment of the $C\overset{\cdot}{O}O$ group is less than that of the $C-F$ group. Thus by oxidation the cancellation of the $C-F$ dipole

moments is broken, resulting in an increase of the net dipole moment of the group which has been oxidized. Similarly the other possible groups are

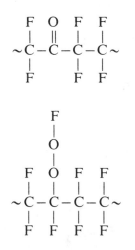

In all cases an increase of the net dipole-moment concentration is expected.

Such an effect can be regarded as a chemical defect: the introduction of a foreign chemical group to the system upsets the dipole balance and an effective dipole moment is formed at the defect site; similarly to polytrifluoro chloroethylene already discussed in section 2.1. In this polymer a fluorine atom is replaced by chlorine

The dipole moment of the C−Cl bond is 1.47 debye. The corresponding dielectric loss in the order of $\varepsilon''_{max} = 0.1$, $\varepsilon_0 - \varepsilon_\infty = 0.6$.

Figure 4.7 shows the transitions in polytetrafluoroethylene schematically. At 19°C there is a crystal–crystal transition. Below 19°C the lattice is triclinic, involving a zigzag chain conformation which is twisted at each 13 CF_2 group. This structure is transformed into a hexagonal lattice in which chains twist by 180° at each 15 CF_2 group (Pierce et al., 1956).

Between 19°C and 30°C the packing of the chains in the hexagonal lattice is somewhat disordered; chain segments have statistical angular displacements about the chain axis, and the segments oscillate about a crystal axis. At 30°C this symmetry is lost; the preferred axis vanishes and segments oscillate at random. This is therefore an order–disorder transition within the crystal while the hexagonal symmetry of the lattice is retained.

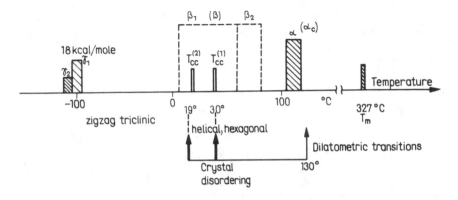

Fig. 4.7 Transitions in polytetrafluoroethylene

The crystal transitions at 19°C and 30°C are detected by dilatometry and by differential scanning calorimetry. They also appear as small jumps of the unrelaxed permittivity $\varepsilon_\infty = n^2$ versus temperature curve, as shown in Fig. 4.4. In the dielectric spectrum this transition is covered by the broad β-peak.

It is difficult to assign the transitions in polytetraflouroethylene to the amorphous and crystalline phase because, as shown in Fig. 4.6, all the transitions are dependent on changes in crystallinity. The temperature dependence of the α- and β-transition frequencies do not obey the Arrhenius law; at 130°C there is a dilatometric transition. On the basis of this, α and β can be regarded as being due to motion in the disordered phase mainly determined by the free volume, similarly to T_g in amorphous polymers.

The γ-transition is Arrhenius-like with an activation energy of 18 kcal/mole. This corresponds to local motion of short-chain segments, most probably a crankshaft-type rotation (see Fig. 2.2) at defects or dislocations. The sensitivity of the γ-transition to changes in crystallinity shown in Fig. 4.6 indicates that this also should correspond to the disordered (defect) phase.

There is a remarkable difference between the dielectric and mechanical relaxation spectra in PTFE, as shown in Fig. 4.6. Especially in the $\alpha - \beta$ transition region, the mechanical band widths are much broader than the dielectric ones. This supports the view that in this range crystallinity, i.e. long-range order, is retained but there is rotational disorder within the crystal.

4.2 Polymers with flexible polar side groups

In this section the dielectric behaviour of polymers which have apolar main chains and polar side chains is discussed. This corresponds to the case illustrated schematically in Fig. 2.18 c. In such polymers the highest dielectric effect is expected at the β_a-transition which corresponds to hindered rotation of the polar side group. An example for this, polymethyl methacrylate, has already been given in Chapter 2. At the glass–rubber transition where the main-chain microbrownian motion is highly activated, the dielectric effect is due to the side groups which move along with the main chain. On the other hand, as a result of the increase in free volume the potential barrier hindering the rotation of side groups is drastically reduced. In this case the α and β-processes are mixed—cf. Fig. 2.19. The dielectric effect at not too low (> 1 kHz) frequencies is uniquely due to side-chain rotation, so it reflects the perturbation of the α-process to β. By comparing the mechanical and dielectric spectra in these systems, $\alpha - \beta$ mixing can thus be easily studied.

In this section the dielectric spectra of three basic groups of polymers having flexible polar side chains will be discussed: polyvinyl acrylates and methacrylates, polyvinyl esters and polyvinyl alkyl ethers.

Polyvinyl acrylates and methacrylates

The general formula of polyvinyl acrylates and methacrylates is

$$
\begin{array}{ccc}
\text{H} & \text{R}_1 & \text{H} \\
| & | & | \\
\sim\text{C} - \text{C} - \text{C} \sim \\
| & | & | \\
\text{H} & \text{C} & \text{H} \\
& \underset{\text{O}}{\overset{}{\diagup}}\;\;\underset{\text{OR}_2}{\overset{}{\diagdown}}
\end{array}
$$

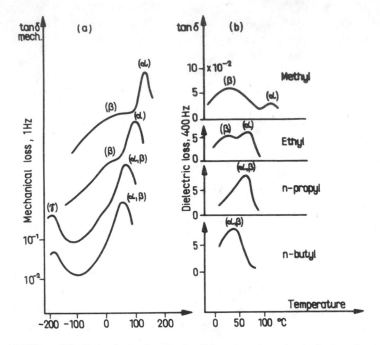

Fig. 4.8 Effect of β-alkyl substitution on the dielectric and mechanical relaxation spectra of polymethacrylates (after Mikhailov, 1958; and Heijboer, 1960)

where R_1 is usually hydrogen (acrylates), CH_3-group (methacrylates), and R_2 is an alkyl group. For polymethyl methacrylate (which has already been discussed in Chapter 2) $R_1 = CH_3$, $R_2 = CH_3$.

The common feature of these polymers is that the polar ester group undergoes hindered rotation about the $C-C$ bond which links it to the main chain. The transition corresponding to this motion is referred to as the β-process. In the glassy state the hindered rotation of this group is fairly well described by the barrier theory. The barrier heights are determined by the chemical composition and by the stereochemical configuration of the molecules.

Effect of substitution at the ester side group R_2

The effect of ester side-chain substitution R_2 with different nonpolar groups on the dielectric and mechanical relaxation spectra is shown in Fig 4.8 after Mikhailov and Borisova (1958) and Heijboer (1960a,b). It is seen that by increasing the size of the side group R_2 the dielectric

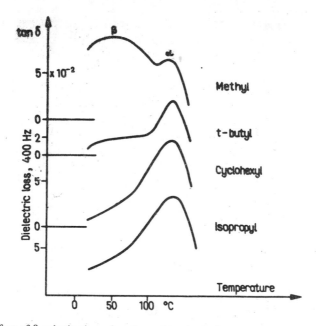

Fig. 4.9 Effect of β-substitution of methacrylates by bulky groups, (after Mikhailov and Borisova, 1958)

α-peaks increase and the β-peaks decrease in amplitude; the α-peaks are shifted to lower temperature until they merge with the β-peak. When the β-peak is resolved the activation energy is found not to be much affected by the substitution. The apparent activation energy of the combined (αβ)-transition was found to increase from 23 kcal/mole (polymethyl methacrylate) to 30 kcal/mole (Poly *n*-butyl methacrylate), 43 kcal/mole (poly *iso*-butyl methacrylate). (Brouckere, Offergeld, 1958).

Figure 4.8 also shows the change of the mechanical spectra by substitution of R_2 with alkyl groups. It is seen that the mechanical α-peaks are shifted to low temperatures by increasing the length of the substituted groups while the β-peaks are practically unchanged.

In the mechanical spectra the α-peaks appear to be more intense than the β-peaks, while in the dielectric spectra this ratio is opposite. The reason is that in the dielectric spectra the α-process is monitored through the α−β interaction because the polar group is located in the side group.

Figure 4.9 shows a set of dielectric spectra obtained by substituting R_2 by bulky types of groups, $R_2 = C - (CH_3)_3$ butyl; $R_2 = CH(CH_3)_2$

*iso*propyl, and *cyclo*hexyl groups. It is seen that this kind of substitution does not affect the position of the α-maximum but increases its height while the intensity of the β-maximum is decreased.

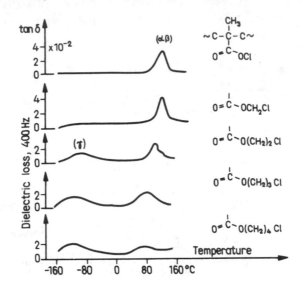

Fig. 4.10 The dielectric spectra of alkyl chloroacrylates (after Borisova *et al.*, 1969)

Borisova *et al.* (1969) investigated the dielectric spectra of such acrylate derivatives by which the group R_2 was substituted by polar groups such as Cl and $C-N$. One group of the studied compounds was

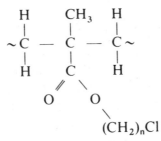

where *n* ranged from 0 to 4. The corresponding dielectric spectra are shown in Fig. 4.10. The α- and β-peaks are not separated, showing that in these compounds the (αβ)-interaction is large; the complex (αβ)-peak is shifted to low temperatures by increasing the length of the substituted alkyl group which is labelled by the polar $C-Cl$ bond. An additional

210

low-temperature peak appears at about $-120°C$, corresponding to the local motion of the substituted $(CH_2)_nCl$ group, which is decoupled from the α and β processes. The spectra correlate well with the mechanical spectra obtained by similar substitution of R by non-polar alkyl groups of increasing length (Fig. 4.7).

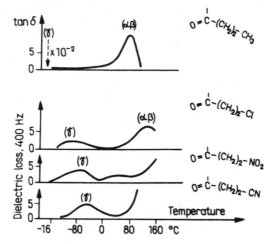

Fig. 4.11 Effect of labelling *n*-propyl methacrylate with groups of different polarity to the dielectric spectra (Borisova *et al.*, 1969)

Another set of spectra shown in Fig. 4.11 corresponds to compounds obtained by substituting the CH_3 group of poly *n*-propyl methacrylate by groups having different dipole moments as

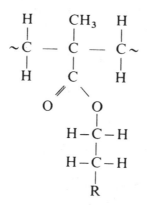

where $R = CH_3$ (0.4 debye), Cl(1.95 debye), NO_2(3.20 debye) and $C-N$(3.8 debye).

It is seen that by increasing the polarity of the labelling groups the γ-relaxation peak and the mixed (αβ)-peak are both shifted to higher temperature. The γ-relaxation peak of poly β−n propyl methacrylate

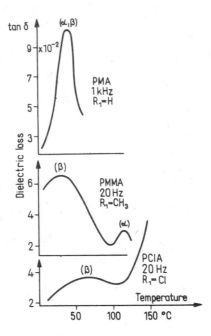

Fig. 4.12 Effect of α-substitution on the dielectric spectra of polymethacrylates

does not appear in the dielectric spectrum because the CH_3 group is not polar. It appears, however, in the mechanical-relaxation spectrum as shown in Fig. 4.7.

Effects of α-substitution R_1

By substituting the R_1 group which is attached directly to the main chain the glass-transition temperature is expected to be influenced mainly. This group is, however, rather effective in the formation of the potential barrier which hinders the rotation of the ester group. As an example for this the dielectric spectra of polymethyl methacrylate (PMMA) $R_1 = CH_3$, $R_2 = CH_3$, polymethyl acrylate (PMA) $R_1 = H$, $R_2 = CH_3$ and polymethyl α-chloro-acrylate (PClA) $R_1 = Cl$, $R_2 = CH_3$ are shown in Fig. 4.12.

212

The peak observed at about 40°C in polymethyl acrylate is probably due to the mixed (αβ) process. Mixing is almost complete in this case because the rotation of the side group is not much hindered. The oscillator strength of this transition is very high, which indicates that the polar side group moves along with the main chain. By substituting an α-methyl group the rotation of the side chain is hindered more and the αβ-mixing becomes incomplete. This means that this substitution has made the chains stiffer.

Polymethyl α-chloroacrylate has considerably higher glass-transition temperature than the methyl acrylate and methyl methacrylate. As the van der Vaals radius of the Cl atom is approximately equal to that of the CH_3 group, this change is attributed to the polarity of the $C-Cl$ bond by which the inter and intramolecular interactions are increased.

Mikhailov and Borisova (1964a, b) labelled the ester side groups in poly α-chloroacrylates by chlorine and studied the dielectric spectra of the corresponding

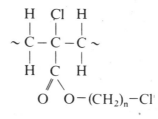

compounds. These compounds have three polar groups; the α-chlorine moves along with the side chain making the dielectric glass transition intense, the ester chlorine labels the motion of the side chain, while the ester carbonyl group indicates ester-group rotation.

By increasing the length of the alkyl chain n the T_g is shifted to lower temperatures, as in the case of methacrylates. The β-transition is clearly observed only in poly methyl chloroacrylate as a broad peak near 60—80°C. In the higher values of n the β-peak merges with α.

The shift of the glass–rubber transition by substitution observed in polyalkyl acrylates and methacrylates means that the chain mobility has increased. This effect is referred to as 'internal plasticization' because it has the same effect on T_g as addition of low molecular weight plasticizers (see Chapter 5). Figure 4.13 shows the T_g-values and the densities of a series of n-alkyl substituted methacrylates as a function of the length of the substituted group. It is seen that the glass transition temperature is greatly reduced by increasing the substituted chain length and that this change runs parallel with the reduction of the density. This is consistent with the picture sketched in Chapter 2 of the glass transition as being mainly governed by the free volume. The packing of the chains is loosened by substitution; the free volume and

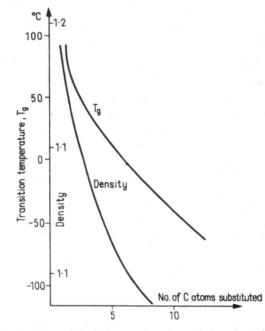

Fig. 4.13 Internal plasticization in poly-*n*-alkyl methacrylates

the density increases, and this results in a decrease of T_g. By substituting less flexible isobutyl, *tert*-butyl or *cyclo*hexyl groups, on the other hand, no internal plasticizing effect occurs (see Fig. 4.9).

For interpreting the effects of substitution on the α-relaxation the change in packing density should be considered. Packing fraction is defined as

$$K = \frac{\varrho \, N_A \sum_i \Delta v_i}{M} \tag{4.3}$$

where N_A is Avogadro's number, Δv_i is the increment of the volume of the repetition unit of the polymer, M is the molecular weight of the repetition unit, ϱ is the density of the polymer.

214

Δv, M, ϱ and K values for the most important groups are tabulated by Slonimsky *et al.* (1970). It is seen there that by substitution of the ester group by alkyl groups of increasing length in polyalkyl methacrylates the packing fraction K is decreased as 0.684 ($R_2 = CH_3$), 0.680 ($R_2 = CH_2CH_3$), 0.668 ($R_2 = (CH_2)_2CH_3$), 0.665 ($R_2 = (CH)_3 CH_3$).

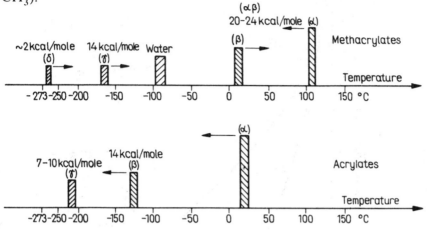

Fig. 4.14 Transitions in alkyl methacrylates and acrylates

In Fig. 4.14 the main transitions in polyalkyl acrylates and methacrylates are summarized schematically. The shifts caused by substitution are also shown. It is seen that these polymers exhibit a glass–rubber transition (α) which is essentially dependent on the packing fraction in accordance with the free-volume theory (cf. Chapter 2). The β-process is due to the rotation of the ester side group about the $C-C$ bond which links it to the main chain. At higher temperatures or at lower packing fractions obtained by substitution, the α and β peaks are merged by the $\alpha\beta$-interaction process.

As has already been discussed in Chapter 2, PMMA exhibits a γ-transition attributed to rotation of the α-methyl group. As this group is apolar this transition is only detected by mechanical resonance or by NMR. Cyrogenic transitions are observed at about 40 K (impurities) and 4 K (δ-process), (see section 2.3). The δ-transition is attributed to the rotation of the ester–alkyl group. This transition has been detected dielectrically by labelling this group by chlorine. In the case of $R_2 = CH_3$ the methyl-group rotation is hindered only at very low temperatures (≤ 4 K).

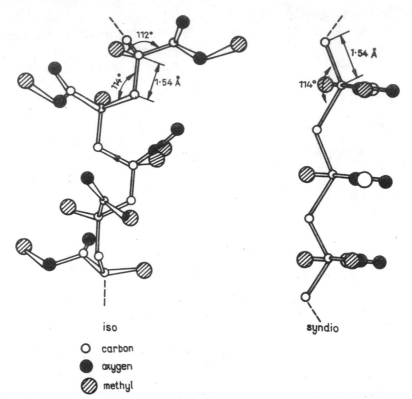

iso   syndio

○ carbon

● oxygen

◎ methyl

Fig. 4.15 Chain configuration in isotactic and syndiotactic polymethyl methacrylate
(after Shindo, 1969)

Effect of tacticity

It was mentioned in section 1.1, that dielectric spectroscopy should in
principle be sensitive to the stereochemical structure of the polymer. One
reason is the dependence of the effective dipole moment on the stereochemical structure; another is that the physical structure, packing, and
molecular mobilities are also highly dependent on the stereochemical
structure. Atactic polymethyl methacrylate is, for example, amorphous,
while the syndiotactic and isotactic polymers are crystalline.

The first evidence that the dielectric spectra of PMMA are sensitive to
change in tacticity was reported by Mikhailov and Borisova (1960) and
by Mikhailov *et al.* (1965); their result was mentioned in Chapter 1.
The difference in the dielectric β-transition of isotactic and syndiotactic
PMMA has been interpreted on the basis of the barrier theory by
Shindo *et al.* (1969).

216

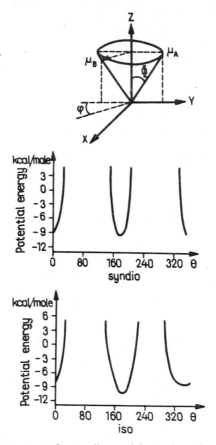

Fig. 4.16 Potential energy maps for syndio- and isotactic polymethyl methacrylates calculated by Shindo (1969)

Shindo *et al.* (1969) calculated the potential barriers which hinder the rotation of the polar ester group in PMMA by using the chain configurations known from X-ray (Stroupe and Hughes, 1958) and infrared spectroscopic (Havriliak and Roman, 1966) data. According to these studies, syndiotactic PMMA exhibits horizontal trans-zigzag chains as shown in Fig. 4.15 while the syndiotactic polymer chains in the crystalline phase are helical. For the isotactic amorphous polymer the configuration is not known; it is assumed to be trans-zigzag as in the isotactic polymer. Values for the main bond angle and bond length values are also indicated in Fig. 4.15.

The calculated potential energy map for syndiotactic and isotactic PMMA is shown in Fig. 4.16.

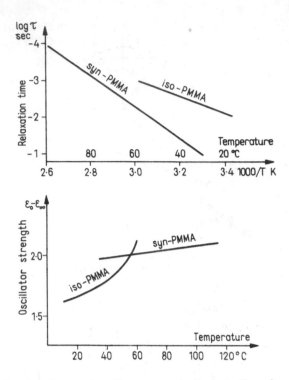

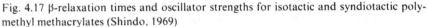

Fig. 4.17 β-relaxation times and oscillator strengths for isotactic and syndiotactic poly-methyl methacrylates (Shindo, 1969)

Figure 4.17 shows the experimental β-transition frequencies (relaxation times) as a function of $1/T$ for both stereoisomers and the measured $\varepsilon_0 - \varepsilon_x$ oscillator strengths as a function of the temperature. It is seen that the activation energies are significantly different for the syndiotactic and isotactic polymers and that the transition frequency at constant temperature is higher for the isotactic polymer than for the syndiotactic one. When measuring the spectrum as a function of the temperature, the β-peak corresponding to the syndiotactic form appears at higher temperature than that corresponding to the isotactic form.

The oscillator strength expressed by the Kirkwood–Fröhlich equation (Eqn. 1.13) should be

$$\varepsilon_0 - \varepsilon_\infty = \frac{3\varepsilon_0}{2\varepsilon_0 + \varepsilon_\infty} \left(\frac{\varepsilon_\infty + 2}{3}\right)^2 \frac{4\pi N_0 \mu_0^2}{3kT} g_r \qquad (4.4)$$

where N_0 is the dipole moment concentration, μ_0 is the permanent dipole moment of the ester group (1.8 debye), and g_r is the Kirkwood reduction factor defined as $g_r = [1 - \langle\mu\rangle_{av/\mu}]^2$ representing the intra- and inter-molecular interactions.

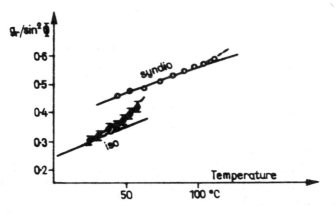

Fig. 4.18 Comparison of measured Kirkwood g_r-values with theory for iso- and syndiotactic polymethyl methacrylates (after Shindo, 1969)

From the potential barrier model the reduction factor is expressed as

$$g_r = \frac{2 \exp\left(\dfrac{-\Delta V}{kT}\right)(1 + \cos\varphi)\sin^2\phi}{\left[1 + \exp\left(-\dfrac{\Delta V}{kT}\right)\right]^2} \qquad (4.5)$$

where the angles φ and ϕ are shown in Fig. 4.16; where μ_A and μ_B are the ester group dipole moments corresponding to sites (A) and (B) respectively. ΔV is the energy difference of the potential minima; its value is $\Delta V = 0.6$ kcal/mole for the syndiotactic and $\Delta V = 1.7$ kcal/mole for the isotactic form. From the measured values of ε_0 and ε_r from Eqn. 4.4 the Kirkwood reduction factor can be determined. It is also possible to calculate g_r from Eqn. 4.5. Figure 4.18 shows the comparison of these calculated and experimental $g_r/\sin^2\phi$ values as a function of the temperature. It is seen that the tendency is clearly reproduced by the barrier calculation, but the absolute values for the isotactic polymer are poor.

219

Polyvinyl esters

Another important family of polymers having apolar main chain and polar side chain are the polyvinyl esters

$$
\begin{array}{ccc}
\text{H} & \text{H} & \text{H} \\
| & | & | \\
\sim\!\text{C} - \text{C} - \text{C} \sim \\
| & | & | \\
\text{H} & \text{O} & \text{H} \\
& | & \\
& \text{O} = \text{C} - \text{R} &
\end{array}
$$

Here R is usually an alkyl group.

The most studied polyvinyl ester is polyvinyl acetate $R = CH_3$ but the higher esters as propionate $R = CH_2CH_3$ and butyrate $R = (CH_2)_2CH_3$ have also been studied.

From the chemical composition it is seen that the structure is similar to that of the acrylates and methacrylates, but the total effective dipole moments of the side group are different because a carbon atom and an oxygen are interchanged. In polyvinyl esters the side group is linked to the main chain by polar $C - O$ bonds, and it is expected to move along the main chain, which makes the dielectric glass transition intenser than that for the acrylates and methacrylates where this link is an apolar $C - C$ bond.

The effect of replacing the R group by alkyl groups of increasing length is essentially the same as with polyacrylates and methacrylates: the glass-transition temperature is shifted to lower temperature and the β-transition temperature corresponding to side group rotation is not changed. This is shown in Fig. 4.19, where the transitions are schematically illustrated. The α-transition is typically a glass–rubber (T_g) transition. In polyvinylacetate this transition has been extensively studied in terms of free-volume theory. In Chapter 2 this polymer was chosen to illustrate volume relaxation (Fig. 2.16). The mechanical-relaxation maxima in the α-transition in PVAc appear at somewhat lower temperatures (higher frequencies) than the dielectric transitions. The position of the α-transition moves to higher temperature as a function of the average molecular weigth, starting from the oligomers (Hendus *et al.*, 1959). This shift levels off at molecular weights in the order of 80,000 and is unchanged in the range of 10,000 and 600,000.

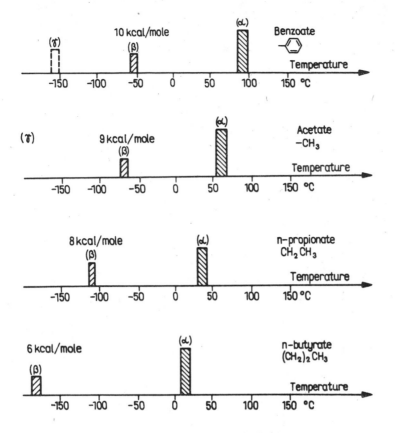

Fig. 4.19 Transitions in polyvinyl esters

The β-relaxation transition is rather weak in the mechanical spectra. In the dielectric spectra the β-peak is well observable; it is attributed to the hindered rotation of the ester side group about the $C-O-C$ bond which links it to the main chain. The corresponding activation energy is 9 kcal/mole.

The γ-relaxation in PVAc has been detected by the ultrasonic method (Thurn and Wolf, 1956) and is believed to be a local mode cryogenic process. This transition has not so far been studied in detail. By replacing the R group with alkyl groups of increasing length the γ-relaxation moves to higher temperature. By labelling with chlorine poly β-chloro vinyl butyral $R = (CH_2)_2 - CH_2 - Cl$ it can be detected by the dielectric method (Mikhailov and Borisova, 1960).

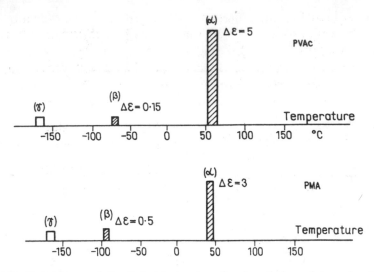

Fig. 4.20 Comparison of the dielectric spectra of polymethyl acrylate and polyvinyl acetate

It is instructive to compare the dielectric properties of polymethyl acrylate (PMA)

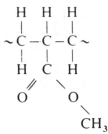

and polyvinyl acetate (PVAc)

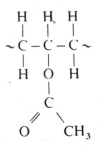

The dielectric spectra of these two polymers as a function of the temperature are shown in Fig. 4.20. Both α- and β-transitions in PVAc

are at higher temperatures than those of PMA. The oscil
of the α-transition $(\varepsilon_0 - \varepsilon_\infty)_\alpha$ is larger in PVAc than that in
facts are consistent with the picture that the dielectric α-peak
to large-scale movement of the main-chain segments which
by the polar group attached to the chain. By this motion th
as a whole moves, and this involves an effective dipole me ...... or the

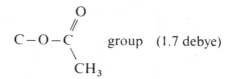

$$C-O-C \quad\quad \text{group} \quad (1.7\ \text{debye})$$

On the other hand, in the β-process the ester side group is rotating
about the $C-O-C$ bond; the effective dipole moment is, correspondingly, due to the

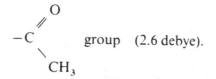

$$-C \quad\quad \text{group} \quad (2.6\ \text{debye}).$$

This is why the oscillator strength of the α-process is increased and that
of the β-process is decreased in PVAc with respect to PMA.

Polyvinyl alkyl ethers

The group of compounds having a general structural formula

$$
\begin{array}{ccc}
H & H & H \\
| & | & | \\
\sim C - C - C \sim \\
| & | & | \\
H & O & H \\
 & | & \\
 & R &
\end{array}
$$

where R is hydrogen (polyvinyl alcohol) or alkyl group, exhibits flexible
polar side groups but in this group the polar bond is the bond of rotation.
It is therefore to be expected that the dielectric α-transition will be
intense and the β-transition weak, which is experimentally observed.

On the other hand, the mechanical β-process should be appreciably
strong. Schell *et al.* (1969) studied a series of polyvinyl alkyl ethers by

223

successively increasing the length of the group R. Figure 4.21 shows, for example, the dielectric and mechanical relaxation spectra of polyvinyl *n*-butyl ether $CR = (CH_2)_2 CH_3$ in comparison with the coefficient of linear thermal expansion measured by dilatometry. It is

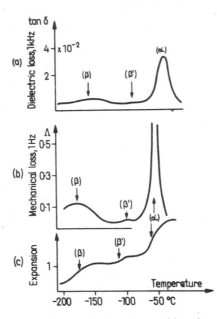
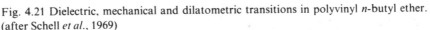

Fig. 4.21 Dielectric, mechanical and dilatometric transitions in polyvinyl *n*-butyl ether. (after Schell *et al.*, 1969)

seen that in this polymer two β-peaks are found (β and β'); the corresponding activation energies are 6.9 and 13.8 kcal/mole respectively.

The dielectric α-peak is strongest, because the $\geqslant C-O-C \leqslant$ dipole is attached rigidly to the main chain. The β-transition, which is attributed to the rotation of the side group, is very weak. In fact it should be absent as the group is rotated about the $C-O$-bond and, so does not contribute to the polarization. Through inductive effect, however, the alkyl group exhibits a certain polarity, which is indeed reflected in the β-loss peak.

The internal plasticizing effect in these polymers is very interesting. By increasing the length of the *n*-paraffin side chain the glass-transition temperature is first decreased and then increased again, as shown in Fig. 4.22 after Schell *et al.* (1969). The β'-transition shows a similar dependence, while β remains unchanged by the substitution. Raising

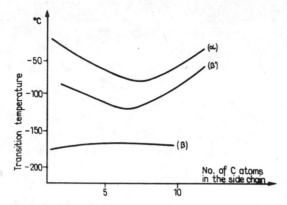

Fig. 4.22 Internal plasticization in polyvinyl alkyl ethers (after Schell *et al.*, 1969)

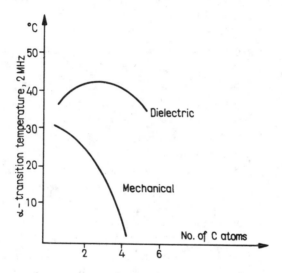

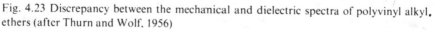

Fig. 4.23 Discrepancy between the mechanical and dielectric spectra of polyvinyl alkyl. ethers (after Thurn and Wolf, 1956)

T_g by increasing the length of the substituted group after C = 8 (*n*-octyl) is explained by crystallization of the side chain which forms a physical crosslink between the main chains and makes them stiffer.

The transition temperatures measured by dielectric spectroscopy in polyvinyl alkyl ethers do not agree well with those measured by the mechanical method, especially in the α-transition region. This is illustrated in Fig. 4.23 for a series of polyvinyl alkyl ethers after Thurn and Wolf (1956). The α-transition temperatures measured at a frequency of 2×10^6 Hz by the dielectric and by the mechanical (ultrasonic) method are plotted against the number of carbon atoms in the alkyl side group.

It is seen that the internal plasticization effect of the flexible alkyl group is shown clearly in the mechanical spectra but not in the dielectric spectra. The dielectric α-transition temperatures at the same frequency are considerably higher than the mechanical ones and do not decrease when chain length is increased. This anomaly might be explained by the fact that the dielectric spectra represent the motion of the $C - O$ dipoles, which are directly linked to the side group, while the mechanical α-transition represents large-scale chain motion.

4.3 Polymers with polar side groups rigidly attached to their apolar main chain

For an important group of polymers the main chain is apolar and the polar groups are directly attached to it. Typical examples are the halogen polymers where chlorine, bromine or fluorine is linked to the carbon–carbon main chain. The $C - Cl$, $C - F$ or $C - Br$ bond which is linked to the chain cannot move independently; it moves along with the chain. The corresponding dielectric spectra indicate the mobility of the main chain. It is therefore to be expected that dielectric transitions will appear at the glass–rubber transition and at transitions where local motion isomerisation of the main chain is involved (cf. section 2.3). In these polymers no β-transition is expected to occur in the sense, defined in the previous section, of being due to side-group rotation. The transitions in the glassy state in these polymers are, however, conventionally denoted by β. A typical polymer having a polar bond attached to its main chain, polychlorotrifluoroethylene, has already been discussed in Chapter 2 to illustrate transitions associated with the crystalline phase.

Polyvinylchloride and related halogen polymers

The most representative polymer belonging to this group is unplasticized polyvinylchloride PVC

$$
\begin{array}{cc}
Cl & H \\
| & | \\
\sim C - C \sim \\
| & | \\
H & H
\end{array}
$$

226

Conventional PVC is mainly amorphous; its crystallinity is about 5 %
and it is prepared above room temperature by a free radical polymerization process. At low temperatures highly crystalline PVC samples can

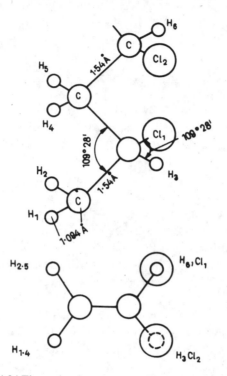

Fig. 4.24 The molecular structure of polyvinylchloride

be prepared (Juhász, 1967). The steric configuration in conventional
PVC is 65 % syndiotactic, having predominantly head-to-tail structure
as shown in Fig. 4.24, where the main bond distances and bond angles
are also given (Natta and Corradini, 1956, Kashivagi and Ward, 1972).

As expected, the strongest transition in PVC is the glass–rubber
transition at about 80°C detected by the mechanical as well as by the
dielectric technique. The oscillator strength of the dielectric transition
is $(\varepsilon_0 - \varepsilon_\infty)_\alpha \approx 8$. The T_g-transition of PVC has already been discussed in
detail in section 2.2, where changes in the mechanical, thermal and
electrical behaviour were illustrated (Fig. 2.9). The transitions in PVC
as detected by the NMR spin-echo method are shown in Fig. 3.39.

Another very broad transition is observed in the temperature range from $-50°C$ to $+40°C$, which is referred to as the β-transition. The oscillator strength of the dielectric β-transition is $(\varepsilon_0 - \varepsilon_\infty)_\beta \approx 0.5$; it increases slightly with increasing temperature. The mechanical β-transition is more pronounced than the dielectric one. Kakutani and Asahina (1969) have found an indication that the β-peak is resolved into two peaks. One is centered at $-50°C$, and is attributed to the amorphous phase (β_a); the other, centered on $0°C$, is attributed to the crystalline phase (β_c).

In the dielectric spectra the β_a and β_c maxima are not resolved; even in the mechanical spectra they are hardly distinguishable.

The β-peak is found to decrease and vanish when small amounts of plasticizers are added (see Chapter 5).

In some cases a γ-relaxation peak can be observed in PVC by the mechanical method at about $-100°C$.

The β-transition has not been clearly explained. As will be shown in section 5.2, this transition vanishes when small amounts of plasticizers ($<10\%$) are added. On the basis of this it is difficult to explain the β-transition in PVC as a local motion of the main chain, as assumed earlier (Boyer, 1963). This problem will be discussed in some detail in Chapter 5.

Figure 4.25 shows the transitions in polyvinylchloride, polyvinylidene chloride, polyvinylfluoride and in polyvinylidene fluoride schematically. The activation energies of the transitions are also indicated. It is clearly seen that by substituting an additional halogen atom, chlorine or fluorine, the glass–rubber transition is shifted to lower temperatures and the oscillator strength is reduced. This shows on one hand that the polarity of the $C-F$ or $C-Cl$ bond has a remarkable influence on the inter- and intramolecular interaction, on the other hand the reduction of the oscillator strength shows that the dipole moments in polyvinylidene chloride and fluoride are partially cancelled.

This cancellation is also observed in the chlorination of PVC. The oscillator strength of the dielectric α-transition is drastically reduced by increasing chlorine content (Reddish, 1965), while the transition is somewhat shifted to higher temperature. A typical series of spectra recorded as a function of the temperature is shown in Fig. 4.26.

On the basis of the cancellation of the dipole moments, Reddish (1965) has suggested that at the glass-transition temperature in PVC a drastic change in the effective dipole moment occurs. This would mean that T_g in PVC could be regarded as a dielectric order–disorder

228

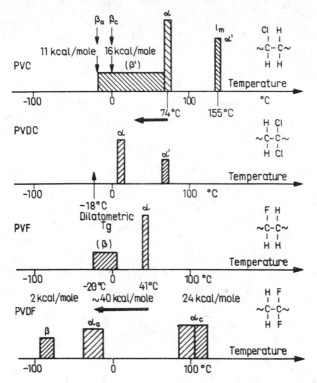

Fig. 4.25 Transitions in halogen polymers PVC : polyvinylchloride, PVDC : polyvinylidene chloride, PVF : polyvinylfluoride, PVDF : polyvinylidene fluoride

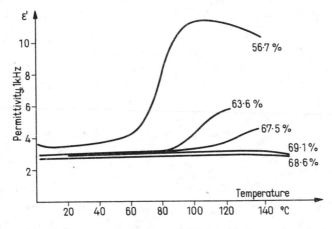

Fig. 4.26 Dielectric spectra of chlorinated PVC with increasing chlorine content. (Reddish, 1962)

transition. Reddish (1965) assumed the presence of an antiferroelectric arrangement in PVC below T_g, in which the dipoles are partially cancelled. At a certain temperature, antiferroelectric Curie temperature, this antiferroelectric order is thought to be broken, resulting in an increase of the effective dipole concentration. This picture is not supported by the analysis of chain conformations in PVC, which would make antiferroelectric orientation of the $C-Cl$ dipoles very improbable.

In order to understand the effect of the polarity of the side group on the glass-transition temperature it is interesting to compare the transitions of the polar halogen polymers with those of their non-polar analogues.

In Fig. 4.27 the transitions in polypropylene and polyvinylchloride, polyisobutylene and polyvinylidene chloride are compared. The $-CH_3$ group has approximately the same van der Vaals radius as the chlorine atom. So the very large shift of the T_g of PVC with respect to PP and that of PVDC with respect to PIB might be attributed to the polarity of the $C-Cl$ bond.

The anomalously high glass-transition temperature of polyvinyl-chloride and of the other halogen polymers with respect to their hydro-carbon analogues is not yet fully explained. On the basis of infrared analysis, Warrier and Krimm (1970a) suggested that the $CH-Cl$ intra-molecular interaction in PVC is analogous to a hydrogen bond. Thus the chains would be linked by the $C-H \ldots Cl-C$ interactions occurring in the syndiotactic chain configuration shown in Fig. 4.24. This is possible because the $C-Cl$ bond of one molecule in the crystalline syndiotactic polymer is collinear with the $Cl-C-H$ bond of the other (Natta and Corradini, 1956). This is strictly true only in the crystalline phase, but as in other polymers in amorphous PVC a short-range order is present which makes such interactions possible. This interaction (weak hydrogen bond) might be effective by plasticization—see Chapter 5 and degradation of PVC (Chapter 7).

Another interesting shift is shown in Fig. 4.27, namely that by alternate symmetric substitution of polar Cl or apolar CH_3 groups the glass-transition temperature is considerably shifted down. The T_g of poly-propylene is near $0°C$, while that of polyisobutylene is $-70°C$. This is explained by Gibbs and di Marzio (1959) by the change in the conformational energy differences between the equilibrium configurations of the main chain.

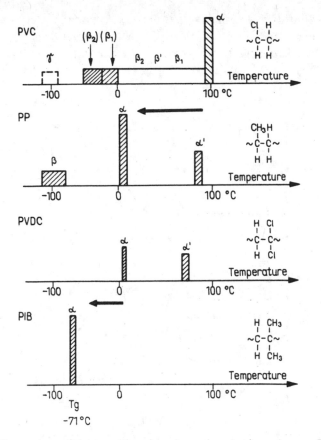

Fig. 4.27 Comparison of the transitions in polar and non-polar polymers: PP: polypropylene, PVC: polyvinylchloride, PIB: polyisobutylene, PVDC: polyvinylidenechloride

The hydrogen-bond structure of halogen polymers might also contribute to the drastic change of the oscillator strength of the dielectric α-transition when the temperature is swept over T_g. According to this view the effective dipole moment of the system is considerably weakened by the hydrogen bonds; correspondingly, the effective dipole moment concentration at $T < T_g$ is relatively small. Near T_g, however, the hydrogen bonds are broken, resulting in an increase of the effective dipole moment. This would offer an alternative explanation to that offered by Reddish (1965) and discussed somewhat earlier in this section. No direct evidence has been found so far for deciding which mechanism is really effective in the dielectric T_g-transition in PVC.

231

The oscillator strength of the dielectric α-transition in halogen polymers measured as a function of the temperature is considerably increased by decreasing the frequency. A series of $\varepsilon'(T)$ spectra indicating this effect has been shown in Fig. 2.11 for unplasticized PVC. From this

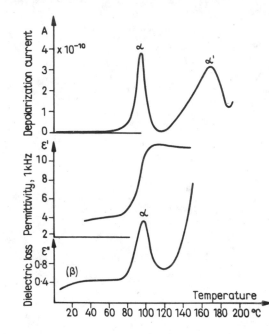

Fig. 4.28 Dielectric depolarization spectrum of polyvinylchloride compared with the a. c. dielectric spectra

behaviour it is assumed that at low frequencies a very high Maxwell–Wagner–Sillars (MWS) polarization (cf. Chapter 5) is developed in this polymer. This is supported by the known aggregate structure of the polymer.

From this it is concluded that at least a part of the observed 'dielectric transition', i.e. sharp increase in the relaxed permittivity at T_g, is due to the MWS mechanism. This is supported by the very high dielectric depolarization peaks observed in halogen polymers. The aggregate structure of PVC and the corresponding MWS-polarization will be discussed in Chapter 7 in connection with physical structural ageing.

The dielectric depolarization spectrum of unplasticized polyvinylchloride is compared in Fig. 4.28 with the 1 kHz dielectric-relaxation spectra. It is seen that the depolarization spectrum exhibits an intense

peak α' near $+150°C$ as well as the $\alpha(T_g)$ and β-peaks
cannot be detected with the conventional a.c. metho
such a high temperature the ohmic conductivity becor

The extremely high MWS polarization in PVC i
permittivity inside the aggregates should differ greatly from
surface. From this it might be concluded that the effective dipole
moment inside the aggregates is much smaller than that at the surfaces
or in the inter-aggregate space. This is rather likely, as the aggregates
exhibit a certain order; they behave like crystallites in which either
ferroelectric cancellation of the moments or cancellation through intermolecular and intramolecular hydrogen bonding is more probable than
at the surfaces.

Polyvinyl acetals

Polyvinyl acetals are amorphous polymers having a general chemical
structure

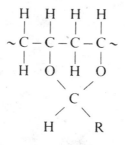

where R is usually an apolar alkyl group. As these polymers are
synthesized from polyvinyl alcohol by condensation with an aldehyde
the polymers contain polyvinylalcohol and polyvinylacetate sequences
in their main chain.

As the polar $C-O$ groups are rigidly attached to the apolar main
chain in these polymers the dielectric spectrum directly reflects the
chain mobility. As in the other polymers belonging to this group, two
main transitions are observed; one is the glass–rubber transition (α),
which is fairly intense, and the other is a very broad transition at about
$-10°C$, corresponding to local motion of the main chain.

By substituting R with apolar alkyl groups of increasing length, the
glass–rubber (α) transition shifts to lower temperatures, as shown in
Fig. 4.29 after Mikhailov (1955). This T_g-shift is due to the internal

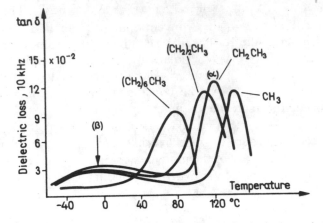

Fig. 4.29 Effect of substitution on the dielectric spectra of polyvinyl acetals (Mikhailov, 1955)

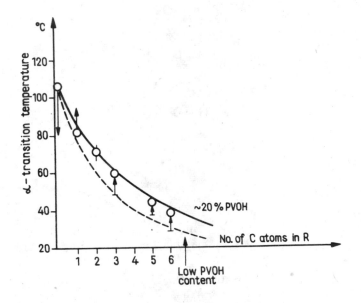

Fig. 4.30 Internal plasticization in polyvinyl acetals

plasticizing effect of the side group; it is similar to that in acrylates and methacrylates discussed in section 4.2.

Figure 4.30 shows the change in the glass–rubber transition temperature as a function of the number of carbon atoms in the substituent group R by normal substitution. The dashed line roughly corre-

sponds to the extrapolated values of no polyvinylalcoho'
solid line shows the dependence for samples having about
alcohol. By increasing polyvinylalcohol content, the T_g ᴠ
formate R$=$H is decreased, that of polyvinyl acetal ,
-propional, R$=$CH$_2$CH$_3$, n-butyral R$=$(CH$_2$)$_2$CH$_3$, n-hexa..
R$=$(CH$_2$)$_4$CH$_3$, n-heptanal R$=$(CH$_2$)$_5$CH$_3$ and n-octanal
R $=$ (CH$_2$)$_6$CH$_3$ is increased (Fitzhugh, Crozier, 1952; Fujino et $al.$,
1962). In Fig. 4.29 it is seen that the β-transition in polyvinyl acetals is
not affected by the substitution.

4.4 Polymers with polar main chain

In one class of polymers the polar group is directly built in the chain.
In these cases the chains contain hetero atoms, as in the oxide polymers
oxygen. The most common polar groups in chain are C$-$O having a
dipole moment of 0.7 debye and Si$-$O 0.95 debye. The inorganic
polyphosphates, -selenates and -arsenates are not so widely used as the
organic polymers or polysylanes silicon rubbers.

   In this section the dielectric spectra of oxide polymers, polyesters and
polycarbonates will be discussed in some detail. Polyamides and polyurethanes will be discussed in section 4.5 in connection with hydrogen
bonds and silicon rubber will be discussed in Chapter 6.

Oxide polymers

Oxide polymers are characterized by the polar C$-$O$-$C bond in their
main chain. As a result of this packing is loosened and the microbrownian mobility of the chains increased; the transitions are
correspondingly shifted to lower temperature. Figure 4.31 shows
schematically the transitions in a series of oxide polymers as polyoxymethylene (POM)

$$\begin{array}{c} H \\ | \\ \sim C-O \sim \\ | \\ H \end{array}$$

Polyethylene oxide (PEO)

$$\begin{array}{c} H \quad H \\ | \quad | \\ \sim C - C - O \sim \\ | \quad | \\ H \quad H \end{array}$$

235

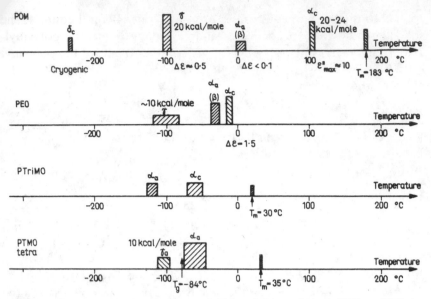

Fig. 4.31 Transitions in oxide polymers; POM: polyoxymethylene, PEO: polyethylene oxide, PTriMO: polytrimethylene oxide, PTMO: polytetramethylene oxide

polytrimethylene oxide (PTriMO)

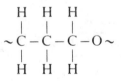

and polytetramethylene oxide (PTMO)

$$\sim C - C - C - C - O \sim$$

The transitions are successively shifted to lower temperatures as the number of CH_2 groupes between the oxygen atoms in the chain is increased.

These oxide polymers are crystallizable and in Fig. 4.31 the transitions attributed to the amorphous and crystalline phases are distinguished.

236

α_c is believed to be essentially the same transition as that in polyethylene, i.e. that corresponding to the motion at the surface of the crystallites. The shift of the α_c-transition, correspondingly, indicates that the crystallite size is reduced. The α_a-transition, usually designated as β-transition in these polymers, is attributed to the amorphous phase, at

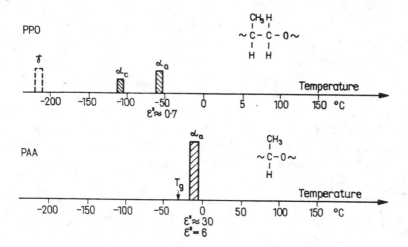

Fig. 4.32 Comparison of the transitions in polyacetaldehyde and polypropylene oxide

least it is highly increased by increasing amorphous content. The shift of α_a is, correspondingly, interpreted as being due to the decrease in the degree of packing, accompanied by higher chain mobility.

Comparing the main transitions in the oxide polymers of Fig. 4.31 with those of polyethylene (Fig. 4.1) it appears that the packing is influenced only when oxygen atoms are separated by more than one CH_2 group. The reason is evidently the influence of oxygen on the chain configuration, i.e. on the twist of the chains.

It is concluded that this substitution mainly affects the crystallinity and not the chain mobility.

It is interesting to compare the intensities of the dielectric transitions in polyoxymethylene (POM) and in polyacetaldehyde. Figure 4.32 shows the transitions observed in polyacetaldehyde

$$\sim \overset{\displaystyle CH_3}{\underset{\displaystyle H}{\overset{|}{\underset{|}{C}}} - O \sim}$$

237

and in polypropylene oxide

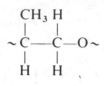

It is seen that by substitution the α_a peak is somewhat reduced and the crystalline transitions become very weak. The increased oscillator strength $\varepsilon_0 - \varepsilon_\infty$ in polyacetaldehyde with respect to POM indicates that the effective dipole moment has been increased by the methyl substitution, as the concentration of the polar $C-O-C$ groups in these two polymers is the same. This is explained by the effect of the substitution on the chain conformation by which the *trans-trans* structure becomes preferred.

The γ-transitions in oxide polymers are observed near $-100°$ C. As discussed in Chapter 2, this peak has been correlated to the motion of the chain-end OH− group in the defects of the crystalline phase in polyoxymethylene

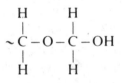

This view is supported by measurement of the γ-transition frequencies over a very wide temperature range up to the crystalline melting temperature (see Fig. 2.37).

In polyethylene oxide single crystals the γ-transition is also present (Ishida *et al.* 1965). This transition has been attributed to local twisting motion in the main chain, but probably it is also due to the hydroxyl end groups

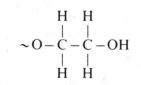

Polyesters and polycarbonates

In polyesters and polycarbonates, oxygen coupled with a benze
is built into the main chain and a carbonyl oxygen is attache
rigidly. The polar groups are found in this class of polymers as

$$\sim R_2 - O - \overset{\displaystyle O}{\overset{\|}{C}} - \underset{\bigcirc}{} - O - R_1 \sim$$

$$\sim R_2 - O - \overset{\displaystyle O}{\overset{\|}{C}} - O - R_1 \sim$$

where R_1 and R_2 may be a phenyl ring or alkyl group.

The benzene ring introduced in the main chain of these polymers makes the chains stiffer and correspondingly the transitions are observed at much higher temperatures than for the oxide polymers. By introducing more phenyl rings a series of polymers having high temperature resistance have been prepared (see, for example, Mark, 1969). The dielectric behaviour of these polymers has not so far been studied in detail.

As a particular example of such polymers, polyethylene terephthalate (PET) will be discussed here.

Polyethylene terephthalate

Polyethylene terephthalate (PET) is a crystalline polymer having a chemical structural formula

$$\sim \overset{\displaystyle O}{\overset{\|}{C}} - \underset{\bigcirc}{} - O - (CH_2)_2 - O \sim$$

It has been extensively studied by dielectric, mechanical, and thermal methods and by NMR because of its technical importance and also because it is a good model for studying the effects of crystallinity on transitions.

The major polar group in this polymer can be considered as rigid in the temperature–frequency range of dielectric spectroscopy, although

239

local rotations, about the $C-O$ bond are likely to be involved. As in the case of the oxide polymers, the effect of the hydroxylic end-groups

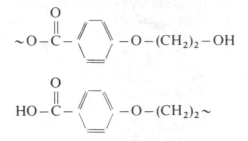

is to be considered in assigning the transitions.

The dielectric and mechanical transitions in PET are shown schematically in Fig. 4.33 for 50 % crystalline polymer.

The crystalline melting temperature is $T_m = 260°C$. At about 100°C the α-relaxation peak can be observed both mechanically and dielectrically. With decreasing crystallinity this peak is reduced in intensity and shifts to somewhat lower temperatures towards the dilatometric T_g of the amorphous polymer $T_g = 67°C$. The α-transition in PET is therefore assigned to the T_g of the amorphous phase.

The dielectric β-transition in this polymer is found at about $-60°C$; its intensity decreases with increasing crystallinity. Reddish (1950) assigned this peak to the hydroxyl end-groups on the basis of the

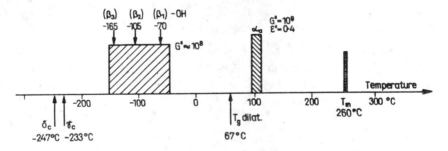

Fig. 4.33 Transitions in 50 % crystalline polyethylene terephthalate

correlation found between the oscillator strength of this transition $\varepsilon_0 - \varepsilon_\infty$ and the $-OH$ content measured by infrared spectroscopy. Illers and Breuer (1963) found three peaks in the β-region, β_1, β_2 and β_3, by mechanical relaxation centered at $-70°C$, $-105°C$ and $-165°C$

respectively. According to these studies the β-process can be considered as being due to the local motion of the main chain, in the amorphous and crystalline phases, and also due to rotation of the end hydroxyl groups.

Crystalline PET exhibits two cryogenic transitions at 40 K (γ_c) and 26 K (δ_c) which are absent in the amorphous polymer and this is attributed to motion in the crystalline state (Armeniades and Baer, 1970; Hiltner and Baer, 1972).

The flexibility of the main chains of polyesters is increased when the methylene chain between the rings is longer:

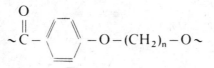

where $n = 2 - 10$.

Figure 4.34 shows the change of the α- and β-transition temperatures as a function of the length of the $(CH_2)_n$ group on the basis of the data of

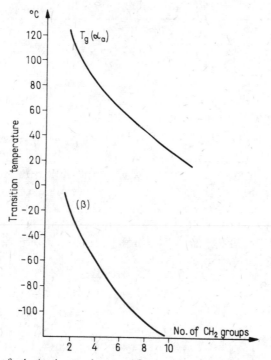

Fig. 4.34 Effect of substitution on the α- and β-transitions of polymethylene terephthalates

Farrow *et al.* (1960). It is seen that both transition temperatures are considerably reduced by increasing the alkyl chain length in the main chain. The shift of the α-transition temperature indicates that in these polymers this transition is the glass–rubber transition; the packing and the large-scale mobility of the main chain are considerably influenced by the substitution. The decrease of the β-transition temperature, on the other hand, indicates that the main contributor to this transition is the local motion of the main chain and not the −OH end group.

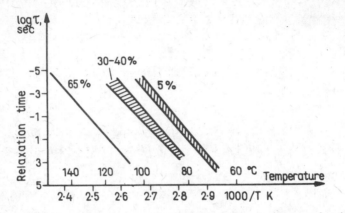

Fig. 4.35 Effect of crystallinity on the dielectric relaxation times of polyethylene terephthalate in the α-transition region

Effects of crystallinity

Polyethylene terephthalate can be prepared in the amorphous state by quenching from the melt to below room temperature. This polymer remains amorphous up to 90°C, where crystallization starts. In the temperature range between 90°C and 170°C polymers having crystallinity up to 30% can be prepared by annealing. The crystallinity proceeds up to a limit at a given temperature and not further. In the temperature range of 170°C to 230°C crystallization proceeds faster, the crystallinity reaching its maximum value of about 50%.

Figure 4.35 shows the effect of crystallinity on the dielectric relaxation times of polyethylene terephthalate in the α-transition region. The transition is shown to be shifted down when crystallinity decreases. There is a considerable difference between the mechanical and dielectric spectra: by increasing crystallinity the mechanical α-peak broadens greatly and the dielectric α-peak practically vanishes.

4.5 Polymers with hydrogen bond

The dielectric properties of an important group of polymers are mainly determined by inter- and intra-molecular hydrogen bonds. Such polymers include polyvinyl alcohol, the polyamides and polyurethanes, cellulose derivatives, proteins, amino-acids and various other polymers of biological interest. The electrical conductivity in the hydrogen-bonded polymers is usually very high with respect to their non-hydrogen-bonded analogues. The conductivity in such systems is considered as being due to proton transfer along hydrogen bond chains (proton-conductivity). Hydrogen-bonded polymers are very sensitive to a small amount of moisture; the electric and mechanical properties are both seriously affected. It seems that a small amount of water is inherent in determining their ultimate properties.

In this section some general features of the dielectric spectra of hydrogen-bonded polymers is discussed and some illustrative examples are presented.

Hydrogen bond chains

In order to form a hydrogen bond which is stronger than a Van der Vaals interaction but weaker than a covalent or ionic chemical bond, polar groups are needed. These are most commonly $-OH$ or $-NH$ groups which would interact through proton exchange. The array of subsequent polar $-NH$ or $-OH$ groups is referred to as a hydrogen bond chain. Such a chain is exhibited intermolecularly for example in polyamides as

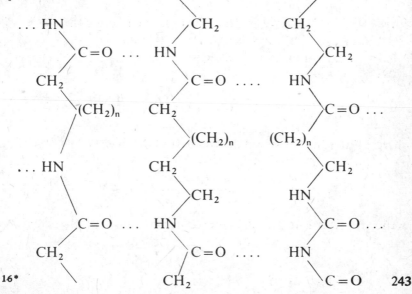

Similar hydrogen-bond chains are formed in polyurethanes, which have similar structure except that oxygen is built in the main chains.

The dielectric polarization of hydrogen-bond chains has been studied by Sack (1952) and Kurosaki and Furumaya (1960) in the case of polyvinyl alcohol. They considered the polarization as being due to orientational defects formed in the hydrogen-bond chains, i.e. in the array of coupled dipoles.

The situation is similar to that of ice studied by Bjerrum (1951), who assumed positive and negative orientational defects in crystalline ice to correspond to two and no protons respectively between two oxygen molecules. Such orientational defects in the hydrogen-bond structure are referred to as Bjerrum defects:

$$\ldots HO \ldots OH \ldots H \ldots O \ldots H \ldots \qquad \text{positive}$$
$$\ldots H-O \ldots HO \ldots OH \ldots OH \ldots OH \ldots \qquad \text{negative}$$

According to Sack (1952), the dielectric polarization of hydrogen-bonded chains is not determined by the individual rotation of the OH bonds but by successive rotations. In the idealized case of having N chains of the same length containing n elementary dipoles (OH groups) each, the Bjerrum defect may have $n+1$ positions for each chain. The probability of finding a chain in the j-th state is

$$\mathscr{P}_j = \mathscr{P}_0 \left(1 + \frac{\mu_j \mu_0}{kT}\right) \mathscr{E}_z \qquad (4.6)$$

where $\mathscr{E}_z$ is the internal electric field component along the chain, $\mathscr{P}_0$ is a constant and μ_0 is the dipole moment of the unit $-OH$.

The polarization per unit volume is

$$\mathscr{P} = \frac{16\,\mu_0^2 (n+1)^3\, N}{\pi^4 kT(1+i\omega\tau_n)} \langle \mathscr{E}_i \cos \vartheta \rangle_{av} \qquad (4.7)$$

where ϑ is the angle between the chains and the field

$$\tau_n = \frac{2\,(n+1)^2}{2}\, \tau_0 \qquad (4.8)$$

where τ_0 is the relaxation time corresponding to the elementary transition.

For random orientation of chains $\langle \cos \vartheta \rangle = 1/3$ and, if the internal field is taken to be equal to the external one, the dielectric permittivity components are expressed as

$$\varepsilon' - \varepsilon_\infty = \frac{64 \mu_0^2 \mathscr{P}_0 N_0 (n+1)^2}{3 \pi^3 kT} \frac{1}{1 + \omega^2 \tau_n^2} \tag{4.9}$$

$$\varepsilon'' = \frac{64 \mu_0^2 \mathscr{P}_0 N_0 (n+1)^2}{3 \pi^3 kT} \frac{\omega \tau_n}{1 + \omega^2 \tau_n^2} \tag{4.10}$$

It is seen that the polarization of hydrogen-bonded chains exhibiting Bjerrum positive or negative defects according to this simple model results in a Debye-type relaxation with an effective relaxation time τ_n related to the chain length n and oscillator strength

$$\varepsilon_0 - \varepsilon_\infty = \frac{64 \mu_0^2 \mathscr{P}_0 N_0 (n+1)^2}{3 \pi^3 kT} \tag{4.11}$$

also related to the chain length n and total concentration of chains N_0.

According to the theory of Sack, the dielectric polarization in hydrogen bonded polymers should depend very strongly on their physical structure. At transitions, especially at the glass–rubber transition, in order to obtain large-scale mobility of the chains hydrogen bonds must be broken. The situation is complicated by the possibility of change in the length of hydrogen-bond chains as a function of the temperature.

Typical transitions in polymers having linear hydrogen-bond chains

There is a certain similarity between the transitions measured by mechanical and dielectric spectroscopy of hydrogen-bonded polymers exhibiting linear (not cyclic) hydrogen-bond chains. Figure 4.36 shows for example the main transitions in polyamides, and in polyurethanes, corresponding to about 10 Hz frequency. It is seen that in these polymers three main transitions are exhibited: the α_a-transition is near 50 C and corresponds to the glass–rubber transition of the amorphous phase. The β-transition is very weak in the dry polymers but is greatly increased by increasing moisture content (see later). The γ-transition is attributed to the local motion of the chain segments located between

245

the hydrogen bonds. These segments mainly consist of apolar CH_2 groups but in polyamides attached polar carbonyl groups, and in polyurethanes in chain polar $C-O$ groups, are also involved, so that transition can be observed dielectrically.

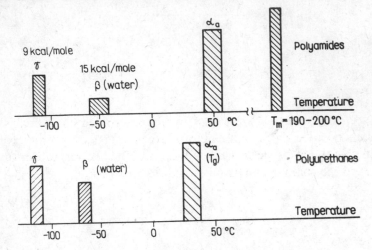

Fig. 4.36 Transitions in polyamides and in polyurethanes

The general formula for one class of polyamides is

$$\sim N-(CH_2)_x-N-\overset{\overset{\displaystyle O}{\|}}{C}-(CH_2)_{y-2}-\overset{\overset{\displaystyle O}{\|}}{C}\sim$$
$$\quad\ |\qquad\qquad\ \ |$$
$$\quad\ H\qquad\qquad\ H$$

where x and y are variable. The compound exhibiting $x = 6$, $y = 6$ is conventionally referred to as 66-nylon, that of $x = 6$, $y = 10$ as 610-nylon, etc.

Another group of polyamides has a general structural formula

$$\sim N-(CH_2)_{x-1}-\overset{\overset{\displaystyle O}{\|}}{C}\sim$$
$$\quad\ |$$
$$\quad\ H$$

These polymer are also termed by the value of x. $x = 6$ is referred to as nylon-6 (poly 6-caprolactam).

246

The general structural formula of polyurethanes is

$$\sim N - (CH_2)_x - N - \overset{\overset{\displaystyle O}{\|}}{C} - O - (CH_2)_y - O - \overset{\overset{\displaystyle O}{\|}}{C} \sim$$
$$\hspace{0.6cm} | \hspace{2.2cm} |$$
$$\hspace{0.6cm} H \hspace{2.2cm} H$$

These polymers are prepared by addition of diisocyanates $OCN - (CH_2)_x - NCO$ and diols $HO - (CH_2)_y - OH$.

The close similarity of the mechanical and dielectric transitions of the various polyamides and polyurethanes suggests that the main factor in determining the transitions is the hydrogen-bond chain structure and not the structure of the main polymer chains. By increasing the number of CH_2 groups in the main chain of polyamides, the transitions would be shifted to slightly lower temperatures. This change is, however, small in comparsion with the effects of crystallinity and the effects of water content.

Figure 4.37 shows the effect of water content on the mechanical relaxation spectrum of a polyurethane based on hexamethylene di-iso-cyanate ($x = 6$) and 1.4 butanediol ($y = 4$) after Jacobs and Jenckel (1961).

For comparison, Fig. 4.38 shows the effect of chemical substitution in the main chain. The polymers are based on hexamethylene diiso-cyanate ($x = 6$), the other components are 1.4 butanediol ($y = 4$), 2.5 hexanediol ($y = 5$) 1.6 hexanediol ($y = 6$) and 1.10 decanediol ($y = 10$).

It is seen that by increasing y the α-peak is slightly shifted to lower temperatures and the β-peak is increased as when water is added but this effect is small.

The behaviour shown in Figs 4.37 and 4.38 is typical for polyamides too. The β-relaxation which is very weak in dry polymers is highly intensified by small moisture content.

Figure 4.39 shows the mechanical and dielectric spectra of 66-Nylon measured as a function of the temperature at frequencies of 1 Hz and 10 kHz respectively. All the main transitions detected by mechanical relaxation spectroscopy are exhibited in the dielectric spectra but there is a continuously increasing loss level indicating the unusually high conductivity. The activation energies calculated from the dielectric spectra in polyamides and in polyurethanes agree fairly well with those obtained from the mechanical spectra.

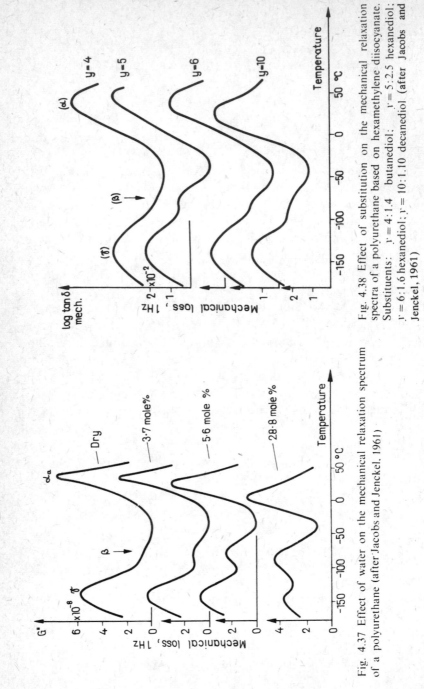

Fig. 4.38 Effect of substitution on the mechanical relaxation spectra of a polyurethane based on hexamethylene diisocyanate. Substituents: $y = 4:1.4$ butanediol; $y = 5:2.5$ hexanediol; $y = 6:1.6$ hexanediol; $y = 10:1.10$ decanediol (after Jacobs and Jenckel, 1961)

Fig. 4.37 Effect of water on the mechanical relaxation spectrum of a polyurethane (after Jacobs and Jenckel, 1961)

248

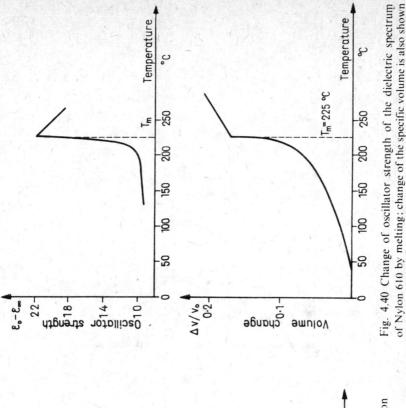

Fig. 4.40 Change of oscillator strength of the dielectric spectrum of Nylon 610 by melting; change of the specific volume is also shown (after Boyd, 1959)

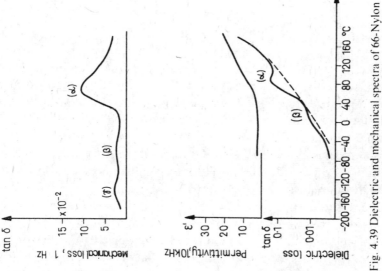

Fig. 4.39 Dielectric and mechanical spectra of 66-Nylon

249

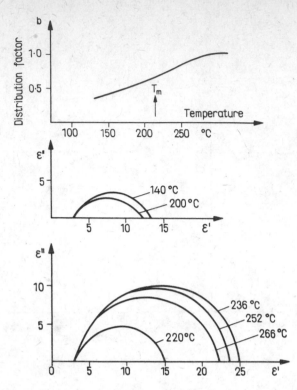

Fig. 4.41 The arc diagrams and relaxation time distribution parameters of Nylon-610 below and above melting temperature (after Boyd and Porter, 1972)

Boyd and Porter (1972) studied the dielectric transition in 610-Nylon by the microwave technique in a wide temperature range up to above the melting temperature. The oscillator strength of the transition as a function of the temperature in comparison with the change in the fractional volume is shown in Fig. 4.40. It is seen that at the crystalline melting temperature of this 50% crystalline polymer a typical 'dielectric transition' is observed: the oscillator strength increases abruptly by passing through the melting temperature.

Figure 4.41 shows a series of Cole–Cole plots measured at different temperatures passing through the melting temperature; the relaxation distribution parameter b is also indicated. It is seen that the distribution parameter increases continuously as a function of the temperature exhibiting no abrupt change at melting.

250

Cellulose derivatives

In cellulose derivatives polar $-C-O-C$ groups are built in the main polymer chain linearly and also in the form of a 6-membered gluco-piranose ring according to the general formula

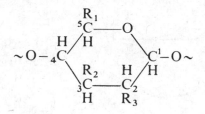

where R_1, R_2, R_3 are usually polar OH, alkyl-ester or ether or nitrogen containing groups.

The rigidity of the structure formed by strong intramolecular hydrogen bonding—see, for example, Forslind, (1971)—makes the glass-transition temperatures of culloluse shift above the decomposition temperature. Correspondingly, in these polymers only T_{gg}-type transitions are observed. The dielectric transitions are strong; the oscillator strengths may be as high as $\varepsilon_0 - \varepsilon_x = 12$. They are attributed to the motion of the polar side groups. This is illustrated in Fig. 4.42 for a series of cellulose derivates after Mikhailov *et al.* (1969b). The first compound, starch,

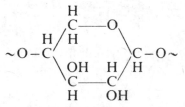

exhibits no dielectric transition up to about $50°$C, where it decomposes. Cellulose hydrate

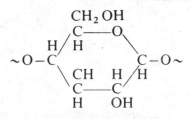

exhibits a distinct β-peak at about $-80°$C at 1 kHz.

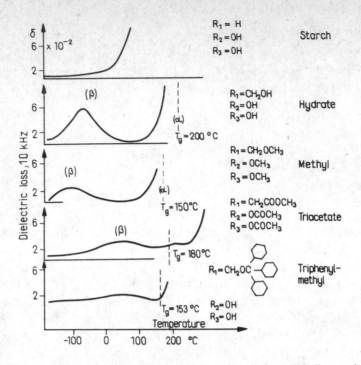

Fig. 4.42 Dielectric spectra of some cellulose derivatives (after Mikhailov *et al.*, 1969)

By substituting the hydroxyl groups in cellulose hydrate by methyl groups

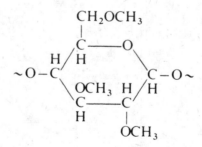

the α- and β-transitions are shifted to lower temperatures and the dielectric peak intensities are reduced.

252

By substitution of more bulky groups as acetate (triacetate cellulose)

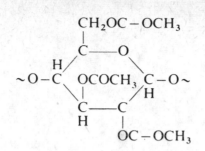

the α- and β-transitions are both considerably shifted to lower temperatures, as shown in Fig. 4.42. It is also seen that the transition intensities are further reduced in comparison with those of hydrate cellulose.

Figure 4.42 also shows the dielectric spectrum of triphenylmethyl-ester cellulose in which R_1 is substituted by a $CH_2OC(Ph)_3$-group

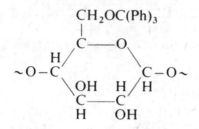

Figure 4.43 shows the proton spin-lattice relaxation times T_1 as a function of the temperature for the cellulose derivates, the dielectric spectra of which are shown in Fig. 4.42. The proton T_1 values exhibit minima at low temperatures (-180 C) for methyl and triacetate celluloses. This low-temperature process has been assigned by Mikhailov et al. (1969 b) to hindered rotation of the methyl group about the $O-C$ bond in the methyl celluloses and about the $C-C$ bond in the triacetate. As the effective frequency of the NMR spin-echo measurement is 20 MHz the dielectric peaks corresponding to this transition should be below the temperature of liquid nitrogen. Although the methyl group is apolar the dielectric γ-transition is expected to be observable as a result of the inductive effect of the oxygen.

That the dielectric β-transition in cellulose derivates is due to the rotation of side groups has been confirmed by the measurements of

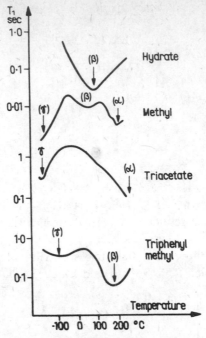

Fig. 4.43 Proton spin lattice relaxation times of some cellulose derivatives (Mikhailov *et al.*, (1969 b)

Mikhailov *et al.* (1967c) for a series of cyanoethyl acetyl celluloses. In these compounds groups containing polar $C-N$ bonds are substituted.

As a particular example, in Fig. 4.44 the dielectric $\varepsilon''(T)$ and $\tan\delta\,(T)$ spectra of cyanoethyl cellulose

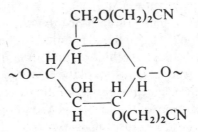

are shown. It is seen that the oscillator strength of the β-transition is considerably higher than that of cellulose hydrate (cf. Fig. 4.42).

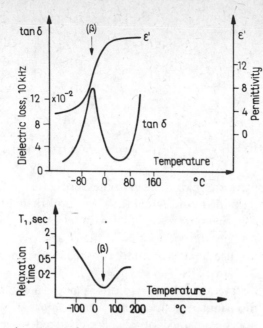

Fig. 4.44 Dielectric spectra of cyanoethyl celluloses in comparison with the proton spin-lattice relaxation curve (after Mikhailov *et al.*, 1969)

The intramolecular hydrogen bonding in cellulose hydrate is assumed to form between the hydroxyl group of the R_1 with that of the R_3 hydroxyl. By this a straight, very rigid chain should be formed (Mann, Marrinan, 1956):

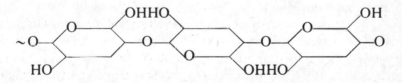

This structure should be broken by substituting the hydroxyl groups by methyl or other groups, as has been discussed previously. Figure 4.42 shows that the transition temperatures are indeed shifted down by methyl substitution. The glass-transition temperature of cellulose is estimated to be 200°C. This is shifted down by about 40—50°C by substitution of apolar methyl or alkyl groups from the hydroxyl group.

CHAPTER 5

SPECTRA OF POLYMER COMPOUNDS

In practice polymers are seldom used in the pure state; to lengthen their life, stabilizers are added, and when they are too brittle plasticizers are often used to make them more flexible. A practical polymeric material often also contains organic or inorganic fillers and dyes. Polymer and copolymer blends are used to improve specific properties such as toughness.

A practical polymeric material is consequently a multi-component system. It is important to study how the physical structure and macroscopic properties of the polymer are changed by low molecular weight or polymeric additives. Dielectric spectroscopy offers a very efficient method, not yet exploited, for studying such problems.

Polymer compounds may be divided into three main groups. In one representative group the additive does not form a separate phase; the resulting compound is homogeneous, at least to such a degree as the pure polymer is. Such an additive which forms a homogeneous phase with the polymer is said to be compatible. Plasticized compounds and polymer solutions will be considered here as quasi-homogeneous systems.

Another group of additives are not compatible with the polymer; they would form a separate phase resulting in an inhomogeneous physical structure. Examples of such additives are fillers, dyes and stabilizers. Traces of impurities would also form structural inhomogeneities. The dielectric behaviour of heterogenous systems is mainly determined by the Maxwell–Wagner–Sillars (MWS) polarization, which will be discussed in detail in section 5.2. In section 5.3 the dielectric spectra of copolymers and polymer blends are discussed.

5.1 Homogeneous systems: plasticized compounds and polymer solutions

Study of the effect of plasticizers on the dielectric properties of polymers is important from the practical point of view because many plasticized polymers are used in the electric industry. From the dielectric behaviour it is possible to deduce information about the mechanism of plasticization. A combination of electrical, thermal and mechanical methods is very useful in such study.

In the present section the effect of some typical plasticizers on the dielectric spectra of some polymers will be illustrated by examples, and some general problems of plasticizing will be discussed. As an extreme case, examples of the behaviour of dilute polymer solutions are also presented.

Plasticizers are low-molecular-weight or polymeric additives which make the glass–rubber transition of the original polymer shift to lower temperatures. By adding 30 % dioctyl phthalate (DOP) to polyvinylchloride, for example, the α-transition temperature (T_g) shifts from 74°C down to −10°C. As a result of this the PVC–DOP mixture at room temperature is a flexible material commonly known as plasticized PVC.

Compatibility

A good plasticizer is expected to be compatible with the polymer. This means that the plasticizer should not form a separate phase; the system at first approximation thermodynamically behaves like a concentrated solution. In a 'compatible' mixture of polymer and plasticizer the intermolecular interactions should be high. Compatibility in general is usually characterized by Flory's equation for the enthalpy change by mixing (Flory, 1953)

$$\Delta H = kT X N_s \bar{v}_s \qquad (5.1)$$

where k is Boltzmann's constant, N_s is the concentration of the solvent molecules, $\bar{v}_s$ is the volume fraction of the solvent with respect to the polymer. X is a parameter characterizing the interaction energy per solvent molecule expressed as a fraction of the thermal energy kT, referred to as Flory's interaction parameter:

$$X = \frac{E_{sp} - E_{ss}}{kT} \qquad (5.2)$$

where E_{sp} is the energy of the solvent molecule surrounded by the polymer solution, E_{ss} is that surrounded by pure solvent.

Flory's interaction parameter X can be determined from osmotic pressure measurements in the case of dilute solutions and from vapour pressure depression measurements in concentrated solutions (see Flory, 1953, Volkenstein, 1963). The interaction parameter X can also be determined by equilibrium swelling of crosslinked polymers (Flory and Rehner, 1943) as

$$X\bar{v}_p^2 = \cdot - N_A(1-\bar{v}_p) - \bar{v}_p - \frac{\varrho_p v_s \bar{v}_p^{1/3}}{M_c} \qquad (5.3)$$

where $\bar{v}_p$ is the volume fraction of the polymer at equilibrium swelling, v_s is the molar volume of the solvent, M_c is the average molecular weight between crosslinks, ϱ_p is the density of the pure polymer. N_A is Avogadro's number.

In the case of plasticized systems the value of X can be more easily determined from the depression of the melting temperature of the polymer by the solvent. According to Anagnostopoulos et al. (1960)

$$\frac{1}{T_m} - \frac{1}{T_m^0} = \frac{R}{\Delta H} \frac{v_p}{v_s} (\bar{v}_s - X\bar{v}_p^2) \qquad (5.4)$$

where T_m^0 is the melting temperature of the pure polymer, T_m is that of the mixture, v_p is the molar volume of the polymer, v_s is the molar volume of the solvent.

Besides the interaction parameter X the compatibility is also characterized by the cohesive energy density expressed as

$$E_{coh} = \frac{Q_e - RT}{v} \qquad (5.5)$$

where Q_e is the latent heat of evaporisation, v is the molar volume, R is the gas constant.

The polymer-plasticizer mixture is said to be compatible if the cohesive energy densities of the components are nearly equal.

Both parameters of compatibility represent roughly the intensity of the intermolecular interactions between the plasticizer and polymer molecules. X involves dipolar interactions; E_{coh} involves dispersion type interactions only.

258

By comparing the X-parameter of various solvents and plasticizers, Doty and Zable (1946) found that for a good plasticizer $X \leq 0.3$. If it is higher, the plasticizer is not retained in the polymer satisfactorily and its rate of exudation will be too high.

Attempts have been made to correlate the X-values of various polymer-plasticizer systems with macroscopic properties such as mechanical strength (Boyer, Spencer, 1947, Jones, 1946) and diffusion (Wartman, Frissel, 1956).

Darby *et al.* (1967) discussed the correlation between the relaxed dielectric permittivity of the plasticizer and the X-values. These correlations, however, have not been found to be decisive. The correlation between the macroscopic properties and the compatibility is much more complicated than can be described by the simple molecular theory of solutions.

The molecular interpretation of the polymer–plasticizer interaction is inadequate because in plasticized polymers, even in solutions, aggregate structure is found (Crugnola, 1968, Kratochwil, 1967). Plasticized PVC is even found to be partially crystalline (Alfrey *et al.*, 1949). Therefore the polymer–plasticizer system is not to be regarded as a true solution; it is more like a colloidal solution in which the aggregate structure (texture) is mainly influenced by the plasticizer and not directly by the intermolecular interaction inside the aggregates. On the basis of this, Kargin *et al.* (1960) assumed two basic plasticization processes—the molecular one in which the plasticizer molecules would directly contact the polymer molecules and the structural, inter-aggregate, inter-bundle one in which the plasticizer cannot penetrate into the aggregates and it is distributed in the inter-aggregate space. This problem has been discussed in detail by Suvarova and Tager (1966).

Efficiency of plasticizers

The efficiency of a plasticizer means its ability to shift the glass–rubber transition of the polymer to low temperature and make the transition broader. As shown in Chapter 2, T_g is principally determined by the free volume. Correspondingly, efficiency of a plasticizer is defined by its ability to increase the free volume of the polymer. By using this concept Kelly and Bueche (1961) expressed the free volume at a temperature $T > T_g$ of a polymer–plasticizer system as

$$v_f(T) = C + \alpha_s [T - T_g^{(s)}] \, \bar{v}_s + \alpha_p [T - T_g^{(p)}] \, \bar{v}_p \qquad (5.6)$$

where C is a constant; according to Fox and Flory $C = 0 \cdot 025$, α_s and α_p are the thermal expansion coefficients of the plasticizer solvent and polymer respectively, $T_g^{(s)}$ and $T_g^{(p)}$ are the corresponding transition temperatures, $\bar{v}_s$ and $\bar{v}_p$ are the volume fractions, $\bar{v}_s = 1 - \bar{v}_p$.

On the basis of Eqn. (5.6) the glass–rubber transition temperature of the plasticized polymer is expressed as

$$T_g = \frac{\alpha_s \bar{v}_s T_g^{(s)} + \alpha_p (1 - \bar{v}_s) T_g^{(p)}}{\alpha_s \bar{v}_s + \alpha_p (1 - \bar{v}_s)} \tag{5.7}$$

This equation is based on the additivity of free volume. It is in agreement with the experiment only when the dipolar interactions between the plasticizer and the polymer are negligible. For the polymethyl-methacrylate-toluene, polymethyl-methacrylate-dioctyl phthalate, styrene-diethyl benzene system Eqn. (5.7) is found to be approximately valid (Kelly, Bueche, 1961) but not for polyvinylchloride. The application of the free-volume theory to plasticization has been studied in detail by Kanig (1963). The reason for this is the aggregation tendency of the polar polymers; PVC, for example would form aggregates even in dilute 3% solution in dioctyl phthalate (Jasse, 1968). Aggregation of polar polymers in dilute solution, even in good solvents, is generally found, and is interpreted as being due to the effect of intermolecular dipole–dipole interaction. For details see Ryshkina and Averianova (1970), Kratochvil (1967), Crugnola (1968), Meerson and Zagraevskaya (1969).

Efficiency is determined not only by the T_g shift but also by the increase of the width of the transition band. According to the general practice those plasticizers are efficient which exhibit broad mechanical or dielectric relaxation bands. This requirement is in contradiction with the thermodynamical requirement of compatibility because if Flory's interaction parameter X decreases, i.e. compatibility increases, the distribution of relaxation times should become narrower. With respect to the efficiency, consequently, it is not desirable for the polymer–plasticizer system to be a real solution. The problem is often solved in practice by using mixtures of plasticizers which would shift T_g to a different extent and thus would broaden the transition region.

The requirement of compatibility implies that in the dielectric spectra of a polar polymer–polar plasticizer system the transitions corresponding to the components are not observed; the compound exhibits a single (common) glass–rubber transition band. In incompatible mixtures the dielectric transitions are separately observed; examples will be presented in sections 5.2 and 5.3. The sensitivity of the dielectric transitions to phase separation makes it possible to study compatibility problems. When, for example, the plasticizer is not properly mixed, the individual transitions of the polymer and plasticizer appear. This is not observed by mechanical relaxation spectroscopy but may be observed by dielectric spectroscopy and by differential scanning calorimetry. The dielectric depolarization spectra are especially sensitive to phase separation. In such spectra, indications of the presence of separate phases of polymer and plasticizer are found even when, according to mechanical and thermal measurements, the compound appears to- be single-phased. An example will be given later in this section for plasticized PVC compounds.

Besides the observation of problems of phase separation, compatibility and T_g shifts, the dielectric method offers a special possibility for measuring polymer–plasticizer dipole–dipole interactions. By increasing the plasticizer concentration, i.e. not only the position but also the oscillator strength of the dielectric spectrum band changes. Figure 5.1 shows, for example, two mechanisms suggested by Würstlin and Klein (1952) and by Leuchs (1956) for explaining the interaction of PVC molecules with polar plasticizers. In both cases the PVC chain dipoles are thought to be partially compensated by the plasticizer dipoles. This suggests that the effective dipole-moment concentration of the compound should so depend on the concentration that for optimum compatibility it is at a minimum. This is a molecular interpretation of plasticization.

Although the picture of plasticizer–polymer interaction in Fig. 5.1 is oversimplified, by analyzing the dielectric oscillator strengths important information can be gained. This method has not so far been utilized in a quantitative way.

As many polymers such as PVC exhibit an aggregate structure, besides the molecular interpretation another model may be effective. According to this view the plasticizer could not penetrate into the aggregates but would fill-up the inter-aggregate inter-bundle space. In

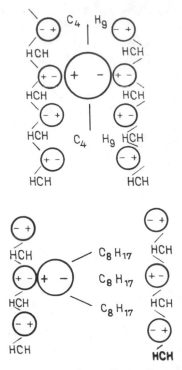

Fig. 5.1 Types of plasticizer–polymer dipole–dipole interactions

this case the aggregate as a whole should be considered as a unit undergoing thermal motion. The aggregate may have a residual dipole moment which would interact with the polar plasticizer. In contrast to the molecular mechanism of plasticization this will be referred to as the textural mechanism.

Another possible method, not yet developed, for investigating by dielectric spectroscopy the mechanism of plasticization is to study low-temperature transitions. The effect of the plasticizer on the α-transitions of PVC and of PMMA has been studied in detail (see later) but the effect on the β and lower temperature cyrogenic transitions is not known. As the low-temperature transitions are usually attributed to local motion of the polymer chains it would be interesting to know how they are affected by the plasticizer. This would help in deciding whether in a particular case the molecular or the textural mechanism of plasticization is effective.

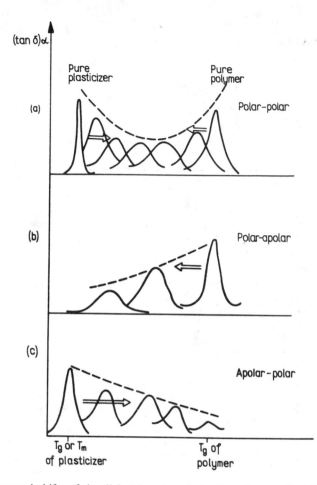

Fig. 5.2 Expected shifts of the dielectric α-transitions for polymer–plasticizer systems of different polarity

Figure 5.2 shows the dielectric α-transition (T_g) bands of various polymer–plasticizer systems when (a) both the polymer and plasticizer are polar, (b) when only the polymer is polar and (c) when only the plasticizer is polar. In the first case the intensity of the dielectric absorption band measured as a function of the composition usually goes through a minimum. In cases (b) and (c) the intensity should be gradually decreased, simply because the concentration of polar groups in the compound decreases. From the T_g-shifts and band widths information

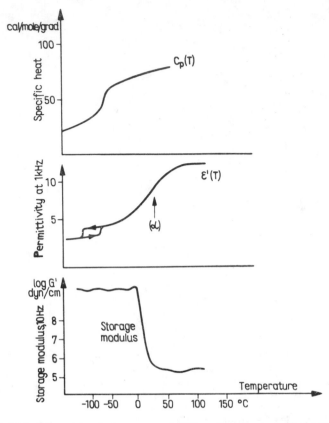

Fig. 5.3 Dielectric, mechanical spectra and differential scanning calorimetric curve of a typical plasticized PVC-compound

about the efficiency of the plasticizer can be gained from the change in the absorption-band intensity and oscillator strength of the transition, the dipolar interaction between polymer and plasticizer can be studied.

Dielectric spectra of plasticized polyvinylchloride

Because of their practical importance, plasticized PVC compounds have been studied by various methods, including mechanical-relaxation spectroscopy, differential scanning calorimetry, NMR and dielectric spectroscopy.

Figure 5.3 shows a series of spectra measured for a conventional plasticized PVC compound containing 30 % (weight) of diisooctyl phthalate as plasticizer and 1 % of tribasic lead sulphate as stabilizer.

The forced oscillation $G'(T)$ spectrum and the differential calorimetric $c_p(T)$ curve at constant load are compared with the $\varepsilon'(T)$ dielectric spectrum. A similar sequence of curves for unplasticized PVC has been shown in Chapter 2, Figs 2.9 and 2.10.

Figure 5.3 shows a correspondence between the mechanical and dielectric relaxation band. The DSC curve shows only a slight increase of the specific heat at the transition; the change is not so abrupt as in unplasticized PVC. A sharp low-temperature transition is indicated by DSC and by the dielectric curve. This transition has been found to be very dependent on the thermal history of the sample; its position could be shifted 40—50°C by changing the rate of cooling from room temperature. This effect is also illustrated in Fig. 5.3 by recording the spectrum with decreasing temperature; the transition appears much lower than by recording with increasing temperature (Kisbényi, 1973).

The sharp low-temperature transition is attributed to the separate plasticizer phase—to a part of the plasticizer which is not compatible with the polymer.

In Fig. 5.4 the dielectric depolarization spectrum (DDS), the mechanical loss spectrum and the dielectric $\sigma(T)$ spectrum of the same plasticized PVC compound are shown. It is seen that T_g is exhibited by well correlated peaks in these spectra. Besides T_g (α-peak) a transition is observed at about 60°C by the depolarization method. This peak (α') has been found to be dependent on the method of processing. When the plasticizer is not sufficiently homogenized during processing the α'-peak becomes very high; it appears even in the mechanical $G''(T)$-spectra. This transition is thought to be due to the PVC-particles which form a separate phase as plasticizer could not penetrate them during the course of processing. The α'-transition has been found to be dependent on long-term storage of the polymer at room temperature. This problem will be discussed in Chapter 7 in connection with physical structural ageing. The α''-peak which appears in the depolarization spectrum is a T_{ll}-type transition corresponding most probably to the destruction of the aggregate structure. This transition has not been studied in detail so far.

Figure 5.5 shows a typical series of dielectric $\varepsilon'(T)$ spectra at 30 Hz for compounds having increasing di-isooctyl phthalate (DIOP) plasticizer content. It is seen that the α-transition temperature T_α is shifted down by increasing plasticizer content and the transition band is broadened. It is reasonable to use the $(\varepsilon_0 - \varepsilon_\infty)_T$, the T_α and the ΔT width of band values for characterizing the compound, as illustrated in Fig. 5.6.

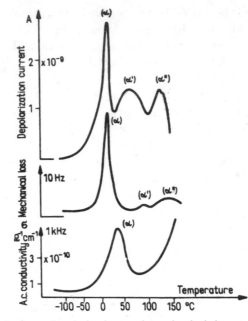

Fig. 5.4 Dielectric loss, depolarization and mechanical loss spectra of a typical plasticized PVC-compound

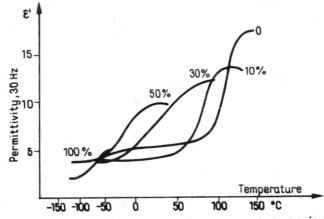

Fig. 5.5 Dependence of the dielectric dispersion spectra of plasticized PVC on the plasticizer content (di-isooctyl phthalate)

These parameters are plotted against plasticizer content in Fig. 5.7. Equation (5.7) is not fulfilled for the PVC–DIOP system.

The oscillator strength of the dielectric β-transition measured as a function of the temperature at 30 Hz drops rather suddenly if a small

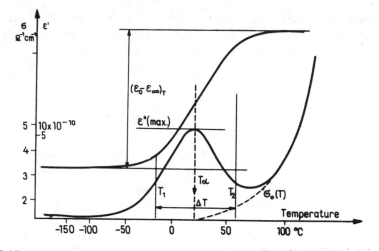

Fig. 5.6 Parameters of the dielectric dispersion and absorption curves of plasticized PVC

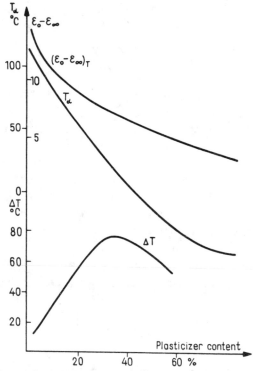

Fig. 5.7 Dependence of the dielectric α-transition temperature T_α the oscillator strength $(\varepsilon_0-\varepsilon_,)_T$ and the spectrum band width ΔT of PVC on the plasticizer content (di-isooctyl phthalate)

amount of plasticizer (up to about 10 %) is added; it then decreases approximately linearly corresponding to the decrease of the PVC dipole concentration in the sample. At high plasticizer contents it approximates the oscillator strength of the pure plasticizer. The PVC–DIOP system represents approximately case (b) in Fig. 5.2, where the plasticizer is apolar. Although DIOP is slightly polar, the oscillator strength of the dielectric transition in it, $(\varepsilon_0 - \varepsilon_\infty)_T \approx 2$, is much smaller than that of pure PVC, $(\varepsilon_0 - \varepsilon_\infty)_T \approx 13$.

The width of the transition ΔT, defined by Fig. 5.6, exhibits a maximum at about 35 % plasticizer content. This concentration is a kind of optimum with respect to efficiency and exudation.

The parameters of the dielectric α-transition band are found to depend strongly on the chemical composition of the plasticizer.

Figure 5.8 compares the mechanical properties and the dielectric spectra for a series of phthalate plasticizers having a gneral structural formula

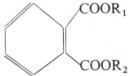

where R_1 is n-butyl or iso-butyl group, R_2 is an alkyl chain of different length having a general formula of C_nH_{2n+1} (Kisbényi, 1970).

In Fig. 5.8 the rupture strain values are plotted against the number of carbon atoms in the group R_2. It is seen that the rupture strain increases as the alkyl group length increases and levels off at about carbon number of 10 for both normal and iso-butyl phthalates. The torsional cold flex temperature measured by the ASTM standard method (ASTM D 1043) shows a minimum at 8 and 10 carbon numbers respectively for the normal and iso-butyl series respectively. Similar minima are exhibited by the a.c. and d.c. dielectric spectra.

The correlation between the main parameters of the dielectric α-transition band and the mechanical and thermomechanical properties are clearly seen.

Antiplasticizing. Effect of small plasticizer content on the β-transition of PVC

Adding small amounts of plasticizers up to about 10 % makes the compound more rigid than the unplasticized polymer; the rupture strain decreases and the rupture stress increases. This behaviour is

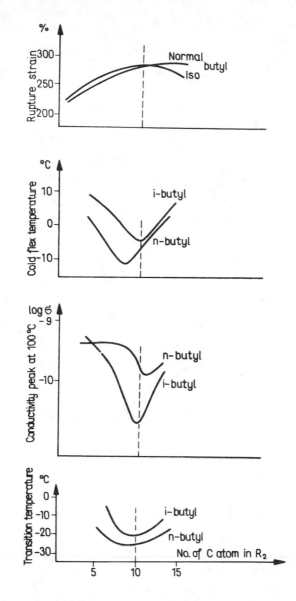

Fig. 5.8 Rupture strain, cold flex temperature and dielectric transition temperature of a series of alkyl phthalate plasticizers as a function of the number of carbon atoms in the alkyl group

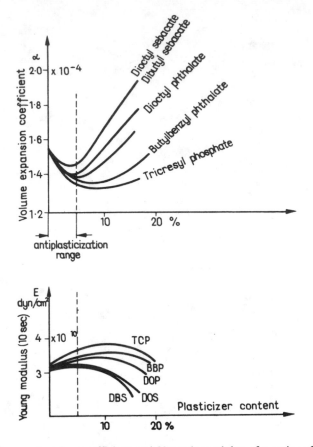

Fig. 5.9 Volume expansion coefficient and Young's modulus of a series of plasticized PVC compounds showing range of antiplasticization (after Kinjo and Nakagawa 1973)

referred to as antiplasticization. For a review see Kinjo and Nakagawa (1973). Antiplasticization is connected with the change of free volume by addition of the plasticizer.

Figure 5.9 shows the volume expansion coefficient of the compounds prepared with different plasticizers at temperatures between −20 and +50°C, which is below the glass–rubber transition. For comparsion, the Young's modulus at −20°C, measured after 10 sec of applying the load, are also shown for the same compounds. It is seen that with low-plasticizer concentration where the volume expansion coefficient decreases by increasing plasticizer content the Young's modulus in-

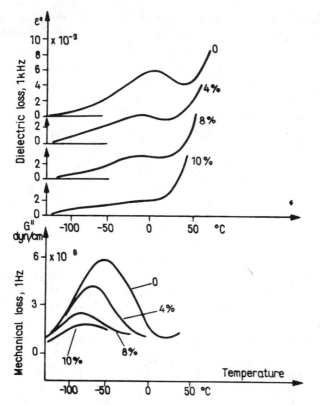

Fig. 5.10 Comparison of the dielectric and mechanical relaxation spectra of the PVC-tricresyl phosphate system at low plasticizer concentrations (Mikhaïlov *et al,* 1967 a)

creases. At higher concentrations, when the volume expansion coefficient of the compound becomes larger than that of the pure polymer the Young's modulus decreases. From this concentration up, plasticization begins to be effective. In the concentration range in which the volume expansion coefficient of the compound is below that of the pure polymer antiplasticization occurs.

In the concentration range of antiplasticization, the dielectric as well as the mechanical β-transitions decrease in intensity with increasing plasticizer content. This is illustrated in Fig. 5.10 for the PVC-tricresyl phosphate (TCP) system after Mikhailov *et al.* (1967a). The ε″-values are shown as a function of the temperature at a frequency of 1 kHz. For comparison, the corresponding mechanical relaxation spectra are also shown.

The dielectric and mechanical β-peaks show similar behaviour: the peaks shift to lower temperatures, decrease, and with more than 10% plasticizer concentration almost vanish. This behaviour is not dependent on the polarity of the plasticizer; the same sequence of spectra has been measured in the PVC-diphenyl system (Fuoss and Kirkwood, 1941). The activation energy of the β-process in PVC is found to decrease by increasing plasticizer content (Mikhailov *et al.*, 1966b). The sharp decrease of the oscillator strength of the β-process with increasing plasticizer content indicates that not only the mobility but also the effective dipole moment has been affected.

The connection between the β-peak in PVC and antiplasticizing is not yet quite clear because the origin of the β-transition is not known (cf. section 4.2). According to Bohn (1963), the β-peak in PVC is connected with the toughness, i.e. brittleness—the higher the β-peak intensity the more brittle the corresponding system. This suggestion has been supported by measurements of correlation of the β-transition of PVC-base polymer blends with the impact strength (see, for example, Heijboer, 1965); the problem will be discussed also in section 5.3 of this chapter in connection with polymer blends.

The disappearance of the β-peak in PVC when small amounts of plasticizer are added is very difficult to interpret by assuming that β is a local mode transition. It is not likely that the plasticizer which increases the free volume of the polymer as indicated by the continuous shift of the α-peak should affect the local motion of the main chain so much. It is more likely that the β-process is connected with the aggregate structure of the polymer, which is evidently influenced by the plasticizer. This assumption has been put forward by Lebedev (1965).

On the basis of the experimental evidence collected so far it may be concluded that in PVC plasticizing is textural rather than molecular.

Plasticized polymethyl acrylate and -methacrylate

As an example of the effect of plasticizers on the dielectric spectra of polymers having flexible polar side groups, polymethyl acrylate (PMA) and methacrylate (PMMA) will be discussed here in some detail. The dielectric spectra of these polymers have been shown in Chapter 4, and the transitions interpreted.

Figure 5.11 shows a typical sequence of dielectric tanδ versus temperature spectra for PMMA plasticized with dibutyl phthalate (DBP) measured at 1 kHz after Lobanov *et al.* (1968). It is seen that by

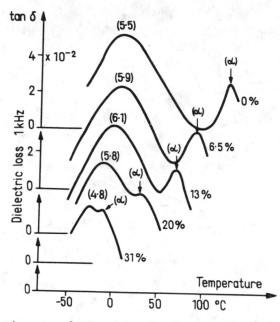

Fig. 5.11 Dielectric spectra of polymethyl methacrylate plasticized by dibutyl phthalate (Lobanov *et al,* 1968)

increasing the plasticizer content the α-peak (T_g) shifts to lower temperatures and is decreased while the β-peak is somewhat increased. As has been shown in Chapter 4, in the dielectric spectrum of PMMA the α-peak appears as a result of the (αβ)-interaction as the main chain motion is monitored through the rotating ester side-group, which is polar. It is seen that in contrast to PVC, discussed previously in this section, the secondary β-peak is not decreased by addition of the plasticizer, although it should be because the ester-dipole moment concentration decreases.

Figure 5.12 shows a series of dielectric spectra of PMMA containing increasing amounts of an apolar plasticizer, toluene. It is seen that in this case again both α- and β-peaks are shifted to low temperatures and the intensity of the α-peak is not appreciably reduced by addition of the plasticizer. As the α-peak shifts more than β above 20 % plasticizer content the α- and β-peaks coalesce. Considering the (αβ)-mixing mechanism discussed in section 2.1, it is concluded that the plasticizer in such systems mainly influences the α − β interaction process.

In Fig. 5.13 the α- and β-transition temperatures are plotted against plasticizer content for the PMMA–toluene and PMMA–DBP systems.

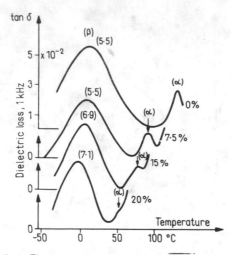

Fig. 5.12 Dielectric spectra of polymethyl methacrylate plasticized by toluene

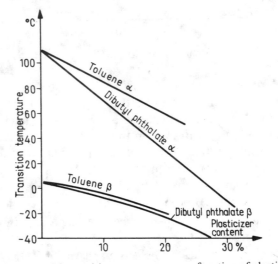

Fig. 5.13 Shift of the α- and β-transition temperature as a function of plasticizer content in polymethyl methacrylate

The polarity of the plasticizer has no appreciable effect on the shift of the transition temperatures.

Figure 5.14 shows the effect of polar and nonpolar plasticizers on the α-transition in polymethyl acrylate PMA after Mikhailov *et al.* (1968a). It is seen that the shift of the α-transition is the same by addition of 20 % polar tricresyl phosphate (TCP) and nonpolar toluene. When 20 %

274

dibutyl phthalate (DBP) is added, the T_g-shift is greater, although ε_0 of DBP (6.5) is smaller than that of TCP (7.2).

Mikhailov *et al.* (1968) also studied the temperature dependence of the relaxation times for the α- and β-processes in PMA up to high

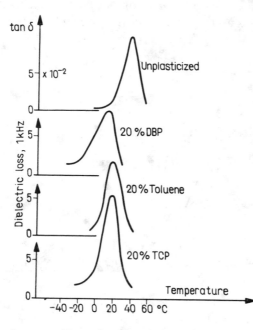

Fig. 5.14 Dielectric α-transition of polymethyl acrylate containing 20%, tricresyl phosphate (TCP), dibutyl phthalate (DBP) and toluene (after Mikhailov *et al.*, 1968)

temperatures. The corresponding relaxation map is shown in Fig. 5.15 for PMA-TCP compounds containing different amounts of plasticizer. It is seen that at temperatures higher than about 120 C the ln τ versus $1/T$ dependence becomes linear and the slope becomes approximately identical with that of the α-process, so these two processes are merged into a common (αβ)-process (cf. section 2.1). With increasing plasticizer content the activation energies for the α, β and (αβ)-processes are found to decrease. Figure 5.15 also shows that the temperature T^* at which the α-relaxation time deviates from the Arrhenius dependence is also shifted to lower temperatures and the virtual activation energy of the α-transition in the range between T^* and dilatometric T_g is decreased by increasing plasticizer concentration. From this it is concluded that the (αβ)-mixing process is made easier by addition of the plasticizer.

Figure 5.16 shows the decrease of the activation energy of the β-transition in the PMA-TCP system as a function of the plasticizer content in comparison with the PVC-TCP system. The change in the dielectric absorption amplitude ε''_{max} is also compared with that to be expected from the decrease in the dipole-moment concentration—dashed line. In Chapters 2 and 5 the β-process of polymers having flexible side

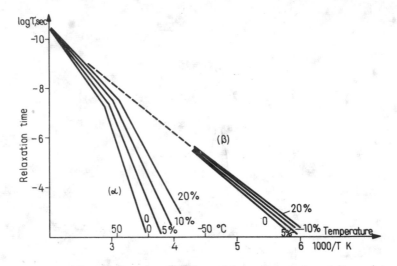

Fig. 5.15 Relaxation map for the α- and β-transitions of the polymethyl acrylate –tricresyl-phosphate system (Mikhailov *et al.* 1968)

groups, such as PMA, has been successfully interpreted in terms of the barrier theory. According to this picture the β-process in these polymers is due to the hindered rotation of the ester side groups. The decrease in the activation energy when the plasticizer content is increased, shows that the potential barrier which hinders the rotation of the ester group has been decreased.

The oscillator strength of the β-transition, i.e. the ε''-maximum, decreases more steeply, as would be expected from the decrease of the ester group dipole moment concentration in the sample 'calculated' dashed line in Fig. 5.16. For comparison, in Fig. 5.16 the corresponding change of the β-oscillator strength in the PVC-TCP system is also shown, for which this drop is even steeper. On the other hand ε''_{max} in polymethyl methacrylate is approximately constant when plasticizer content is increased, which means that this transition is in fact increased.

276

From these experimental data it is concluded that plasticization in acrylates and methacrylates is due to the molecular rather than the textural mechanism.

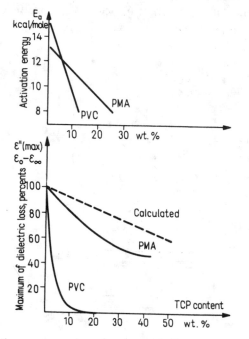

Fig. 5.16 β-activation energy and maximum dielectric loss of the polymethyl acrylate tricresyl-phosphate and PVC-tricresyl-phosphate systems as a function of the plasticizer content

Polymer solutions

In dilute solution of polar polymers with nonpolar solvents the polymer molecule as a whole may exhibit thermal motion. The corresponding dielectric relaxation is determined by the effective dipole moment of the molecule as a whole. On the other hand the thermal motion of the gaussian chain segments results in another dipole relaxation process which is analogous with the α-process (T_g) of pure polymers. Dipole relaxation due to the orientation of side groups also may result in a separate dielectric spectrum band in solution. Consequently, in dilute solution all the basic processes discussed in Chapter 2 may be observed. The dielectric behaviour of polymers having an apolar main chain but flexible polar side chains, such as the acrylates and methacrylates, are expected to exhibit dielectric behaviour rather different from that of those having polar groups built in their main chain (oxide polymers, polyesters and polycarbonates).

As the relaxation of the polymer molecules as a whole is a rather slow process in the frequency range covered by dielectric spectroscopy,

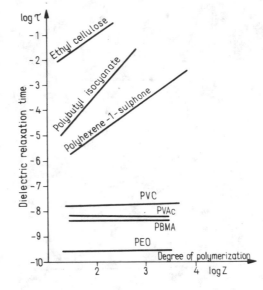

Fig. 5.17 Dependence of the dielectric relaxation time on the degree of polymerization for different polymer solutions

usually the dipole relaxation processes of the segmental or side-groups are measured. This is shown by the fact that the dielectric relaxation time of most polymers is independent of the molecular weight, i.e. of the degree of polymerization. In some cases, when the dipoles are built rigidly into the chains and the chains are not flexible the relaxation of the polymer molecule as a whole can be detected. This is illustrated in Fig. 5.17, where the dipole relaxation frequencies of some polymers in dilute solution are plotted against polymerization degree in a logarithmic scale. It is seen that the most common flexible, chain polymers exhibit a dipole relaxation time in the order of 10^{-8} sec in solution near to room temperature, while ethyl cellulose and some other rigid chain polymers exhibit a molecular weight dependence and the relaxation times are in the order of $10^{-1} - 10^{-5}$ sec.

278

In solution the dielectric permittivity of flexible polar polymer molecules in the main relaxation region can be calculated by considering the normal modes of vibrations of polymer chains. This method has been developed by Rouse (1953), Bueche (1954) and Zimm et al. (1956); it is referred to as the normal mode theory. For details see Volkenstein (1963), Flory (1953) and McCrum et al. (1967).

For solutions of polymers having polar groups built in their main chain according to Zimm's theory

$$\varepsilon^* - \varepsilon_\chi = \frac{4\pi N_A^2}{27\,RTM}\,(\varepsilon_\vee + 2)^2 \sum_k P_k^2\,(1 + i\omega\tau_k)$$

$$P_k = [2\,\langle l^2\rangle]^{1/2}\,(k\pi)^{-1} \int_0^1 \cos k\pi s\,\varrho(s)\,ds$$

where N_A is Avogadro's number, R is the gas constant, T is the temperature, M is the molecular weight of the polymer, $\varepsilon_\vee$ is the permittivity of the solvent, ω is the angular frequency, $\langle l^2\rangle$ is the mean square end-to-end distance for the polymer chains, $\varrho(s)$ is the charge density along the chain, k is the mode of vibriation of the chains, τ_k is the relaxation time corresponding to vibrational mode k.

The dielectric relaxation time is expressed as

$$\tau_k = 3.42\,\frac{M\,[\eta]\,\eta_\vee}{RT\lambda_k} \tag{5.8}$$

where $\eta_\vee$ is the viscosity of the solvent, $[\eta]$ is the intrinsic viscosity of the solution, λ_k is a parameter depending on the draining conditions; values for λ_k are found in Zimm et al. (1956).

In polymer solutions the rotational–diffusional mode ($k = 1$) is considered to be dominant. The relaxation time in this case is

$$\tau_1 = 0.85\,\frac{M\,[\eta]\,\eta_\vee}{RT} \tag{5.9}$$

Most of the investigations in polymer solutions have been performed in order to study the motion of the parts of polymer molecules in the solvent, so the experiments have been made usually in dilute solution of polar polymers in nonpolar solvents. Another interesting possibility is to study the motion of the solvent molecules in the polymer matrix,

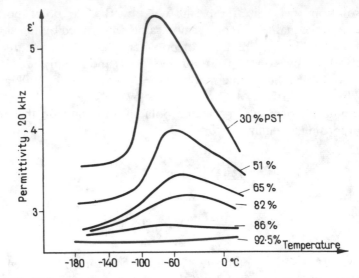

Fig. 5.18 Dielectric dispersion spectra of the polystyrene-anisole system (after Borisova *et al.*, 1972)

using concentrated solutions of nonpolar polymers in polar solvents. Figure 5.18 shows a series of $\varepsilon'(T)$ spectra of anisole

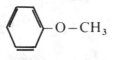

in polystyrene after Borisova *et al.* (1972); the dielectric transition is considerably broadened by increasing the polystyrene concentration. Figure 5.19 shows the corresponding $\tan\delta$ versus temperature spectra; at low and intermediate polystyrene concentrations the solvent spectrum exhibit three peaks labelled by α', α, β. The highest intensity α-loss peak corresponds to the sharp change in $\varepsilon'(T)$ as seen by comparing Fig. 5.19 with Fig. 5.18. With increasing PST-concentration the α'-peak shifts to higher temperatures and is increased. The α-peak is also somewhat shifted but decreased; the β-peak is unchanged. Borisova *et al.* (1972) measured the transitions also by the NMR spin echo method. For solutions containing 75 % polymer, minima in the spin–lattice relaxation time versus $1/T$ curve have been found at about -90 C, -10 C and 120 C respectively, corresponding to the β-, α- and α'-transitions.

280

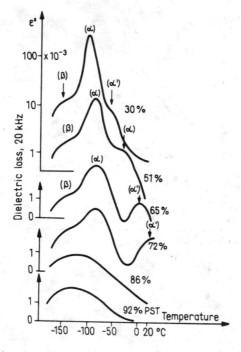

Fig. 5.19 Dielectric absorption spectra of the polystyrene-anisole system (after Borisova *et al.*, 1972)

At high PST-concentration the solvent α- and β-transitions are smeared out and the α'-transition is detected at about 140 C.

The solvent α-transition is attributed to the motion of the entire anisole molecule, the β-transition is interpreted as being due to the intramolecular rotation of the OCH_3 group. The α'-peak is attributed to the α-transition of polystyrene shifted down by the solvent and monitored through the motion of the polar solvent molecules. Figure 5.20 shows the transition temperatures and activation energies calculated from the temperature dependence of the dielectric relaxation time for the solvent α- and β-transitions as a function of the polystyrene concentration. It is seen that the activation energy corresponding to the motion of the entire solvent molecule decreases with increasing polymer content, while that corresponding to internal rotation (β) is not affected. The α-transition temperature exhibits a maximum at high ($\sim 70°_0$) polymer concentration.

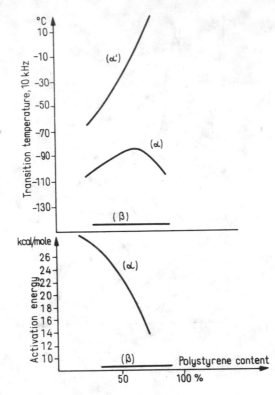

Fig. 5.20 Activation energies and transition temperatures of the polystyrene-anisole system as a function of polystyrene concentration

5.2 Heterogeneous systems: Effects of fillers and other non-compatible additives

Practical polymeric systems are almost always heterogeneous. The stabilizers which are added to prevent ageing processes are usually not mixed but dispersed in the polymer. Fillers and dyestuff also would form separate phase. Impurities as traces of monomer, solvent or water also make the practical polymeric system heterogeneous. The effect of such non-compatible additives to the dielectric properties of the compound is not well understood. The most remarkable effect is expected from the Maxwell–Wagner–Sillars (MWS) polarization which always develops in heterogeneous dielectrics. This problem is discussed in the present section in some detail.

282

The Maxwell–Wagner–Sillars (MWS) polarization

It has long been known (Maxwell, 1892) that in heterogeneous dielectrics a polarization occurs as a result of the accumulation of virtual charge at the interface of two media having different permittivities and/or conductivities. The simplest model of this is a medium having two layers of permittivities ε_1' and ε_2' and conductivities σ_1 and σ_2 respectively. Simple calculation '(see van Beek, 1967) shows that the frequency dependence of the permittivity components of the composite system are expressed by:

$$\varepsilon'(\omega) = \bar\varepsilon_\gamma + \frac{\bar\varepsilon_0 - \bar\varepsilon_\gamma}{1 + \omega^2 \tau_{MW}^2} \tag{5.10}$$

$$\varepsilon''(\omega) = \frac{(\bar\varepsilon_0 - \bar\varepsilon_\gamma)\,\omega\tau_{MW}}{1 + \omega^2 \tau_{MW}^2} + \frac{\bar\sigma}{\omega} \tag{5.11}$$

where

$$\bar\varepsilon_0 = \frac{d(\varepsilon_1' d_1 \sigma_2^2 + \varepsilon_2' d_2 \sigma_2^2)}{(\sigma_1 d_2 + \sigma_2 d_1)^2} \tag{5.12}$$

$$\bar\varepsilon_\gamma = \frac{d\varepsilon_1' \varepsilon_2'}{\sigma_1 d_2 + \sigma_2 d_1} \tag{5.13}$$

$$\tau_{MW} = \frac{\varepsilon_1' d_2 + \varepsilon_2' d_1}{\sigma_1 d_2 + \sigma_2 d_1} \tag{5.14}$$

$$\bar\sigma = \frac{d\sigma_1 \sigma_2}{\sigma_1 d_2 + \sigma_2 d_1} \tag{5.15}$$

d_1, d_2 are the thicknesses for layers 1 and 2, $d = d_1 + d_2$.

Equations (5.10) and (5.11) are similar to those corresponding to the Debye single-relaxation-time approximation for orientational dipole polarization in homogeneous media—Eqn. (1.55). The only difference is the appearance of the conductivity term in Eqn. (5.11) which would shift the baseline of the MWS absorption band.

283

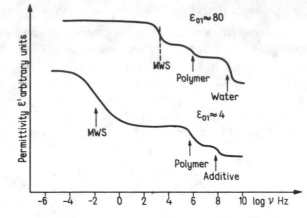

Fig. 5.21 Schematic representation of the Maxwell–Wagner–Sillars transition for highly polar (water) inclusions ($\varepsilon_{01} \approx 80$) and for slightly polar ones ($\varepsilon_{01} \approx 4$)

In general, the individual permittivity ε_i' and conductivity σ_i components are themselves frequency dependent

$$\varepsilon_i'(\omega, T) = \varepsilon_{i0} + \frac{\varepsilon_{i0} - \varepsilon_{i\infty}}{1 + \omega^2 \tau_i^2} \qquad (5.16)$$

$$\sigma_i(\omega, T) = \frac{\omega}{4\pi} \frac{(\varepsilon_{i0} - \varepsilon_{i\infty})\,\omega\tau_i}{1 + \omega^2 \tau_i^2} \qquad (5.17)$$

where τ_i is the relaxation time corresponding to the orientational dipole polarization in medium i.

At first approximation it can be assumed that the individual dielectric media exhibit no dispersion in the frequency range on which the MWS transition is observed, so ε_i' and σ_i are constant. In this case the MWS relaxation time τ_{MW} is independent of the frequency.

The assumption that the dispersions of the individual components are not represented in the MWS-transition is not unreasonable when the effect of fillers on the dielectric spectra of polymers is considered. Fillers are usually inorganic materials having no dielectric transitions at the range where the host polymer does. On the other hand, if water or other compounds of low molecular weight are dispersed in the polymer, the frequency range of their dispersion might coincide with that of the host polymer. This is illustrated schematically in Fig. 5.21, where the

284

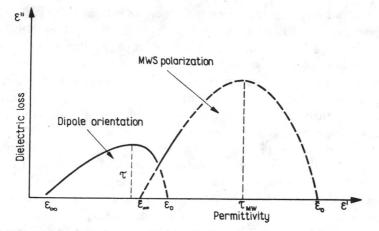

Fig. 5.22 Arc plot in the presence of Maxwell–Wagner–Sillars polarization. Schematic

dielectric transitions of heterogeneous mixtures of polymeric and low-molecular-weight materials (e.g. water) are represented. The transition of the compound of low molecular weight is expected at the highest frequency, lowest temperature, range. The MWS transition is expected at the low frequency range, while the transition of the polymer is usually in between them. Unfortunately the MWS transition is not always separated from the transitions of the individual components, resulting in a distortion of the orientational dipole transition.

The MWS transition of heterogeneous systems having high permittivity inclusions (e.g. water) are in the kHz range depending on the size and shape of the heterogeneities. If the permittivity of the additive is reduced, the MWS transition band shifts to quite low frequencies. This is also shown schematically in Fig. 5.21 for an $\varepsilon_{01} = 4$ additive (lubricant or monomer).

Figure 5.22 shows schematically the arc plot of a heterogeneous system in which dipole orientation and MWS polarization takes place. It is seen that the ε_0 and ε_∞ values of the dipole transition are different from those of the MWS transition, which usually occurs at the low frequency range and involves rather high ε' and ε'' values.

From this simplified treatment it is apparent that dielectric MWS transition may be detected in polymers having structural inhomogeneities even when there is no transition due to the orientational polarization of the polar inclusions, or even if the inclusions are nonpolar. In any case, when structural inhomogeneities of different permittivity and conductivity from the base material are present, MWS polarization is expected to occur. This principally influences the low frequency $10^{-5} \ldots 10^2$ Hz dielectric properties, as the MWS polarization usually decreases with increasing frequency. As even pure polymers are never homogeneous, the very low frequency dielectric permittivities are expected, and found to be, much higher than the relaxed permittivities obtained from extrapolation of Cole–Cole diagrams measured at intermediate and high frequencies.

Equation (5.10) and (5.11) for the MWS transition have been found to be valid for ellipsoidal, spherical, and cylindrical inclusions. In such cases the limiting permittivities $\bar{\varepsilon}_0$ and $\bar{\varepsilon}_\alpha$ are dependent on a so-called depolarization factor, which, in turn, depends on the shape of the inclusions. For a general treatment and formulae see van Beek (1967). Here only some simplified cases will be discussed which are more closely related to the practical polymeric systems.

For the special case when spheres or ellipsoids of conductivity σ_2 and permittivity ε_2' are dispersed in a homogeneous medium $\varepsilon_1'\ \sigma_1$, the following relations are found to be valid

$$\tau_{MW} = \frac{\varepsilon_1' + A_a(1-v_2)\,(\varepsilon_2'-\varepsilon_1')}{\sigma_1 + A_a(1-v_2)\,(\sigma_2-\sigma_1)} \tag{5.18}$$

$$\bar{\varepsilon}_0 = \varepsilon_1'\ \frac{\sigma_1 + [A_a(1-v_2)+v_2]\,(\sigma_2-\sigma_1)}{\sigma_1 + A_a(1-v_2)\,(\sigma_2-\sigma_1)} +$$

$$+ v_2\sigma_1\ \frac{\sigma_1 + A_a(\sigma_2-\sigma_1)\,(\varepsilon_2'-\varepsilon_1') - [\varepsilon_1' + A_a(\varepsilon_2'-\varepsilon_1')]\,(\sigma_2-\sigma_1)}{[\sigma_1 + A_a(1-v_2)\,(\sigma_2-\sigma_1)]^2} \tag{5.19}$$

$$\bar{\varepsilon}_\infty = \frac{\varepsilon_1' + [A_a(1-v_2)+v_2]\,(\varepsilon_2'-\varepsilon_1')}{\varepsilon_1' + A_a(1-v_2)\,(\varepsilon_2'-\varepsilon_1')}\ \varepsilon_1' \tag{5.20}$$

where A_a is the depolarizing factor along the a axis of the ellipsoid expressed as

$$A_a = \frac{-1}{(a/b)^2-1} + \frac{a/b}{[(a/b)^2-1]^{3/2}}\ \ln(a/b) + [(a/b)^2-1]^{1/2} \tag{5.21}$$

286

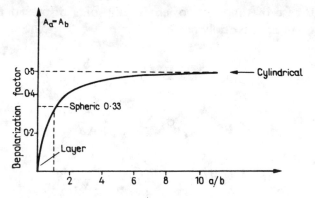

Fig. 5.23 Depolarizing factor of the Maxwell–Wagner–Sillars polarization of spheroid inclusions

for prolate ellipsoids, $a > b$, where a and b are the principal axes and

$$A_a = \frac{1}{1 - (a/b)^2} - \frac{a/b}{[1 - (a/b)^2]^{3/2}} \text{ arc } \cos(a/b) \qquad (5.22)$$

for oblate ellipsoids, $a < b$ and

$$A_a = 1/3 \quad \text{for spheres,} \quad a = b$$

Tables for the values of the depolarization factors are found in O'Konski (1960). A diagram for $A_a = A_b$ as a function of the a/b ratio is shown in Fig. 5.23.

In many practical cases in polymeric insulators, lossy material is dispersed i.e. $\sigma_2 \gg \sigma_1$. In such cases, when $v_2 \ll v_1$, i.e. the concentration of the additive is low,

$$\tau_{MW} \approx \frac{\varepsilon_1' + A_a(\varepsilon_2' - \varepsilon_1')}{A_a \sigma_2} \qquad (5.23)$$

$$\bar{\varepsilon}_0 \approx \varepsilon_1' \left[1 + \frac{v_2}{A_a(1 - v_2)} \right] \qquad (5.24)$$

$$\bar{\varepsilon}_\infty \approx \varepsilon_1' \left[1 + v_2 \frac{\varepsilon_2' - \varepsilon_1'}{\varepsilon_1' + A_a(\varepsilon_2' - \varepsilon_1')} \right] \qquad (5.25)$$

287

In the case of randomly oriented ellipsoids dispersed in low concentration in the base polymer

$$\bar{\varepsilon}_0 = \varepsilon_1' + \frac{1}{3} v_2 \sum_{i=1}^{3} \frac{\sigma_2 - \sigma_1}{\sigma_1 + A_i(\sigma_2 - \sigma_1)} +$$
$$+ \sigma_1 \sum_{i=1}^{3} \frac{\sigma_1 \varepsilon_2' - \sigma_2 \varepsilon_1'}{[\sigma_1 + A_i(\sigma_2 - \sigma_1)]^2} \qquad (5.26)$$

$$\bar{\varepsilon}_\infty = \varepsilon_1' \left(1 + \frac{1}{3} v_2 \sum_{i=1}^{3} \frac{\varepsilon_2' - \varepsilon_1'}{\varepsilon_1' + A_i(\varepsilon_2' - \varepsilon_1')} \right) \qquad (5.27)$$

where A_i are the depolarization factors along the principal axes of the ellipsoids a, b, c.

Hanai (1961) derived equations for the limiting permittivities and conductivities for such cases when spheroids are distributed in a homogeneous medium at large concentration. The Hanai (1961) equation for $\bar{\varepsilon}_0$ is

$$\bar{\varepsilon}_0 \left(\frac{3}{\bar{\sigma} - \sigma_2} - \frac{1}{\bar{\sigma}} \right) = 3 \left[\frac{\varepsilon_2' - \varepsilon_1'}{\sigma_2 - \sigma_1} + \frac{\varepsilon_2'}{\bar{\sigma} - \sigma_2} \right] - \frac{\varepsilon_1'}{\sigma_1} \qquad (5.28)$$

where the conductivity $\bar{\sigma}$ is determined by

$$\frac{\bar{\sigma} - \sigma_2}{\sigma_1 - \sigma_2} \left(\frac{\sigma_1}{\bar{\sigma}} \right)^{1/3} = 1 - v_2 \qquad (5.29)$$

When lossy material is distributed in an insulator, i.e. $\sigma_1 \ll \sigma_2$, Eqn. (5.28) simplifies to

$$\bar{\varepsilon}_0 = \varepsilon_1' \frac{\bar{\sigma}(\sigma_2 - \bar{\sigma})}{\sigma_1(2\sigma_1 + \sigma_2)} \qquad (5.30)$$

$$\frac{\sigma_2 - \bar{\sigma}}{\sigma_2} \left(\frac{\sigma_1}{\bar{\sigma}} \right)^{1/3} = 1 - v_2 \qquad (5.31)$$

In many practical cases the conductivity of the compound is smaller than that of the inclusions, $\bar{\sigma} \ll \sigma_2$; in this case

$$\bar{\varepsilon}_0 = \frac{\varepsilon_1'}{(1 - v_2)^3} \qquad (5.32)$$

The unrelaxed permittivity in the general case is expressed by the following equation

$$\frac{\bar{\varepsilon}_\infty - \varepsilon_2'}{\varepsilon_1' - \varepsilon_2'} \left(\frac{\varepsilon_1'}{\bar{\varepsilon}_\infty} \right)^{1/3} = 1 - v_2 \qquad (5.33)$$

The validity of the Hanai equations has been verified for several models in which artificial inclusions have been introduced (van Beek, 1962; Kharadly and Jackson, 1953) and also for various dispersions such as oil–water (Hanai, 1961), nitrobenzene–water (Hanai et al., 1962), woolwax–water (Dryden and Meakins, 1957). Colloidal solutions have been studied by Miles and Robertson (1932) and O'Konski (1960).

For heterogeneous polymer systems the relaxation time of the MWS polarization is below the frequency range usually studied ($10^{-5} \ldots 10^{+6}$ sec). Correspondingly, the presence of the MWS effect is manifested by the presence of the $\bar{\varepsilon}_\infty$-value as a background corresponding to the MWS process. When dipole orientation spectra are measured at low frequency as a function of the temperature, the variation of the dipole orientation $\varepsilon'(T)$ or $\varepsilon''(T)$ is added to the $\bar{\varepsilon}_\infty$-value which is not, or slightly, temperature dependent. This is why in heterogeneous systems extremely high permittivity values are measured.

Expressions for $\bar{\varepsilon}_\infty$ for some representative cases are shown in Table 5.1. It is seen that $\bar{\varepsilon}_\infty$ is high when the permittivity difference between the host material and the additive $\varepsilon_1' - \varepsilon_2'$ is large and when the depolarization factor A_i is small, i.e. the inclusions deviate strongly from spherical, for which $A_i = 1/3$.

By measuring dielectric spectra as a function of the frequency the MWS polarization appears usually as a background by increasing the absolute values of the permittivities. By choosing the temperature as a variable, the presence of the MWS effect is observed by the rapid increase of the oscillator strength $(\varepsilon_0 - \varepsilon_\infty)_T$ when frequency decreases. An example of this has been presented in Chapter 2 for PVC—Fig. 2.11. As $\bar{\varepsilon}_\infty$ depends on the conductivity of the host polymer and of the inclusions and the conductivites are temperature dependent, peaks due to MWS, polarization might be observed even if the inclusions have no transition in the temperature range studied.

When the hetero-phase additive exhibits a transition in the temperature range studied, for example the included monomer or water melts, the $\bar{\varepsilon}_\infty$ background exhibits a sharp change as a result of the change in the permittivity difference $\varepsilon_1' - \varepsilon_2'$. An example of this has been given in

Table 5.1 Maxwell–Wagner–Sillars–formulae for various heterogeneous systems

Dispersed particle	$\bar{\varepsilon}$,
Spheres $r_2 < 0.2r_1$	$\varepsilon_1' \dfrac{2\varepsilon_1' + \varepsilon_2' + 2v_2(\varepsilon_2' - \varepsilon_1')}{2\varepsilon_1' + \varepsilon_2' - v_2(\varepsilon_2' - \varepsilon_1')}$
Spherical voids $\varepsilon_1' = 1$	$\dfrac{1 + 2v_2}{1 - v_2}$
Conducting spheres $\varepsilon' \gg \varepsilon_1'$ $\sigma_2 \gg \sigma_1$	$\varepsilon_1'\left(\dfrac{1 + 2v_2}{1 - v_2}\right)$
Spheroids $r_2 < 0.1r_1$ along field	$\varepsilon_1'\left(1 + \dfrac{v_2(\varepsilon_2' - \varepsilon_1')}{\varepsilon_1' + A_i(\varepsilon_2' - \varepsilon_1')}\right)$
Prolate ellipsoids rods, cylinders along field	$\varepsilon_1' + \dfrac{v_2}{3}\dfrac{(\varepsilon_2' - \varepsilon_1') + (5\varepsilon_1' + \varepsilon_2')}{\varepsilon_1' + \varepsilon_2'}$
Randomly oriented ellipsoids	$\varepsilon_1\left[1 + \dfrac{v_2}{3}\sum_i \dfrac{\varepsilon_2' - \varepsilon_1'}{\varepsilon_1' + A_i(\varepsilon_2' - \varepsilon_1')}\right]$
Regularly arranged cylinders $\pm$ field. $r_2 < 0.1r_1$	$\varepsilon_1' \dfrac{\varepsilon_1' + \varepsilon_2' + v_2(\varepsilon_2' - \varepsilon_1')}{\varepsilon_1' + \varepsilon_2' - v_2(\varepsilon_2' - \varepsilon_1')}$
Oblate spheroids (lamellae) $r_2 < 0.1r_1$	$\varepsilon_1' + \dfrac{v_2}{3}\dfrac{(\varepsilon_1' + 2\varepsilon_2')(\varepsilon_2' - \varepsilon_1')}{\varepsilon_2'}$
Randomly oriented conducting ellipsoids $\varepsilon_2 \gg \varepsilon_1$ $\sigma_2 \gg \sigma_1$	$\dfrac{\varepsilon_1'}{1 - (v_2/3)\sum_i 1/A_i}$
Conducting disks $\varepsilon_2 \gg \varepsilon_1$ $\sigma_2 \gg \sigma_1$	$\varepsilon_1' + \dfrac{2v_2}{3}\varepsilon_2'$
Conducting needles $\varepsilon_2 \gg \varepsilon_1$ $\sigma_2 \gg \sigma_1$	$\varepsilon_1' + \dfrac{v_2}{3}\varepsilon_2'$
Arbitrary shaped inclusions $\phi = $ experimental constant	$\varepsilon_1' + \dfrac{v_2(\varepsilon_2' - \varepsilon_1')(1 - \phi)}{1 - v_2\,\phi}$

section 5.1 (Fig. 5.3) for plasticized polyvinylchloride. The sharp increase in ε' at about $-50°\text{C}$, depending on the thermal history of the sample, is interpreted as being due to inclusions of plasticizer which is incompatible with the polymer.

The MWS polarization is most clearly exhibited in polymeric compounds which contain non-polar fillers such as glass fibre, silicon sand, clay, minerals, oil, etc. Such fillers are very often used for rubbers and for thermosetting resins. The most suitable techniques for studying such systems seem to be the Fourier transform and depolarization methods. Unfortunately very little work has been done in this field so far. The effect of nonpolar additives on the a.c. dielectric spectra is indirect; it is exhibited through change of free-volume (α-processes) and even more indirectly through modifying the potential barriers hindering side-group rotations, (β-processes) or local motions of the main chain (γ-processes). Some fillers are known to affect the crystalline structure, especially, e.g., the size of crystalline mineral oxides. Such fillers are expected to change the crystalline transitions.

Lipatov and Fabulyak (1972) studied the effect of some non-polar fillers on the dielectric spectra and on the NMR relaxation times of polymethyl methacrylate (PMMA), polystyrene (PST), cellulose acetate and an MMA-ST copolymer.

Figure 5.24 shows a typical sequence of spectra for polymethyl methacrylate filled with aerosil. The average diameters of the filler particles were in the order of 1μ. In this case the filler is non-polar and exhibits no transition in the temperature range between $0°C$ and $150°C$, where the α- and β-transitions in PMMA are observed. It is seen that the intensity of the dielectric α-peak is decreased while that of the β-peak is increased by increasing filler concentration. Figure 5.24 also shows that the α- and β-transitions are shifted the opposite way by increasing filler concentration; the α-peak (T_g) is shifted to higher temperature, the β-peak (ester side-group rotation) is shifted to lower temperature.

In the dielectric depolarization spectra of polymers containing non-polar inclusions, fillers, air-inclusions, microcracks etc. an individual MWS-peak should develop as a result of the interfacial polarization at the inclusions. Evidence for this has been obtained by studying depolarization currents in an arrangement with an air gap between the electrode and the polymer (van Turnhout, 1971). Figure 5.25b shows how the observed depolarization spectrum of a PVC foil metallized on one side and having an air gap between the other side and the electrode in comparison with that measured without air gap (a). The observed depolarization current is composed of the dipole depolarization part and the MWS part from polarization of the air gap. The origin of the

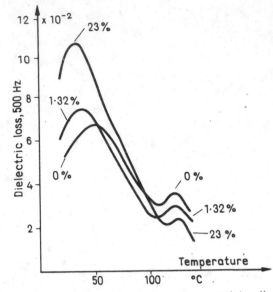

Fig. 5.24 Dielectric spectra of polymethyl methacrylate containing dispersed aerosil filler (after Lipatov and Fabulyak, 1972)

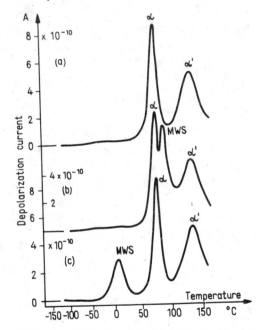

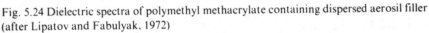

Fig. 5.25 Maxwell–Wagner–Sillars peak in the dielectric depolarization spectrum of PVC (a) PVC without air-gap, (b) with air-gap, (c) PVC-polyethylene blend without air-gap

292

MWS-peak in such systems is explained by depolarization by conduction; the conductivity of the host polymer PVC above T_g is rapidly increased. On the other hand, Fig. 5.25c shows the depolarization current spectrum without air gap of a PVC-polyethylene blend containing 10 % polyethylene. By introducing the non-compatible non-polar polyethylene component into PVC an MWS peak is observed below the T_g of PVC.

Effect of conductive fillers

Fillers of high conductivity, especially carbon black, are often used for preparing thermoplastic or thermosetting compounds. Carbon black is used, for example, to prevent ultraviolet degradation of polyethylene. Rubber filled with carbon black will be discussed in detail in Chapter 6. Thermosetting resins are also often filled with metallic powders or with carbon blacks. These systems are usually insulators, but by special technique conductive resins can be prepared (see, e.g., in Goul *et al.*, 1968).

As the conductivity difference between the host polymer and the filler is extremely great, very strong MWS-effect is expected and found in such systems. The dielectric permittivity of metal or carbon black filled polymer may be as high as 1000 at low frequencies.

Effects of water and solvent traces

Water may be present in polymeric systems caged in the structural defects or bound by hydrogen bonds. As water is polar it exhibits individual dielectric transitions with a relaxation time depending on the method of bonding with the polymer matrix. Free water has a relaxation time in the order of $10^{-10} - 10^{-11}$ sec between 0°C and 100°C. The dielectric relaxation time of ice, on the other hand is 8 orders of magnitude longer. Figure 5.26 shows the temperature dependence of the dielectric relaxation times of water in different environments. It is seen that in the sorbed state it may have dielectric relaxation times of the order of $10^{-2} - 10^{-8}$ sec, which covers the whole range where the relaxation of the host polymer is expected to occur. Consequently water caged in the defects of polymers would produce loss peaks which are separated from those of the host polymer or peaks which would coincide with them. Some examples of water peaks have already been mentioned in Chapters 2 and 4. By measuring the temperature dependence of the

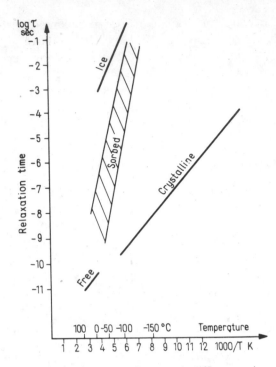

Fig. 5.26 Relaxation map for water in different environments

dielectric permittivities water peaks usually are observed at low temperatures $-50°C$ to $-100°C$ at intermediate frequencies. Especially in such polymers are found individual water peaks which can form hydrogen bonds, as polyamides, polyurethanes, cellulose. In polyesters, polycarbonates and in oxide polymers the chain end hydroxyl groups may interact with water. The appearance of water peaks in polyalkyl acrylates and methacrylates has also been mentioned already in Chapters 2 and 4.

As a particular example Fig. 5.27 shows the water peak in poly-sulphone

$$\sim O - \langle\!\!\!\langle\bigcirc\rangle\!\!\!\rangle - SO_2 - \langle\!\!\!\langle\bigcirc\rangle\!\!\!\rangle \sim$$

The T_g of this polymer is at $214°C$ where a large dielectric peak is observed as a result of the polar chain dipoles. Another dielectric

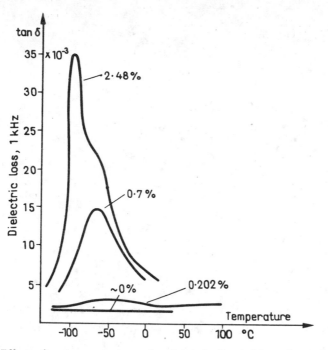

Fig. 5.27 Effect of water content on the dielectric β-transition of a polysulphone (after Allen, 1971)

transition is found near to $-50°C$, assigned as the β-transition, in the polymer unless specially dried. By careful drying, this peak is completely removed, indicating that it is due to water. In Fig. 5.27 this peak is shown at different water contents. It is seen that by increasing water content the dielectric $\tan\delta$ value at $-50°C$ sharply increases. Above 1% water content the peak is doubled. Allen *et al.* (1971) interpreted this as being due to the appearance of a second water phase (lower temperature peak) which is less strongly bound to the polymer. The sites where water is bound to polysulphone are most probably the dislocations created at the chain ends where hydroxyl end groups are present. The situation is similar to that of polyoxymethylene discussed in Chapter 4. Water also affects the mechanical properties, the water peak appears in the mechanical relaxation spectra too.

The water peaks in polycarbonates, polymethylene terephthalates, polysulphone, polyoxymethylene and in polyamides appear approximately in the same temperature range: between $-100°C$ and $-50°C$

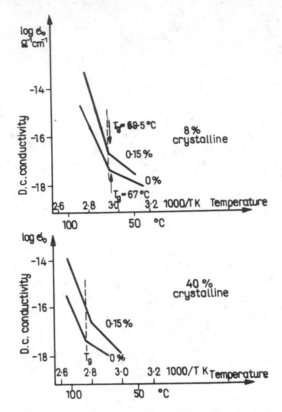

Fig. 5.28 Arrhenius plot of the d.c. conductivities of polyethylene terephthalate containing 0.15 % $Ca(OAc)_2 \cdot H_2O$ (after Sasabe *et al.*, 1969)

in the kHz frequency range. The peak is shifted to a higher temperature by increasing the frequency, the relaxation process is of Arrhenius-type. The activation energy values are very close together, suggesting that the mechanism of the interaction of water with the polymer matrix is similar.

Effects of impurities: catalysts, stabilizers, inhibitors

Small quantities of inorganic impurities remaining from the catalyst residues or added deliberately as stabilizers have no considerable effect on the a.c. dielectric spectra but through the MWS process influence d.c. conductivities and hence polarization currents appreciably. An example of this is Fig. 5.28 where the temperature dependence of the

296

d.c. conductivities of two pairs of polyethylene terephthalate samples of different crystallinity are shown after Sasabe *et al.* (1971). One of the samples has been specially purified and is believed to contain 0 % impurity. The other samples contained 0.15 % by weight of $Ca(OAc)_2 \cdot H_2O$ and 0.025 % by weight of Sb_2O_3.

Table 5.2 WLF constants of polyethylene terephthalate (Sasabe *et al.*, 1969)

Polymer	From d.c. measurements		From dielectric spectra	
	A	B	A	B
Pure	24.39	167.3	19.61	40.5
0.15 %	13.33	55.3	19.61	40.5

In Fig. 5.28 the breaks exhibited by the $\log \sigma$ versus $1/T$ curves are due to the glass–rubber transition. It is seen that the d.c. conductivities and the activation energies for the conductance are appreciably increased by the presence of the impurity. The a.c. dielectric spectra are not so much influenced by the impurity: the shape of the α-peak is unaffected, its position is somewhat shifted to higher temperatures. Sasabe *et al.* (1971) analysed the d.c. conductivity σ_0 versus $1/T$ curves, and the dielectric relaxation time (τ) versus $1/T$ curves, on the basis of the WLF-equation (Eqn. 2.1 in Chapter 2) and found that the WLF constants are dependent on the impurity content only in the $\log \sigma$ versus $1/T$ case. The WLF constants for a pure 0 % and impure (0.15 % additive) polymer are shown in Table 5.2. From these results it is concluded that non-compatible impurities, such as traces of catalysts and stabilizers, would not directly affect molecular motion of the system, but would affect very strongly its texture.

As with fillers, it is concluded that, rather than the intermediate and high frequency technique, the very low frequency Fourier trans-formation dielectric method and dielectric depolarization spectroscopy is to be preferred for such problems.

5.3 Dielectric spectra of copolymers and polymer blends

Copolymers are prepared by polymerization of mixtures of two or more monomers, so in their polymer chains different monomer units are represented at random (random copolymers) or regularly alternating (regular copolymers). In a type of copolymers sequences of like units blocks alternate in the chain (block copolymers). By copolymerization linear or crosslinked systems can be formed. A special case of copolymerization is grafting, i. e. polymerization of a monomer in the presence of a polymer so that the second polymer formed is chemically bonded to the host polymer.

(a)   . . . A B A B A B A B . . .

(b)   . . . A A B A B B A A A B · · ·

(c)   . . . A A A A A B B B B A A A A A B B . . .

                          B
                          B
                          B
(d)   . . . A A A A A A A A A A A · · ·

Fig. 5.29 Schematic representation of copolymer types (a) regular (b) random (c) block (d) graft

The main types of copolymers are schematically represented in Fig. 5.29 where A and B signify different monomer units.

Polymer blends are mechanical mixtures of different polymers without formation of chemical bonds between them. As the dielectric relaxation properties are mainly determined by the motion of parts of the molecules, the transitions of the block copolymers and graft copolymers are not expected to be much different from those of the corresponding mechanical mixtures. In the random and regular copolymers, significant differences may be found.

Dielectric spectroscopy of copolymers and polymer blends is not highly developed. These systems have been studied mostly by mechanical methods and by differential thermal analysis. It is surprising that the dielectric method has been so neglected in studying multi-component polymeric systems. As dielectric transitions are only exhibited by polar polymers, comparison of the transitions measured by dielectric and mechanical relaxation methods is of much help in assigning the transitions. On the other hand, changes in the oscillator strength of the dielectric transitions (effective dipole moments) might reveal the presence of dipolar inter and intramolecular interactions or changes in molecular configurations. Finally, by dielectric depolarization spectroscopy, important information about the phase-heterogeneity of the compositions may be gained. These possibilities are so far unexploited.

298

The glass–rubber transition of copolymers and polymer blends

The T_g of copolymers and compatible polymer blends is usually intermediate between those of the individual components. The problem is very similar to the effect of plasticization discussed in section 5.1. T_g of the copolymer is considered as being determined mainly by the weighted average of the T_g's of the components. According to Wood (1958)

$$A_1 M_1 m_1 (T_g - T_{g1}) + A_2 m_2 M_2 (T_g - T_{g2}) = 0 \qquad (5.34)$$

where m_1 and m_2 are the mole fractions of the components, M_1, M_2 are the molecular weights, T_{g1} and T_{g2} are the corresponding glass transition temperatures, T_g is the glass-transition temperature of the copolymer, the constants A_1 and A_2 are related to the thermal expansion coefficients of the components. Gordon and Taylor (1952) assumed that these constants are simply the differences of the thermal expansion coefficients below and above T_g for the individual components, Mandelkern et al. (1957) used the following expression for the constants A_i

$$A_i = \alpha_i (T > T_g) - \alpha_i^* - 0.025 \alpha_i (T < T_g) \qquad (5.35)$$

where α_i^* is the thermal expansion coefficient of the occupied volume.
Fox (1956) derived a simplified equation by assuming that

$$\frac{A_2 M_2 T_{g2}}{A_1 M_1 T_{g1}} \approx 1 \qquad (5.36)$$

The Fox equation is

$$\frac{1}{T_g} = \frac{m_1}{T_{g1}} + \frac{m_2}{T_{g2}} \qquad (5.37)$$

or by using the weight fractions w_1 and w_2 and considering that $w_2 = 1 - w_1$

$$\frac{1}{T_g} = w_1 \left(\frac{1}{T_{g1}} - \frac{1}{T_{g2}} \right) + \frac{1}{T_{g2}} \qquad (5.38)$$

299

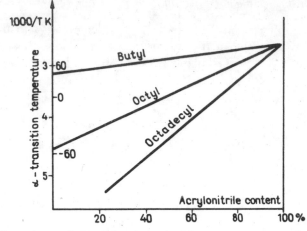

Fig. 5.30 Dependence of the dielectric α-transition temperature on the acrylonitrile content in acrylonitrile-alkylamide copolymer (after Jordan, 1966)

di Marzio and Gibbs (1959) expressed the constants A_i in terms of the number of single bonds in the monomer unit in the copolymer n_i as

$$A_i = \frac{n_i}{M_i} \qquad (5.39)$$

where M_i is the molecular weight of the unit. By using this assumption

$$\frac{m_1 n_1}{M_1} (T_g - T_{g1}) + \frac{m_2 n_2}{M_2} (T_g - T_{g2}) = 0 \qquad (5.40)$$

From Eqns (5.38) and (5.40) it is seen that the reciprocal glass–rubber temperature is expected to be approximately proportional to the volume fraction or weight fraction of the second component, provided the free volume effect is dominant in determining the process, and free volume is additive. An example of this is shown in Fig. 5.30, for a series of acrylonitrile-alkylamide copolymers after Jordan *et al.* (1966). The $10^3/T$ versus acrylamide concentration curves are fairly linear. Similar behaviour is observed for blends of ethylene-vinylacetate copolymer with polyvinyl chloride (Hammer, 1971) and in methyl methacrylate-ethyl acrylate copolymer (Sperling, 1970). The curves are shown in Fig. 5.31, where the linearity of the concentration versus $1/T$ curve is

300

exhibited by the copolymer and blends as well in a fairly wide concentration range.

In another group of copolymers the glass–rubber transition exhibits a minimum as a function of the composition. Examples of this are presented in Fig. 5.32 for ethylene-propylene copolymer (Boyer, 1968) where the α_a- and α_c-transitions are both shown.

The deviation from the composition dependence of T_g predicted by Eqn. (5.40) is interpreted as being due to the change of the free volume by mixing. In order to account for this, Jenckel and Hensch (1953) introduced parameters $k(g)$ and $k(l)$ defined as

$$v(g) = w_1 v_1(g) + w_2 v_2(g) + k(g) w_1 w_2$$
$$v(l) = w_1 v_1(l) + w_2 v_2(l) + k(l) w_1 w_2 \qquad (5.41)$$

where $v(g)$ $v(l)$ are the specific volumes of the copolymer blend in the glassy and in the liquid state respectively, $v_1(g)$, $v_1(l)$, $v_2(g)$, $v_2(l)$ are those of the corresponding components, w_1, w_2 are the weight fractions. By introducing

$$\Delta k = k(g) - k(l) \qquad (5.42)$$

the T_g of the copolymer is expressed as

$$T_g = \frac{w_1 A_1 T_{g1} + w_2 A_2 T_{g2}}{w_1 A_1 + w_2 A_2} + \frac{w_1 w_2 k}{w_1 A_1 + w_2 A_2} \qquad (5.43)$$

considering that $w_1 = 1 - w_2$

$$T_g = \frac{w_1(A_1 T_{g1} - A_2 T_{g2}) + \Delta k - w_1^2 \Delta k + A_2 T_{g2}}{w_1(A_1 - A_2) + A_2} \qquad (5.44)$$

Jenckel and Hensch (1953) used the Gordon–Taylor (1952) approximation for constants A_1 and A_2, i.e.

$$A_1 = \alpha_1(l) - \alpha_1(g) = \Delta\alpha_1$$
$$A_2 = \alpha_2(l) - \alpha_2(g) = \Delta\alpha_2 \qquad (5.45)$$

where $\alpha_i(l)$ and $\alpha_i(g)$ are the thermal volume expansion coefficients in the liquid and glassy state respectively for the individual components.

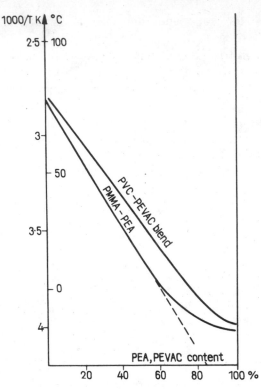

Fig. 5.31 Linearity of the $1\,T$ versus concentration plot for a copolymer methyl-methacrylate-ethylacrylate (PMMA-PEA) and for a blend of PVC with ethylene-vinylacetate copolymer (PVC-PEVAC)

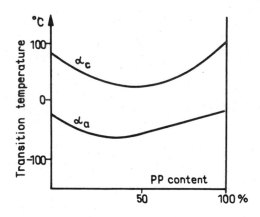

Fig. 5.32 Dependence of the α_c and α_a transition temperature on the composition of the ethylene-propylene copolymer (after Boyer, 1968)

302

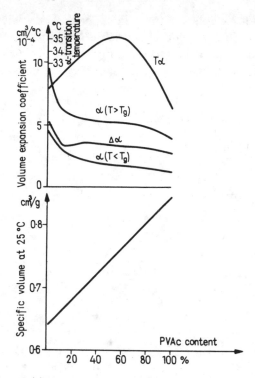

Fig. 5.33 Dielectric transition temperature (T_α), volume expansion coefficient (α), at $T > T_g$ and $T < T_g$, their difference $\Delta \alpha$ and the specific volume, as a function of the composition for blends of polyvinyl nitrate and polyvinyl acetate (after Akiyama, 1972)

According to Eqn (5.44), the glass-transition temperature of the copolymer or blend may be lower or higher than those of the individual components, depending on the relative values of the parameters Δk, $A(\Delta \alpha)$ and those of T_{g1} and T_{g2}.

As an example of this Fig. 5.33 shows the glass-rubber transition temperature, the specific volume, and the volume expansion coefficients measured below and above T_g for a blend of polyvinyl nitrate (PVNi) and polyvinyl acetate (PVAc) after Akiyama (1972). It is seen that the glass-rubber transition temperature of the blend is higher than that of the individual components and the T_g versus PVAc content exhibits a maximum. This behaviour is also satisfactorily described by the Jenckel–Hensch equation (5.44). The maximum of the T_g versus composition curve is explained by the sharp drop of the thermal volume expansion coefficients as the PVAc-concentration is increased.

The parameter Δk obtained from the experimental values of the specific volume is $\Delta k = 8.7 \times 10^{-3} \, \mathrm{cm}^3/\mathrm{g}$. By introducing this value to Eqn. (5.44), together with the $\Delta\alpha$ values shown in Fig. 5.33, the shape of the experimental T_g-versus PVAc concentration curve is satisfactorily represented.

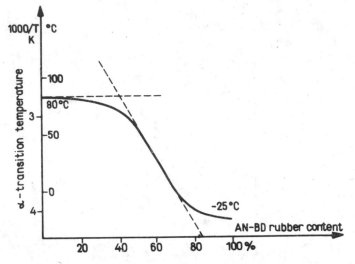

Fig. 5.34 Dependence of the α-transition temperature on the composition of a blend of PVC with acrylonitrile butadiene rubber (Aibasov *et al.*, 1970)

A peculiar dependence of the α-transition temperature is exhibited for acrylonitrile-butadiene copolymer (AN-BD rubber) blended with polyvinylchloride. Figure 5.34 shows the α-transition temperature (T_g) as a function of the AN-BD rubber content in the $1000°/T$ versus concentration representation. It is seen that by increasing the rubber phase content the T_g of PVC is only slightly shifted up to concentrations of about 50%. For concentrations between about 50% and 75%, the T_g is shifted sharply to lower temperatures to reach almost the T_g-value of the copolymer, $-25°C$. In the low rubber concentration range, up to about 40%, the mixture is not compatible, the individual α-transitions of the components are separately observed. In the compatibility range the α-peaks coalesce into a single one. In this range the linear dependence of $1000°/T$ on concentration is exhibited (Aibasov *et al.*, 1970).

In many practical compounds the components are not compatible; they would form separate phases. In such systems the main transitions

304

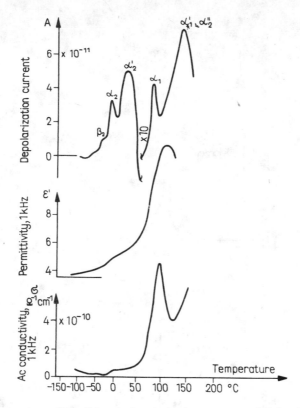

Fig. 5.35 Comparison of the dielectric depolarization spectra and dielectric spectra of a blend of PVC with chlorinated polyethylene

of the individual components appear separately and the position of the maxima are not notably shifted. As shown in section 5.2, in heterogeneous mixtures a very significant MWS polarization develops, dependent on the shape and size of the heterogeneities, i.e. on the texture of the compound. Thus by dielectric depolarization measurements performed in comparison with a.c. dielectric, and with mechanical methods, important information can be gained.

Figure 5.35 shows the dielectric depolarization spectrum of a blend of PVC with chlorinated polyethylene in comparison with the a.c. dielectric spectra recorded as a function of the temperature at a constant frequency of 1 kHz. Transitions α_1 and α_1' correspond to the PVC-phase, those of α_2, α_2', α_2'' and β_2 to the chlorinated polymer phase. Figure 5.36 shows the depolarization spectrum of the rubber phase alone. By com-

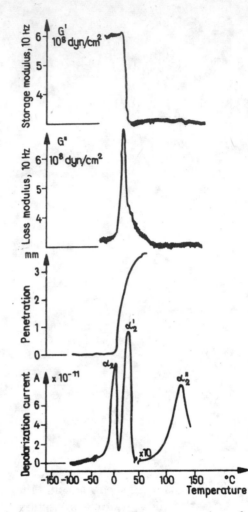

Fig. 5.36 Dielectric depolarization spectra of the elastomer phase of toughened polyvinyl-chloride in comparison with the mechanical spectra and with the thermomechanical curve

parison of the depolarization spectra with thermomechanical and mechanical relaxation curves, Hedvig and Földes (1974) have shown that the smallest peak in the doublet labelled by α_2 corresponds to the glass-rubber transition of chlorinated polyethylene.

The peak α'_2 appearing in the depolarization spectra is attributed to the MWS polarization. This peak appears also in a blend of PVC with unchlorinated pure polyethylene, as shown in Fig. 5.25c.

306

From the experimental evidence obtained so far on the shift of the glass–rubber transition temperature in copolymers and compatible polymer blends it is concluded that this effect is mainly governed by the change in free volume. This is another support of the general conclusion obtained in section 2.2, that the glass–rubber transition in polymeric systems is mainly governed by the free volume.

Secondary transitions

As shown in Chapter 2, secondary transitions, i.e. transitions in the glassy state in pure polymers, arise from hindered rotation of side groups, chain-end groups or from local motion of short segments of the main chain. Evidently, such transitions will not be greatly affected by copolymerization, especially in block-copolymers, graft-copolymers and polymer blends. Some effects may, however, be expected from the interaction of the α- and β-processes, discussed in section 2.1. In regular and in random copolymers the potential barrier which hinders the local motion might be significantly influenced by the presence of the second component.

The effective dipole moment of copolymers

It has been shown in Chapter 1 that the oscillator strength of the dielectric transition $\varepsilon_0 - \varepsilon_\infty$ or the area of the dielectric loss versus frequency curve are related to the concentration of the polar groups involved and to their effective dipole moment (cf. Eqn. 1.175). It is extremely interesting to know how the effective dipole moment of a certain polar group of the host polymer is influenced by other polar or apolar groups introduced by copolymerization or by blending another polymeric component.

Mikhailov *et al.* (1967a, 1969b) studied the effect of an apolar co-polymer (styrene) on the effective dipole moment of acrylates (methyl methacrylate and methylacrylate), and of *p*-chlorostyrene.

In dilute solution the polarization of the copolymer is expressed by Borisova *et al.* (1962) as

$$\mathscr{P}_{cop} = \frac{3 v_0}{(\varepsilon_0 + 2)^2} \frac{d\varepsilon}{dw} + \left(v_0 + \frac{dv}{dw}\right) \frac{\varepsilon_0 - 1}{\varepsilon_0 + 2} \tag{5.46}$$

where ε_0 is the permittivity, v_0 is the specific volume extrapolated to infinite dilution, w is the weight fraction of the copolymer with respect to the solvent; and $\varepsilon(w)$, $v(w)$ are respectively the permittivity and specific volume measured as a function of the concentration w.

The polarization of the copolymer is expressed in terms of that of the individual components as

$$\mathscr{P}_{cop} = (1 - w_2)\,\mathscr{P}_1 + w_2\mathscr{P}_2 \tag{5.47}$$

where w_2 is the weight fraction of the polar component to the apolar one.

The effective dipole moment of the monomer unit is expressed as

$$\mu_{eff}^2 = \frac{9\,kT}{4\,\pi\,N_A}\,N_2\left[\frac{\mathscr{P}_{cop} - (1 - w_2)\,\mathscr{P}_1}{w_2}\,M_2 - R_2\right] \tag{5.48}$$

where N_2 is the number of the polar groups in the chain, $\mathscr{P}_1$ is the polarization of the apolar component, w_2 is the weight fraction of the polar component, M_2 is the molecular weight of the comonomer, R_2 is molar refraction.

For random copolymers of methyl methacrylate with styrene according to Simha (1962) and Mikhailov $et\ al.$ (1967 b)

$$\mu_{eff} = 0.0128\,[(\mathscr{P}_2\,M_2 + 24.40)T]^{1/2} \tag{5.49}$$

where 24.40 is the molar refraction of PMMA

$$\mathscr{P}_2 = \frac{\mathscr{P}_{cop} - (1 - w_2)\,\mathscr{P}_1}{w_2} \tag{5.50}$$

where $\mathscr{P}_1 = 0.3415$ is the polarization of polystyrene.

Figure 5.37 shows the dependence of the effective dipole moment of styrene-methyl methacrylate random and block copolymers on the styrene content, after Mikhailov $et\ al.$ (1968 a). It is seen that with increasing styrene content the effective dipole moment of methyl-methacrylate is increased for the random copolymer, but is unchanged for the block-copolymer. This shows that by building-in the styrene units the interaction among the MMA ester side group dipoles is decreased, resulting in an increase of the effective dipole moment. For the block-copolymer such an effect is not expected as the range of dipole-dipole interaction is smaller than the length of the polar blocks in the chain.

The effective dipole moment can be expressed also in terms of the Kirkwood reduction factor introduced in Chapter 1 (cf. Eqn. 1.176). If the dipole moment of the polar group is μ_0

$$\mu_{eff} \propto (N_2 \, m_2 \, g_r)^{1/2} \mu_0 \tag{5.51}$$

where m_2 is the molar fraction of the comonomer, N_2 is the number of polar sequences in the chain.

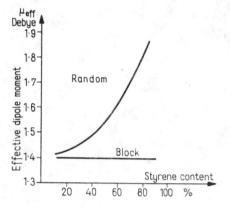

Fig. 5.37 Effective dipole moment of a random and block copolymer of styrene with methyl methacrylate, as a function of the composition (Mikhailov $et\ al.$, 1968)

As the maximum of the dielectric loss curve ε''_{max} is related to the area under it

$$\varepsilon''_{(max)} \propto N_2 \, \mu_0^2 \, g_r \tag{5.52}$$

where N_2 is the concentration of the polar groups. This means that the variation of the effective dipole moment or the Kirkwood correlation factor as a function of the copolymer composition can be studied simply by measuring the loss maxima as a function of the weight fraction of the apolar component.

To keep the number of the polar groups the same by increasing the apolar fraction, i.e. decreasing the polar group concentration, the concentration of the solution is suitably increased. Figure 5.38 shows a typical series of such spectra for the methyl methacrylate-styrene random copolymer, measured at 2 MHz as a function of the

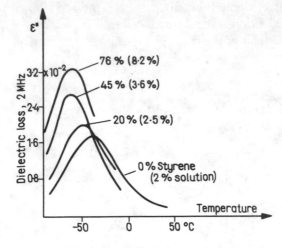

Fig. 5.38 Dielectric spectra of styrene-methylmethacrylate random copolymer in toluene solution of increasing concentration to keep that of the polar groups constant by increasing styrene content (after Mikhailov *et al.*, 1967)

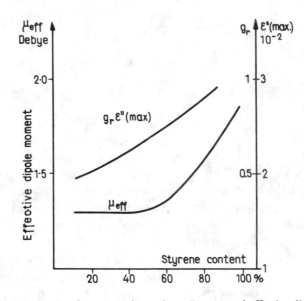

Fig. 5.39 The Kirkwood g_r-factor, maximum loss ε''(max) and effective dipole moment μ_{eff} of the styrene methyl methacrylate copolymer as a function of the composition (Mikhailov *et al.*, 1967)

temperature. Besides the styrene content of the copolymer the concentrations of toluene solution are given in parentheses. It is seen that the ε''(max) values are increased by increasing styrene content.

The temperature dependence of the dipole relaxation time of the ester group of PMMA is found to be of Arrhenius type for the MMA-STY random copolymer. The energy of activation of the process decreases with increasing styrene content. This means that the intramolecular barrier which hinders the rotation of the ester group is decreased by introducing styrene in the chain. The relatively high value of the activation energy in solution (7 kcal/mole) indicates strong intramolecular interaction. In Fig. 5.39 the g_r-values and the ε'' (max) values are shown as a function of the styrene content in the random copolymer.

CHAPTER 6

THE STUDY OF CROSSLINKING PROCESSES BY
DIELECTRIC SPECTROSCOPY

In this chapter the application of dielectric spectroscopy for studying such crosslinking processes as vulcanization of rubber, curing of epoxy, polyester and phenolformaldehyde resins will be discussed. Although there are quite early works, not many efforts have been made to apply the dielectric method to the study of crosslinking processes in these systems. This was probably caused by the fact that, with the classical non-recording dielectric spectrometers, it was too time-consuming to measure sets of spectra at different stages of curing. Recording spectrometers have opened up new possibilities. A very great advantage of the dielectric method over the more usual thermomechanical and chemical methods is especially apparent in studying epoxy and polyester resins. In these systems the reaction can be followed up to the end, when the resin becomes completely rigid, the embedded monomer cannot be extracted and the thermomechanical properties exhibit no change. Another yet unutilized possibility is that of measuring under pressure in the actual conditions of processing. It is not unduly difficult to measure thin layers of lacquers. It is also possible to follow crosslinking induced by ultraviolet illumination or by high energy irradiation.

In the following sections the main possibilities will be discussed and some illustrative examples presented. As in our earlier discussions, d.c. methods are also included because, as in most polymeric systems, in resins and in rubbers, d.c. conductivities measured in the usual conditions are mainly due to polarization effects and not to ohmic conductivity involving charge transfer.

6.1 Possibilities of the method. Experimental aspects

As discussed in Chapters 1 and 2, for measuring dielectric spectra it is essential for the polymer to contain polar groups. Pure apolar hydro-carbons should not exhibit dielectric spectra at all. In fact they do, although the spectrum bands are much less intense than those of polar polymers. Plenty of examples of that have been discussed in the previous chapters. The main sources of the appearance of dipole loss peaks in apolar polymers are the polar groups formed by oxidation (e.g. carbonyl groups), those formed by polar or ionic impurities by trapped electrons or ions and by the interfacial Maxwell–Wagner polarization, which is found to correlate with the defect concentration, i.e. free volume of the polymer (cf. Chapters 2 and 7). These factors make it possible to observe the main transitions of the structure and changes in molecular mobility in apolar polymers. The transition regions and the activation energies calculated from the temperature dependence of the mean dielectric relaxation time are found to agree fairly well with those measured by mechanical relaxation spectroscopy, dilatometry, differential scanning calorimetry and by the nuclear magnetic spin echo experiments. Hence molecular mobilities and transitions of structure can be investigated in apolar as well as in polar thermosetting resins. Very important information can be gained by measuring dielectric dispersion curves and calculating the total concentration of the polar groups in the system.

Crosslinking is a process of bonding linear macromolecules together directly or by introducing crosslinking atoms, molecules or chains. The main possibilities are represented in Fig. 6.1. Case (a) represents direct crosslinking; an example is crosslinking of polyethylene by chemical peroxide type initiators or by high energy irradiation. The macro-molecular chains are linked together randomly by chemical bonds, the average number of crosslinks per molecule is 1—10. This loose network is, however, enough to change the mechanical properties dramatically; the crosslinked polymer does not actually melt at the crystalline melting temperature although crystallinity vanishes.

Case (b) represents such a system in which crosslinking is due to a polyfunctional atom A, which links the chains together. The classical example for this is rubber vulcanized by sulphur. Rubber can be cross-linked by peroxide initiators and also by high energy radiation. The effect of vulcanization on the elastic properties of rubber is very well known.

Case (c) represents such a system in which crosslinking is due to a second macromolecular component. In this way, a kind of copolymer is formed which exhibits a three dimensional network with the host polymer. Typical examples are the unsaturated polyester resins which are usually crosslinked by styrene monomer which copolymerizes with the base resin to form a crosslinked network.

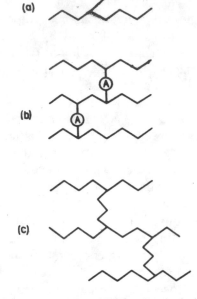

(a)

(b)

(c)

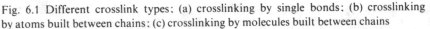

Fig. 6.1 Different crosslink types: (a) crosslinking by single bonds; (b) crosslinking by atoms built between chains; (c) crosslinking by molecules built between chains

From the dielectric spectra it can be deduced how the molecular mobility of the polymer molecules, submolecules and groups is influenced by crosslinking. The change in the mobility is generally observed as a shift of the glass-transition temperature of the polymer to higher temperatures. This process can be easily followed by dielectric spectroscopy.

In many cases besides the T_g-transition several T_{gg}-type transitions can be observed which are correlated with rotations or local motion of short chain segments (cf. Chapter 2). The important possibility presents itself of studying how these transitions are affected by crosslinking.

When the initial polymer or the crosslinking agent is polar, the total concentration of polar groups might change during the course of the reaction. This is reflected in the dielectric spectra as a decrease of the dispersion amplitudes. The concentration of the polar groups can be approximately calculated by using the molecular theories of dielectric relaxation discussed in Chapter 2.

The crosslinking reaction involves not only changes in molecular mobilities but also some redistribution of the structure; hence the total defect concentration can be expected to change during the course of the reaction. As discussed in Chapter 2, changes in the defect concentration are connected with changes in free volume and consequently with changes in the Maxwell–Wagner interfacial polarization. This effect is also reflected in the dielectric dispersion spectra. Examples have already been discussed in Chapters 4 and 5. The ideas used there might be useful to study crosslinked systems too, a possibility not utilized so far.

In some thermosetting systems, polar molecules or groups may be formed during the course of reaction, as in phenol formaldehyde resins, for example, where water is split off. In such cases the dielectric signals due to the polar products can be used to monitor the reaction.

For following crosslinking reactions the most straightforward way is to record dielectric spectra as a function of the frequency, at the fixed reaction temperature. Such an experiment can be performed under high pressure, ultraviolet illumination, or high energy irradiation. The only problem is that there are but few recording dielectric spectrometers available in which the frequency is scanned in a sufficiently wide range (cf. Chapter 3).

In many systems, however, the reaction temperature is much higher than the glass-transition temperature. In these cases, instead of the frequency scan, the more convenient temperature scan can be used with a linear 'up and down' temperature program so that the system stays at the reaction temperature for preset time intervals and the spectrum is repeatedly recorded up to this temperature. A set of spectra is thus obtained in which a shift of the transition temperatures and a change in the dielectric dispersion amplitudes are observed. Such a scheme, constructed in the author's laboratory, is shown in Fig. 6.2. The sample is thermostated in a copper block as shown in Chapter 3. The temperature of this block can be raised and lowered linearly in time at preset rates and can be kept constant within $\pm 0.5°C$ at any temperature in the range between $-150°C$ and $+200°C$. For sample holders coaxial cells made of fused silica have been used. The outer electrode is the copper block itself, the inner one is a silver mirror deposited on the inside wall of the silica tube. In this construction the electrodes do not directly contact the sample. The fused silica walls produce negligible background in the dielectric spectrum. For epoxy, polyester and phenol formaldehyde resins Pyrex glass sample holders have also proved to be useful.

Dielectric spectra can be detected simply by separating the resistive and capacitive current-components as described in Chapter 3. At fixed frequency the temperature dependence of the real part of the dielectric permittivity $\varepsilon'(\omega, T)$ and the total a.c. conductivity $\sigma(\omega, T)$ is

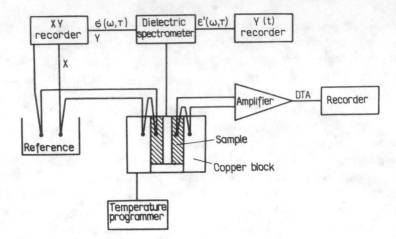

Fig. 6.2 Simplified scheme of a dielectric spectrometer for studying crosslinking reactions

recorded. The total a.c. conductivity is expressed as

$$\sigma(\omega, T) = \sigma_0(T) + \frac{2\pi}{\omega} \varepsilon''(\omega, T) \tag{6.1}$$

where $\sigma_0(T)$ is the ohmic conductivity which is an exponential function of temperature,

$$\sigma_0(T) = \sigma_0^0 \exp\left(-\frac{E_0}{kT}\right) \tag{6.2}$$

where E_0 is the activation energy for the conduction, σ_0^0 is a constant, $\varepsilon''(\omega, T)$ is the dielectric loss factor which exhibits maxima at the transition temperatures, ω is the fixed angular frequency. By repeating the measurement at different frequencies the activation energy of the transition can be determined. In practice, from the recorded curves the dielectric loss $\varepsilon''(\omega, T)$ can be separated from the ohmic conductivity $\sigma_0(T)$, which supplies additional information about the crosslinking process.

With the experimental arrangement shown in Fig. 6.2 it is possible to perform differential thermal analysis simultaneously with dielectric analysis.

In Fig. 6.3 it is schematically shown how the dielectric dispersion curve is expected to change by repeated scanning in a system in which the concentration of the polar groups is changing during the course of crosslinking. The sample is first cooled down to $-150°C$ and the first spectrum is scanned by raising the temperature. At the preset reaction temperature T_r the program is stopped and the $\varepsilon'(T_r)$ level is recorded as a function of reaction time (t). After a reaction time-interval Δt the temperature is lowered linearly and the spectrum is again recorded. If the reaction has proceeded during the time interval Δt a shift of the T_g

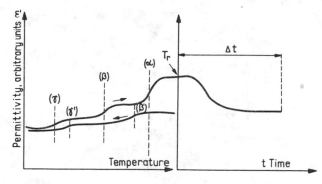

Fig. 6.3 Expected change of the dielectric spectrum during crosslinking. Temperature program: linear heating-up; crosslinking at constant temperature T_r, linear cooling

and a decrease of the corresponding $\varepsilon_0 - \varepsilon_\infty$ value is observed. The spectrum is recorded on raising the temperature. This procedure can be fully programmed beforehand and repeated up to high conversions. Simultaneously the $\sigma(\omega, T)$ curves are recorded. In such experiments it is convenient to use an XY recorder for recording $\sigma(\omega, T)$ and a simple pen-recorder to plot $\varepsilon'(\omega, T)$. When the temperature is stopped the pen-recorder continues to record the permittivity at the reaction temperature as a function of time, $\varepsilon'(T_r, t)$. Contacts deposited on the X axis of the XY recorder are used to introduce temperature marker signals to the pen-recorder.

Measurement during ultraviolet or high energy irradiation

For studying curing of resins or vulcanization of rubber by ultraviolet illumination or high energy irradiation the scheme shown in Fig. 6.4 can be used. A relatively thin layer of the resin (1 mm) is placed in a flat cell, the upper electrode of which is translucent. Such an electrode can be prepared by vacuum evaporating thin metallic layers in a fused silica or glass plate. For irradiation by X-rays, thin aluminium foil electrode can be used.

317

The flat cell is thermostated as described in **Chapter 3**. Spectra can be measured under irradiation, after irradiation, or intermittently by using a mechanical chopper or shutter to produce radiation pulses. The intermittent technique is useful for reaction kinetic studies.

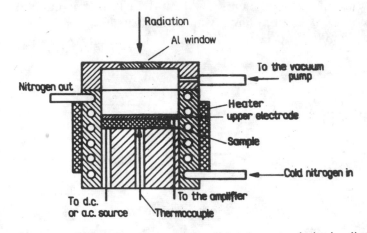

Fig. 6.4 Experimental arrangement for measuring dielectric spectra during irradiation by X-rays or by ultraviolet light

Transient d.c. conductivity and depolarization methods

It was shown in Chapter 3 that the charge and discharge curves measured as a function of time under the action of a step voltage can be transformed to a dielectric loss versus frequency scale by a Fourier-transformation procedure. The transient conductivity $\sigma(t)$ obtained this way is then proportional to the dielectric loss factor approximately as follows

$$\varepsilon''(\omega) = t\,\sigma(t)\,(\Omega^{-1}\,cm^{-1})\,1.8\,10^{13} \tag{6.3}$$

$$\omega = \frac{0.63}{t} \tag{6.4}$$

By the usual technique the frequency range from 10^{-4} Hz to 1 Hz can be covered.

At such low frequencies the glass-transition temperature of the resin is shifted to low temperatures, correspondingly the T_g-shift as a function of curing is easy to follow. In practice it is possible to measure at the fixed reaction temperature and obtain a part of the dielectric absorption spectrum as a function of frequency according to Eqns (6.3) and (6.4). It is also possible to scan the temperature linearly at a low rate and measure the transient conductivity as a function of temperature by measuring transient conductivities at a preset time. In this case the

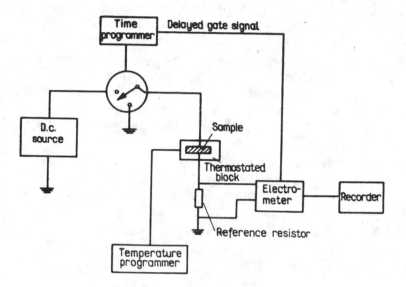

Fig. 6.5 Scheme of an apparatus for measuring transient conductivities during cross-linking

sample must be depolarized before each measurement. A possible solution is to use polarizing and depolarizing cycles alternately. In the polarizing cycle a step voltage is introduced to the sample and the conductivity after a time t_0 is measured by a recording electrometer. The electrometer is gated by a delayed signal in order to avoid saturation caused by the very high initial polarization current. In the next, depolarization cycle the voltage is removed from the sample which is grounded and the depolarization current after a time t_0 is recorded. Then a step voltage of opposite polarity is switched on and the process is repeated. The scheme of such a device is shown in Fig. 6.5.

A simplified variation of this technique is to measure the conductivity of the sample under a constant voltage at a relatively high constant heating rate. In this experiment the glass-transition temperature appears as a peak in the exponential conductivity versus temperature curve. The peak is shifted to higher temperatures when the heating rate is increased. Another useful procedure is to cool down the resin to the temperature of liquid nitrogen under a d.c. voltage, and by removing the external voltage record the thermo-depolarization currents (cf. Chapter 3) as a function of the temperature. At the transition temperatures current peaks are observed which are shifted to higher temperatures when the rate of heating is increased. In these experiments the heating rate is connected with the relaxation frequency. The thermodepolarization peaks are shifted to higher temperatures when the curing reaction proceeds analoguously to that observed by the a.c. methods. As discussed in Chapter 3 the physical basis of all these methods is the same: the change in the dielectric polarization at the transitions where molecular mobilities change.

It is reasonable to use the dielectric methods along with infrared spectroscopy and differential thermal analysis in order to obtain a fair picture of the process of curing. Very few such complex studies have been carried out so far.

6.2 The vulcanization of rubber

The dielectric method has been proved to be useful for studying vulcanization of various kinds of rubber. The earliest investigations in this field were performed by Scott et al. (1933) by measuring the change in the dielectric constant and loss tangent values as a function of sulphur content of the vulcanizates. Subsequent works have dealt with peroxide vulcanizates, with effects of accelerators, fillers and other additives. Considerable work has been done on the dielectric properties of the base polymers in comparison with their mechanical relaxation properties. Although rubbers are extremely important materials and very many new rubber types have been developed, the dielectric spectra of many of them are not known. Even in those cases where the constituent polymers have polar groups, and consequently strong dielectric effects are expected, few if any data are available.

The dielectric properties of unvulcanized rubber

For studying the effects of crosslinking it is essential to know the dielectric spectra of the base materials. These are mostly non-polar

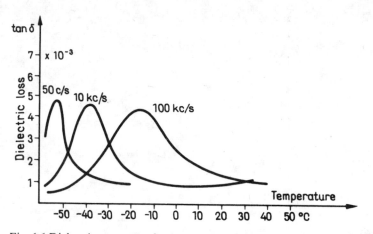

Fig. 6.6 Dielectric spectrum of poly-*cis* 1.4 isoprene (after Norman, 1953)

polymers. The chemical structure of natural rubber (*Hevea*-rubber) poly-*cis* 1,4 isoprene is, for example.

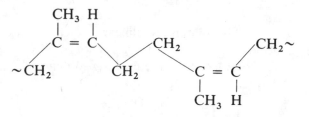

This polymer can be prepared by Ziegler catalysts or by metallic lithium catalysts.

The dielectric loss in this polymer is mainly due to the carbonyl groups formed by oxidation. Norman (1953) showed that the dielectric loss peaks are increased by oxidation. As with other apolar polymers discussed in Chapter 4, here again the polar C=O bonds formed by oxidation act as probes to indicate molecular motion. Even in the purest, most carfully prepared polymers the concentration of polar groups is high enough to produce measurable losses in the order of $\varepsilon'' = 2 - 4 \times 10^{-3}$. A set of dielectric absorption curves as a function of temperature for pure poly-*cis*-1,4-isoprene is shown in Fig. 6.6 at different frequencies after Norman (1953).

It is seen that although the losses are small the transition temperatures are easy to determine. The Arrhenius plot in the T_g-relaxation region is

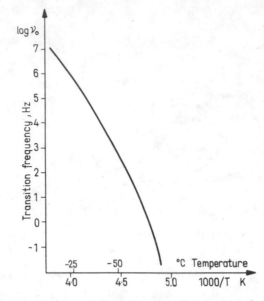

Fig. 6.7 Arrhenius diagram of the relaxation frequencies of polyisoprene (after Norman, 1953)

shown in Fig. 6.7. The plot is not linear, which indicates that the Arrhenius equation of the relaxation times is not fulfilled.

The data can be fitted to the following Williams–Landel–Ferry equation (cf. Chapter 2)

$$\log \left(\frac{\tau}{\tau_0}\right) = - \frac{8.86\,(T - T_g)}{101.6 + T - T_g} \tag{6.5}$$

the extrapolated glass-transition temperature value is found equal to $T_g = 248$ K $-25°$C. The data derived from dielectric spectra agree fairly well with those obtained by mechanical relaxation measurements (Payne, 1958).

Another important base material for rubber is polyisobutylene

This polymer is usually mixed with 1—2 percent of polyisoprene or with styrene to form butyl rubber by crosslinking.

322

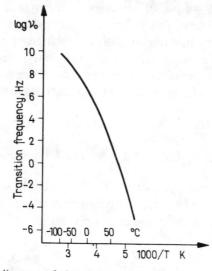

Fig. 6.8 Arrhenius diagram of the relaxation frequencies of polyisobutylene (after Kabin and Mikhailov, 1956)

The dielectric loss in polyisobutylene has been studied by Kabin and Mikhailov (1956). It is found to arise from the carbonyl groups formed by oxidation. The relaxation frequency versus reciprocal temperature (Arrhenius plot) of this polymer is shown in Fig. 6.8. The WLF equation which fits the data is the following

$$\log \left(\frac{\tau}{\tau_0}\right) = -\frac{16.5 \, (T - T_g)}{104 + T - T_g} \tag{6.6}$$

with $T_g = 201.8$ K $-71.2°$C. For comparison of the mechanical and dielectric data see McCrum *et al.* (1967).

For dielectric spectroscopy, those systems are easy to investigate which contain strongly polar groups. In rubbers the chlorinated and brominated base materials are of special interest. 1.4 polychloroprene, the basis of neoprene is, for example,

$$\sim CH_2 - \overset{\displaystyle Cl}{\underset{\displaystyle |}{C}} = CH - CH_2 \sim$$

The dielectric relaxation spectrum of this polymer is rather intense owing to the high dipole moment of the $C - Cl$ bond. The dispersion curve amplitude is about $\varepsilon_0 - \varepsilon_\infty = 10$, the loss factor is in the order of 0.1. The dielectric behaviour of this polymer is somewhat similar to that of PVC but the glass-transition temperature is much lower. Similarly strong dielectric spectra are expected from chlorobutyl and bromobutyl rubbers and from butadiene nitrile rubber. This latter is a copolymer of butadiene and acrylonitrile, the dielectric spectrum is due to the polar $C \equiv N$ bonds

$$
\begin{array}{c}
CN \\
| \\
\sim CH-CH_2-CH_2=CH-CH_2\sim
\end{array}
$$

Very few data have been published about the dielectric spectra of these systems. Another, yet not sufficiently studied, group of elastomers is based on polyurethanes. A typical base polymer is

$$
\sim N-(CH_2)_n-N-\overset{\overset{\displaystyle O}{\|}}{C}-O-(CH_2)_n-O-\overset{\overset{\displaystyle O}{\|}}{C}\sim
$$
$$
\underset{H}{|} \qquad \underset{H}{|}
$$

The adjacent molecules in these polymers are linked by hydrogen bonds $N-H\ldots O=C$. (Jacobs, Jenckel, 1961).

It is seen that this polymer chain is also well marked with polar groups, which makes these systems suitable for dielectric studies. In polyurethanes three transition regions are observed. The α-relaxation is near $40°C$ at low frequencies; it is interpreted as being the glass transition in the amorphous phase. The β-relaxation is found near $-60°C$; it is the effect of the absorbed water, which is always present in these systems. The γ-peak is found near $-150°C$, it is associated with crankshaft motion of the $(CH_2)_n$ chain. (Jacobs, Jenckel, 1961; Flocke, 1963.)

Another important group of rubbers is based on siloxane polymers exhibiting the general formula

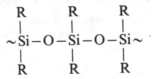

with $R = CH_3$, $CH_2=CH-$, C_6H_5, $N CCH_2CH_2-$ and other groups.

An example of the dielectric spectrum of silicon rubber will be presented later in this section.

324

Vulcanization with sulphur

The most common method of vulcanizing rubber is to mix it with sulphur and heat it up to 100—150°C. This procedure results in crosslinking of the polymer chains by mono- or polysulphide bridges. A side reaction is also known to proceed along crosslinking, leading to formation of heterocyclic groups in the polymer chains. In sulphur-vulcanizates correspondingly the following polar groups are found (Waring, 1951). (a) The carbonyl groups present in the unvulcanized rubber and those formed during vulcanization. (b) Mono- or polysulphide crosslinks as

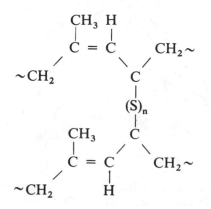

With $n = 1$ to 8. (c) S—C bonds in the heterocyclic groups in the main chains.

At low crosslink densities the average length between crosslinks is higher than the average segment length involved in the T_g-relaxation process. As discussed in Chapter 3, the average segment which becomes mobile at T_g, contains about 50—100 C—C bonds. Up to a sulphur content of about 10—15 % the average number of C—C bonds between subsequent crosslinks is estimated as 100—200. This means that vulcanizates containing less sulphur than 10 % should exhibit but a small T_g-shift and the dispersion amplitudes are expected to be unchanged as no polar groups are introduced to the mobile segments. Experiments show, however, that in sulphur vulcanizates, the glass—rubber transition is appreciably shifted and the dispersion amplitudes are increased. This indicates that the side reaction resulting in heterocyclic groups is effective; they would hinder the segmental motion which results in a T_g-shift and would increase the total number of polar groups involved, which results in an increase of the dispersion amplitude or loss maximum.

From this it follows that at low sulphur contents the dielectric spectra are mainly determined by the side reaction, while at high sulphur content above 15% the crosslink density becomes effective.

As discussed later in this section, in peroxide or in radiation vulcanization such an effect does not take place: the T_g-shift is uniquely

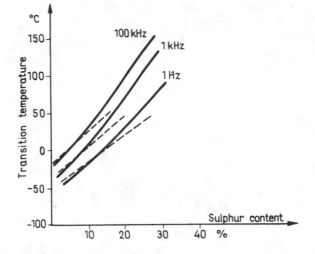

Fig. 6.9 The dependence of the dielectric α-relaxation temperature on the degree of sulphur vulcanization in polyisoprene (after Scott *et al.*, 1933)

due to crosslinking. Comparison of the dielectric spectra of sulphur vulcanizates and peroxy (or radiation) vulcanizates provides important information about the special side-reaction leading to formation of heterocyclic groups in sulphur vulcanizates. Scott *et al.* (1933) found an increase of the dielectric α-relaxation temperature of sulphur-vulcanizates with increasing sulphur content. In Fig. 6.9 this is illustrated at different frequencies. Similar results were obtained by using the transient d.c. conductivity method and by mechanical relaxation. (Schmieder and Wolf 1953). It is seen that by increasing the sulphur content the glass–rubber relaxation temperature is shifted from about $-50°C$ up to $+150°C$ at a frequency of 1 kHz. Up to about 10 per cent sulphur content this shift proceeds at a lower rate, then it is accelerated. The change of the slope of the T_g-shift curve can be interpreted as a result of the increasing contribution of the crosslink density above 10 per cent sulphur content, where the average segmental length is of the same order as the average distance between crosslinks. In the low sulphur content range 1—10 %, correspondingly, the T_g-shift is mainly determined by the hindrance of the heterocyclic groups built in the polyisoprene chains; at high concentrations it is determined by the hindrance of the crosslinks themselves.

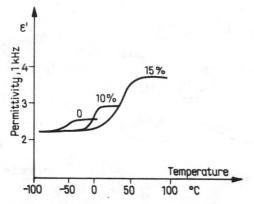

Fig. 6.10 Change of the dielectric dispersion spectrum of natural rubber by vulcanization with sulphur

The change of the dielectric dispersion amplitudes with increasing sulphur content is illustrated in Fig. 6.10. In this experiment the dispersion spectra of different vulcanizates were recorded, as a function of temperature at a fixed frequency of 1 kHz. It is seen that the glass temperature is shifted and the dispersion amplitude is increased with increasing sulphur content.

The effect of accelerators on the process of crosslinking by sulphur has been studied by Schallamach (1951) by the dielectric method. He used tetramethyl thiuramsulphide and diethyl dithiocarbamide accelerators and found that the α-relaxation temperatures were dependent on the accelerators used. On accelerating the crosslinking process with tetramethyl thiuramsulphide (TTMT) the products exhibited higher α-relaxation temperatures and higher dispersion amplitudes (loss maxima) than the corresponding vulcanizates accelerated by diethyl dithiocarbamate (DDTC) at the same sulphur content. This observation indicates that by using DDTC the crosslinking reaction is favoured against formation of heterocyclic polar groups in the chain. Correspondingly the T_g of the products obtained by this accelerator are shifted to higher temperatures than those obtained by using TTMT with the same sulphur content.

327

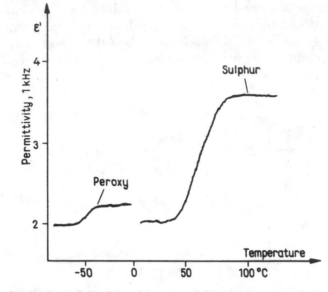

Fig. 6.11 Comparison of the dielectric spectra of polyisoprene vulcanized with sulphur and with peroxide

Vulcanization with peroxides and by irradiation

The crosslinking process in rubber can be initiated by peroxides at sufficient high temperatures and by radiation at low temperatures. A typical initiator is dicumyl peroxide which decomposes to benzoiloxy radicals above 60°C. The radicals thus formed initiate the crosslinking process. An essential difference with respect to sulphur-vulcanization is that no polar groups are built into the chains. Correspondingly the shift of the relaxation maxima to higher temperatures is uniquely due to the hindrance of the motion of the polymer chain segments by the crosslinks. This is illustrated in Fig. 6.11 where the dielectric dispersion spectra of peroxy and sulphur vulcanizates of polyisoprene are compared. The fixed frequency was 1 kHz, the samples cured in different ways contained the same 30% crosslink density. It is seen that at the same crosslink density the sulphur vulcanizate exhibits much higher transition temperature than the peroxide vulcanizate and the dispersion amplitude is much higher.

In the case of crosslinking by irradiation, the chains are linked by direct chemical bonds as with polyethylene discussed in Chapter 4.

328

On irradiation of polyisoprene at low temperatures, the following radical is observed by electron spin resonance

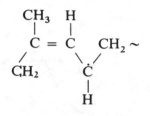

The crosslinking reaction is interpreted as being a result of recombination of such chain radicals to form a crosslinked network

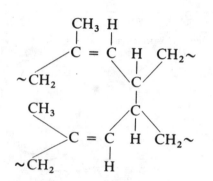

In this structure, evidently, no polar groups are introduced if the polymer is irradiated *in vacuo*. Correspondingly the dielectric dispersion amplitudes and the loss maximum values do not change as a function of crosslink density; only the transition temperature is shifted as the motion of the segments is hindered by the crosslinks. This is illustrated in Fig. 6.12, where the dielectric dispersion spectra of a natural rubber are shown after irradiation with X-rays of different doses. The sample was prepared from a latex as a film of thickness 3 mm. Irradiations were made with a 60 kV copper anticathode X-ray set-up at a dose rate of 100 Mrad/hr.

It is seen that with increasing irradiation dose, the dielectric transition is continuously shifted to high temperatures and the oscillator strength is somewhat increased. This increase is due to the radiation-induced oxidation of the rubber, because irradiations were performed in the presence of air and thus the oxidative degradation process proceeded along crosslinking.

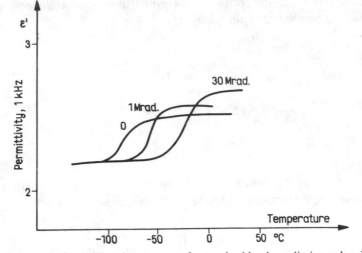

Fig. 6.12 Change of the dielectric spectrum of natural rubber by radiation vulcanization: radiation dose rate 100 Mrad/hr

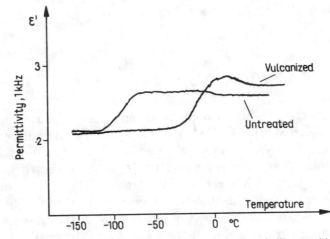

Fig. 6.13 Change of the dielectric dispersion spectrum of silicon rubber by peroxide vulcanization at 200°C

As an example of the dielectric spectra of silicon rubbers, Fig. 6.13 shows the effect of vulcanization on the dielectric dispersion spectra of Rhodorsil MM50 type silicon rubber. The vulcanization has been performed within the sample holder of the dielectric spectrometer, at a temperature of 200°C. The spectra of the unvulcanized rubber and the

330

vulcanized one are shown. It is seen that the transition temperature is shifted considerably to high temperatures and the dispersion amplitude is practically unchanged.

Effects of fillers

Rubber is seldom used in a pure state; it is usually loaded with carbon black and sometimes reinforced by some kind of cord.

Carbon black loaded rubber vulcanizates exhibit additional dielectric properties to those of the base material. Carbon black particles exhibit fairly high conductivities but the loaded rubber is usually an insulator. At high carbon black concentrations, by special treatment it is possible to prepare electrically conductive rubber vulcanizates exhibiting specific resistivities in the order of 10—1000 Ω cm. (Goul *et al.*, 1968). Usually the carbon black particles are separated by insulating polymer layers. In these systems, however, the Maxwell–Wagner interfacial polarization becomes extremely large. This appears in the dielectric spectra as an additional maximum at low frequencies, i. e. high temperatures (Thirion and Chasset, 1951; Chasset *et al.*, 1959). This maximum is increased by decreasing the frequency (Carter *et al.*, 1946). At high carbon black concentrations the dielectric permittivity increases rapidly to reach values above $\varepsilon' = 100$ near 30–40 % content.

Besides the frequency and temperature dependent Maxwell–Wagner loss, Lukomskaya and Dogadkin (1960) have found that in the carbon black vulcanizates a frequency and temperature independent dielectric polarization exists, which depends only on the concentration and shape of the carbon black particles. The high frequency dielectric permittivity of the loaded system is expressed as

$$\varepsilon_\infty = \varepsilon_\infty (u) [1 + 3 \phi C] \tag{6.7}$$

where ε_∞ (u) is the high frequency (low temperature) permittivity of the unloaded vulcanizate, C is the weight concentration of the carbon black, ϕ is a shape factor, which has a value of 1 for spherical particles and $\phi > 1$ when the carbon black particles form non-spherical aggregates.

The frequency dependence of the dielectric permittivities in carbon black filled rubber vulcanizates has been studied by Carter *et al.*, (1946), and by Lukomskaya and Dogadkin (1960). From the spectra, important

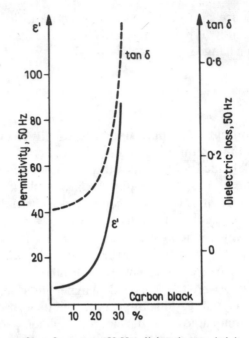

Fig. 6.14 Dependence of low frequency (50 Hz) dielectric permittivity and loss of rubber on the concentration of carbon black (Groul *et al.*, 1968)

information about the distribution of the carbon black particles can be derived.

In Fig. 6.14 the dielectric permittivity and loss tangent of a carbon black loaded rubber is shown as a function of the carbon black concentration at a frequency of 50 Hz. It is seen that ε' reaches extremely high values as a result of the interfacial polarization between the carbon black particles and the polymer (Goul *et al.*, 1968).

6.3 Curing of epoxy and phenol formaldehyde resins

Epoxy resins exhibit fairly intense dielectric spectra as a result of the polar $C-O$ and $C-C-O$ groups in the chain and the OH groups attached to the polymer chain. The structural formula of a typical epoxy resin is the following. (See for example Paken, 1962.)

332

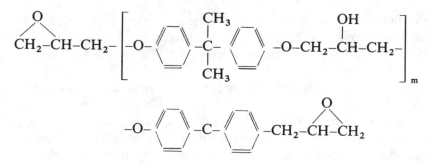

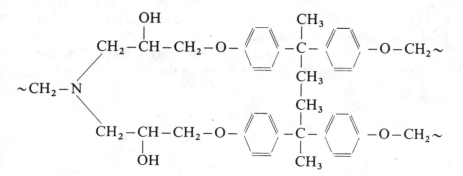

By curing these resins the epoxy ring is opened and a crosslinked network is formed.

For crosslinking agents mainly amines are used, such as diethylene triamine. A typical structure of an amine cured resin is the following

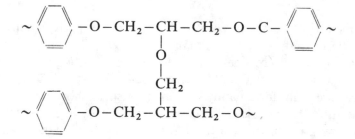

Anhydrides are also used for curing agents; in this case the following links are formed

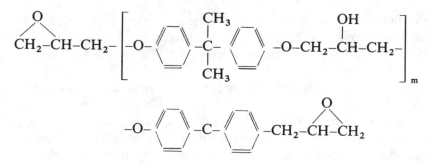

The main polar groups which can contribute to the dielectric spectra are the $C-O-C$ bonds in the epoxy ring. The concentration of these groups is expected to change by crosslinking, while the $C-O$ bonds in the chain and the OH groups attached to the chain are unaffected. The $C-O$ bonds

in the chain are the same as found in polymethylene terephthalates and in different polyesters and polycarbonates discussed in Chapter 4. See also McCrum *et al.* (1967). The dielectric loss maxima in these systems are in the order of $\varepsilon'' = 0.1 - 0.5$, the dispersion amplitudes are $\varepsilon_0 - \varepsilon_\infty = 1 - 4$.

Similarly, polyethylene oxide, and polypropylene oxide and other linear polymers having $C - O$ groups in their chain exhibit dielectric loss maxima of the same order; see Chapter 4. Comparison of the dielectric spectra of uncured and cured epoxy resins with those of linear polymers containing the same polar groups might be of help in assigning the loss peaks.

The dielectric spectra of uncured epoxy resins exhibit three main relaxation regions. The highest temperature (α) transition is observed at -10 to $-60°$C depending on the molecular weight of the resin; this is attributed to the glass transition. Another glass–glass transition is usually observed at somewhat lower temperature β-transition. The third, γ, transition is observed near -120 to $-80°$C.

The γ-relaxation region in the amine cured resins has been attributed by Pogány (1969) to the rotation of the group

$$\sim CH_2 - \underset{\underset{\displaystyle OH}{|}}{CH} - CH_2 - O \sim$$

The rotation of this group is hindered by the hydroxyl groups. This makes the γ-relaxation somewhat shifted to higher temperatures from the usual one for such groups, $-150°$C. This peak is not much affected by the cross-linking.

The dielectric spectra of some anhydride cured epoxy resins have been recently studied by Blyachman *et al.* (1970). In the $\tan\delta$ versus temperature spectra, two peaks were found, one at a high temperature between $160°$C and $200°$C and another broad, peak at $-70°$C at 1 kHz. A typical spectrum for an anhydride cured resin is shown in Fig. 6.15. It is seen that the α-peak in the cured resin is fairly intense, while the β-peak is weak and broad. The temperature dependence of the relaxation times corresponding to the β-process is shown in Fig. 6.16 in Arrhenius representation. The corresponding activation energy is 14 kcal/mole. This process has been interpreted by Blyachman *et al.* (1970) as a dipole-group relaxation analogous to that observed in polyvinylchloride and in polyvinylacetate.

334

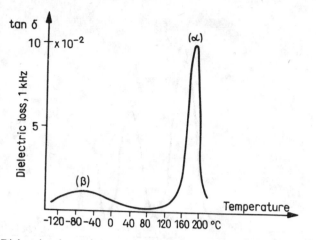

Fig. 6.15 Dielectric absorption spectrum of an anhydride cured epoxy resin (after Blyachman *et al.*, 1970)

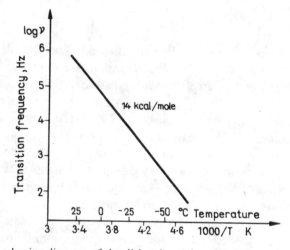

Fig. 6.16 Arrhenius diagram of the dielectric γ-relaxation frequencies of an anhydride cured epoxy resin (after Blyachman *et al.*, 1970)

The effect of the molecular weight of the epoxy resin on the dielectric spectrum has been studied by Antonov *et al.* (1966). The resin has been prepared by condensation of diphenylolpropane and epichloropropane. The dielectric absorption spectra of three resins of different molecular weights are shown in Fig. 6.17. It is seen that the α-relaxation peak is shifted to higher temperatures as the molecular weight is increased.

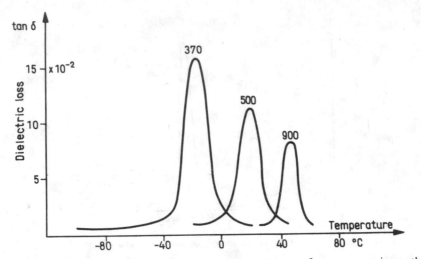

Fig. 6.17 Dependence of the dielectric absorption spectrum of an epoxy resin on the molecular weight (after Antonov *et al.*, 1965)

The β-relaxation peak is practically not affected. The β and γ-peaks are not shown in this scale. The loss maxima are decreased by increasing molecular weights, an effect evidently due to the decrease of the concentration of the epoxy groups in the system. This concentration in the resin with molecular weight 900 is about 50% of that with a molecular weight of 370. This is reflected in the dielectric spectra: the $\varepsilon_0 - \varepsilon_\infty$ values and loss maxima are decreased accordingly. This supports the view that the α-relaxation peak is due to the epoxy groups, which move along with long chain segments of the polymer. This motion is hindered by crosslinking and by the reaction the concentration of polar groups is decreased. In the cured resins the α-peak arises from the uncured parts which are strongly embedded in the three dimensional crosslinked network. This opens up a possibility of studying the completeness of the cure: in completely cured resins no α-peak should be found. Indeed the α-peak is found to decrease when the sample is stored at a high temperature (after cure).

Measurement of the dielectric loss factor as a function of the cure time

Haran *et al.* (1965) measured the change of the dielectric loss tangent of epoxy resins as a function of the cure time at fixed frequency and temperature. In these experiments a flat cell containing a thin, 0.5 – 1 mm,

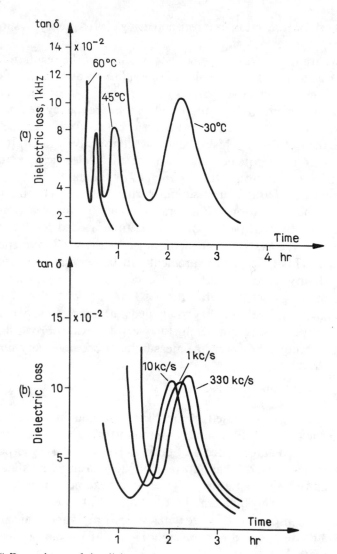

Fig. 6.18 Dependence of the dielectric loss tangent on the cure time of an epoxy resin (a) at different temperatures and (b) at different frequencies (after Haran *et al.* 1965)

layer of resin was used in order to approximate isothermal conditions during the course of the reaction. A typical set of curves measured at 1 kHz at different temperatures are shown in Fig. 6.18. It is seen that as the curing reaction proceeds the loss passes through a maximum, the position of which depends on the curing temperature. This maximum is shifted to shorter reaction times on increasing the frequency, as shown in Fig. 6.18 b.

The appearance of the loss peaks as a function of the reaction time is evidently due to the shift of a relaxation maximum as a result of cross-linking. In general, in these systems the dielectric loss or the dielectric permittivity is dependent on three variables: the frequency, temperature and reaction time. It would be most desirable to fix the temperature and sweep the frequency. This is, however, technically difficult. It is also possible to fix the frequency and measure the dielectric spectra as a function of temperature at different stages of the cure, i.e. to use the reaction time as a parameter, as in our earlier examples. In the present experiment the relaxation maximum is swept as the reaction proceeds and the observed peak indicates the time needed for the system to reach a preset relaxation state, determined by the fixed frequency and temperature. This type of measurement can be very easily performed automatically by recording the total a.c. conductivity as a function of reaction time at preset temperatures. As Haran *et al.* (1965) used the conventional bridge method, they concluded that this method of studying the curing process was promising but very laborious. By using the recording technique, however, the measurement becomes very simple.

Curing of phenol-formaldehyde resins

The thermosetting resins based on phenol and formaldehyde are prepared either by a one stage process or more commonly from a novolak type linear polymer by a two stage process of curing. The novolak type polymer contains uncrosslinked phenolformaldehyde chains. Crosslinking is usually achieved by mixing the polymer with hexamethylene tetramine and curing at relatively high temperature, $100-120°C$.

Novolak can also be crosslinked with cyclic formyls and acid catalysts, as with 1,3 dioxalane and phenolsulphonic acid for example (Heslings and Schors, 1964).

The dielectric permittivity of the uncured phenol formaldehyde resins is rather low; typical values for ε' and ε'' are 4 and 0.04 respectively. The pure uncrosslinked resin exhibits a dielectric transition at the melting temperature, the change in ε' is in the order of 3. Commercial resins usually contain some water, in this case the dielectric peak at the melting temperature is somewhat increased and another peak at $110°C - 120°C$ is observed, which disappears after annealing the sample. In Fig. 6.19 dielectric dispersion spectra of a commercial novolak type

338

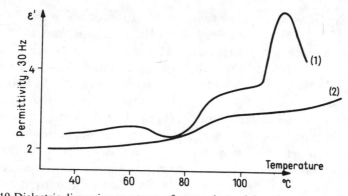

Fig. 6.19 Dielectric dispersion spectrum of uncured novolak resin: two successive runs

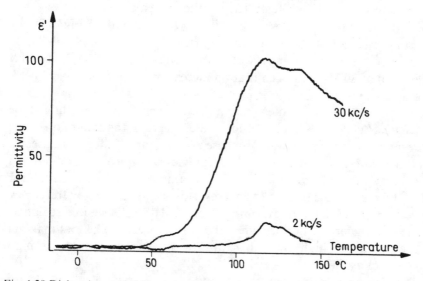

Fig. 6.20 Dielectric permittivity ε' recorded as a function of temperature in the novolak-hexamethylenetetramine system at 30 Hz and at 2 kHz

resin containing no crosslinking agent are shown. Two successive spectra were recorded, the second after 20 min annealing at 140°C. It is seen that the second dielectric dispersion region vanishes completely upon annealing.

The ε'-value of novolak drastically increases upon curing. This is illustrated in Fig. 6.20. In this experiment novolak resin containing 15 percent of hexamethylene-tetramine has been heated up from room temperature and the permittivity ε' is recorded as a function of temperature at a fixed frequency of 30 Hz. It is seen that the ε'-value increases to reach a maximum near $100°C$ and then decreases again. The maximum is found irreversible; repeated scanning results in a shift of the maximum to higher and higher temperatures while the maximum value of ε' decreases. As shown in Fig. 6.21, in pure novolak the dispersion amplitude at the melting temperature is near 4, while during crosslinking it may change by 100—150 units. This enormous effect is interpreted as being due to water which is split off by the reaction. The water droplets are strongly trapped in the crosslinked network and can be expelled only at high temperatures. The extremely high dielectric permittivity can be explained by the strong interfacial Maxwell-Wagner-type polarization of trapped water. This is illustrated by the extremely strong frequency dependence of the peak observed in the $\varepsilon'(T)$ spectrum. In Fig. 6.20 for comparison the dielectric dispersion spectrum of the novolak hexamethylene tetramine is shown at a frequency of 2 kHz besides that measured at 30 Hz. It is seen that on increasing the frequency the peak is drastically reduced.

Van Beek (1964) studied the dielectric spectra of Heslings-type (Heslings, Schors, 1964) phenol-formaldehyde resins as a function of the cure time at fixed temperature. In these experiments the dielectric spectra were measured as a function of the frequency. A typical set of $\varepsilon'(\omega)$ and $\varepsilon''(\omega)$ spectra are shown in Fig. 6.21. The samples were cured at $20°C$ for several days as indicated on the curves. It is seen that the low frequency dielectric permittivity is extremely high and the dispersion region is shifted to higher frequencies when the cure proceeds. The loss maxima are also rather high and the transition peak is shifted to high frequencies as the cure proceeds.

The extremely high permittivities and losses in these experiments are due to the effect of water captured in the crosslinked network. This has been shown in our earlier example by measuring the observed temperature dependence of the permittivity (Fig. 6.20). When the measurement is performed at a fixed, low temperature as a function of frequency (Fig. 6.21) the water remains captured in the crosslinked network and correspondingly the low frequency ε'-value is not appreciably changed as the reaction proceeds.

These experiments show that the crosslinking reaction can be monitored by the split-off water, which indicates the local molecular motion of the system because it is trapped in its physical structural traps. This opens up a new, yet undeveloped possibility of studying the crosslinking

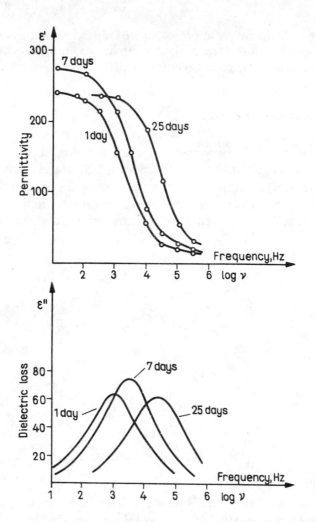

Fig. 6.21 Dielectric dispersion $\varepsilon'(T)$ and absorption $\varepsilon''(T)$ spectra of a Heslings-type phenol formaldehyde resin after different 20°C cure times (after Van Beek, 1964)

reactions in these systems, and simultaneously determining the remaining trapped water in the system cured under defined conditions. Some preliminary work has been done in the author's laboratory with compounds containing mixtures of phenol-formaldehyde resin (novolak) with linear polymers containing wood as reinforcing material. Such systems could be easily investigated by recording the dielectric permittivity as a function of temperature at preset rates of repeated heating and cooling. From the position and shape of the $\varepsilon''(T)$ maxima, important information about the crosslinking reaction could be deduced.

6.4 Curing of polyester resins

The base materials of polyester resins are prepared from saturated or unsaturated dibasic acids or anhydrides which are esterized by some alcohols (glycols); in this way various unsaturated polyester resins have been synthesized. Esterification is a condensation reaction. A typical example is the reaction of phthalic anhydride with ethylene glycol

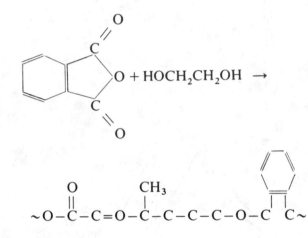

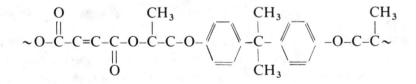

An unsaturated polyester is formed, which, containing double bonds in its main chain, is capable of crosslinking. Another typical polyester resin based on a bisphenol is the following

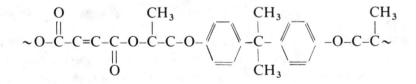

Here, besides the carboxyl groups, polar $C-O$ bonds are found in the main chain attached to a benzene ring.

342

The polar groups are differently placed in the resin based on iso-phthalic acid

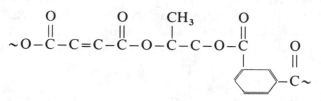

The molecular weights of these resins are between 2000 and 4000.

Besides these main types other acids and anhydrides with other alcohols are suitable for synthesizing polyester resins. For special purposes, chlorinated products, with their highly polar molecules, are also used.

The polyesters are usually crosslinked with such monomers as styrene or with methyl methacrylate, or divinyl benzene. During this process the monomer units are built between the chains to form a crosslinked network. The reaction is performed in such a way that no homopolymer is formed; only single monomer units are built in the network.

A typical polyester-system crosslinked with styrene is the following

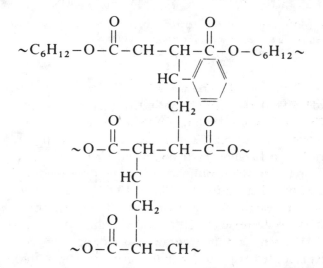

This system is prepared from maleic acid hexane diole polyester. The crosslinking of such systems is correspondingly a kind of copolymeriza-tion reaction, the unsaturated resin copolymerizing with styrene. The reaction can be initiated by peroxides, which decompose to peroxy radicals, by ultraviolet illumination or by high energy irradiation.

Change of conductivity during curing

The curing process in polyester resins seems specially suitable for studying by electrical methods, as the resin exhibits relatively high conductivity, while the cured product is a good insulator. Indeed the d.c. conductivity has been found to increase by 4—5 orders of magnitude

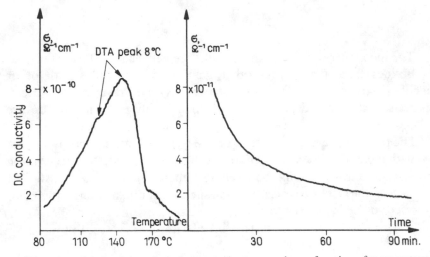

Fig. 6.22 D.c. conductivity of a polyester resin measured as a function of temperature, and later, by stopping the temperature program, as a function of cure time

upon curing. This effect has been used by Warfield and Petree (1962) and by Judd (1965) as a semi-quantitative method for following the reaction. The effect of crosslinking on the electrical conductivity of the resin is illustrated in Fig. 6.22. The d.c. conductivity in this experiment was recorded as a function of temperature by using a linear rate of heating. The reaction was simultaneously monitored by the differential thermal method. Between $20°C$ and $100°C$ the conductivity is exponentially increased, the reaction has not yet started. The conductivity starts to decrease at $130°C$ and simultaneously a DTA peak is observed. The temperature program was stopped at $170°C$ and the conductivity recorded as a function of the reaction temperature. It is seen that the conductivity continues to decrease for a long time. In well cured systems the conductivity reaches a level of $10^{-13}\ \Omega^{-1}\ cm^{-1}$ from the original $10^{-9}\ \Omega^{-1}\ cm^{-1}$ of the uncured resin.

Unfortunately the change of the d.c. resistivity cannot be rigorously correlated with the curing reaction, measured by infrared spectroscopy

(Demmler, 1969) or by other methods. The reason for this is probably that by measuring d.c. conductivities one necessarily would mix polarization and ohmic currents. Evidently the relative contribution of the ohmic and polarization currents to the total conductivity is continuously changing as the crosslinked network is being formed. For this reason the d.c. technique might be useful to check the degree of cure approximately and to study after-cure effects, but is not useful for following the reaction kinetically.

Although the d.c. conductivities during curing do not change exactly parallel with the conversion, some empirical correlations can be established. The pseudo-first-order rate constant of the reaction has been empirically correlated with the d.c. resistivity by Learmonth and Pritchard (1969 a, b) as

$$k = 0.280 \frac{d \log \varrho}{dt}$$

where ϱ is the d.c. resistivity in ohm cm, t is the reaction time. The rate constants obtained from resistivity measurements have been compared with those determined by refractive index measurement (Learmont and Pritchard, 1967) and for those determined by viscometry (Dominic, 1967). An approximate correlation with the infrared data has been also observed.

In Fig. 6.23 the Arrhenius diagram of the rate constants determined by refractometry, viscometry and by the resistivity method are shown. It is seen that the apparent activation energies are not very different.

These examples show that the d.c. resistivity or conductivity measurements are useful for following the reaction semi-empirically for a given system when other methods can be used for calibration.

Dielectric spectra

Much more information can be gained by measuring dielectric spectra during the course of the cure. By using the recording method the ohmic currents are separated from polarization currents and thus not only the changes of the physical structure of the resin can be followed, but changes in its chemical structure as well. The situation is similar to that observed by the degradation of plasticized polyvinylchloride, where the glass temperature is shifted as a result of the loss of plasticizer and the a.c. ohmic conductivity is increased as a result of the conjugation of the chains. This will be discussed in Chapter 7.

In polyester resins, curing results in a formation of a crosslinked network, causing a shift of the glass temperature to higher temperatures. This process can be measured by dielectric spectroscopy. On the other hand the unsaturated resin becomes saturated by crosslinking. This results in a decrease of the ohmic conductivity level, which can be separated from the

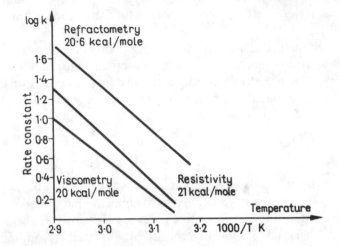

Fig. 6.23 Comparison of the Arrhenius diagrams of the rate constant of crosslinking in styrene-cured polyester resin, determined by refractometry, viscometry and by the resistivity method (after Learmonth and Pritchard, 1967)

polarization current by recording the total conductivity as a function of the temperature. Thus, recording the dielectric permittivity and the total conductivity as a function of temperature at different stages of cure supplies one with fair information of the process. This method has not yet been fully developed; an example is shown in Fig. 6.24.

From the point of view of dielectric spectroscopy the co-monomer styrene is inert, the measured peaks are thus exclusively due to the resin. This is favourable, because it is easier to interpret the peaks. According to the mechanical relaxation studies of Pohl and Kästner (1968) the styrene bridges are too short to appear in the main relaxation regions, which are mainly determined by the motion of different segments of the polyester chains.

The dielectric method has been applied by Learmonth and Pritchard (1969 a, b) to the cure of unsaturated polyester resins. In these experiments the frequency and the temperature were kept constant and the permittivity

ε' and loss tangent, tan δ, were measured as a function of the cure time. As in the case of epoxy resins discussed in section 6.3, the dielectric loss exhibits a maximum as a function of the reaction time. Figure 6.25 gives an example. The resin tested was Beetle 4116 type crosslinked with

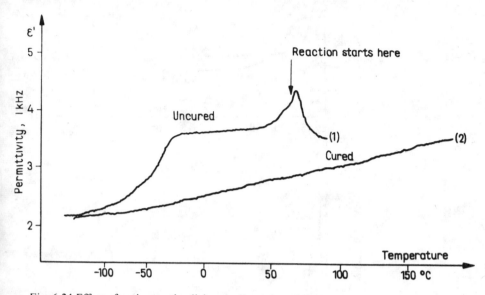

Fig. 6.24 Effect of curing on the dielectric dispersion $\varepsilon'(T)$ spectrum of a polyester resin

styrene by using a benzoil peroxide–tertiary amine initiating system. Two curves are shown in Fig. 6.25, measured at 60°C and 45°C at a fixed frequency of 10 kHz. The dielectric loss peak appears at lower reaction time when the reaction temperature is increased. The corresponding changes in the permittivity are also shown.

When the frequency is increased the peak observed as a function of reaction time is shifted to shorter times. Therefore it is possible to choose such a high frequency as would shift the peak to unmeasurably short reaction times. In this case the dielectric loss is continously decreased as the cure proceeds. This is shown in Fig. 6.26, where the loss tangent of a resin is plotted against reaction time. In this experiment a pure resin containing no initiator was crosslinked by using X-irradiation. The irradiation was made with a copper anticathode 60 kV 30 mA X-ray set up at 20°C. The cure time here corresponds to increasing radiation dose, which is also illustrated in the figure. The loss factor was measured by a resonance method (cf. Chapter 3) at a frequency of 55 MHz.

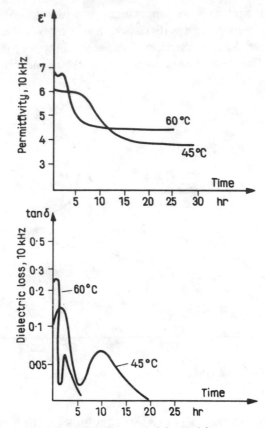

Fig. 6.25 Dependence of the dielectric permittivity and loss tangent on the cure time in a polyester resin (after Learmonth and Pritchard, 1969)

Curing of wood-plastic combinations

Wood-plastic combinations are prepared by impregnating wood with resin *in vacuo* and by crosslinking it within the wood matrix. See for example Czvikovszky (1968). The crosslinking reaction in this case is very difficult to follow. Viscometry, refractometry and the optical spectroscopic methods are evidently useless. Although wood itself exhibits a dielectric spectrum especially as a result of its moisture content, the cure in such systems can be followed by dielectric spectroscopy. The reason for this is that the dielectric effects of the unsaturated polyester are higher than those of wood especially when the moisture is removed from wood during the evacuation preceding impregnation. In the first

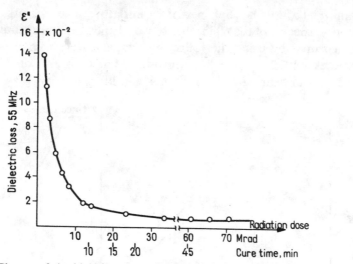

Fig. 6.26 Change of the high frequency dielectric loss by radiation curing of a pure polyester resin; radiation dose rate 80 Mrad/hr, X-rays; temperature 25°C

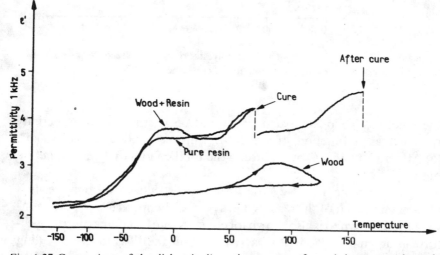

Fig. 6.27 Comparison of the dielectric dispersion spectra of wood, impregnated wood and pure polyester resin

stage of cure, at least, the dielectric peaks observed in impregnated wood are exclusively due to the polyester resin. As an example, in Fig. 6.27 the dielectric dispersion spectra of a wood-polyester system is shown. Coaxial cylinders of maple having a thickness of 1 mm were evacuated to remove moisture and impregnated with a conventional polyester resin containing 1% of lauroyl peroxide initiator. The dispersion and absorption spectra of the resin, that of wood and the impregnated wood were measured by using the following temperature program (Hedvig, Czvikovszky, 1972). (1) Cooling down to −150°C from room temperature. (2) Linear heating with a rate of 4°C/min up to 65°C by recording

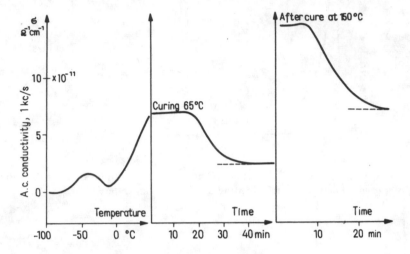

Fig. 6.28 Total a.c. conductivity of a wood-polyester resin system recorded as a function of temperature and successively as a function of cure time

the spectra. (3) Constant temperature at 65°C, the total conductivity recorded as a function of time. (4) Further heating up to 150°C. (5) Aftercure at 150°C by recording the total conductivity as a function of time.

As is shown in Fig. 6.27 the dispersion spectrum of pure resin is identical with that of the resin-wood system. The spectrum of pure, non-vacuum-dried wood is also shown for comparison. By comparing the spectra of impregnated wood with that of pure resin, the difference in the filling factors was taken into account; the gain of the spectrometer was increased by the ratio of the weight of the pure resin to that in wood when the spectrum of the impregnated wood was recorded.

In Fig. 6.28 a typical recording of the total a.c. conductivity is shown according to the given temperature program. The total conductivity was recorded by a simple pen-recorder as a function of time. In the heating-up period the temperature scale was transferred from the XY recorder in the form of marker signals. The peak observed somewhat above −50°C

350

corresponds to the glass-transition temperature of the resin. The temperature program was stopped at 65°C and the decrease of the total a.c. conductivity at a frequency of 1 kHz was recorded as a function of time. It is seen that the conductivity levels off after an elapsed cure time of about 40 minutes. Then the temperature was linearly raised up to 150°C which resulted in an exponential increase of the total conductivity, which was recorded but is not shown in the figure. At 150°C an aftercure was followed by recording again the total conductivity.

6.5 Radiation crosslinking of polyvinylchloride

On irradiating linear polymers with X-rays, γ-rays or by accelerated electrons, crosslinking processes are known to proceed parallel with degradation. On irradiating polyethylene, for example, in the solid state, excluding oxygen, crosslinking is found to proceed faster than degradation (see, for example, Charlesby, 1960). In polyvinylchloride the situation is reversed. Even at low temperatures, down to $-196°C$, the elimination-type degradation reaction is found to proceed at a considerably high rate resulting in the formation of polyenyl radicals

$$\sim \left\{ \begin{matrix} H & & H \\ | & & | \\ C = \dot{C} - C = C \end{matrix} \right\}_n - \begin{matrix} H \\ | \\ C \\ | \\ H \end{matrix} \sim$$

The average value of n is found to be near 7. During formation of polyenyl radicals HCl is split off. This problem will be discussed in some detail in Chapter 7. The radiation crosslinking reaction in pure PVC is very slow. It can be considerably increased by addition of a poly-functional monomer which polymerizes under irradiation and is built in between the PVC chains. This process is very similar to that of vulcanization of rubber. Suitable polyfunctional monomers are n-butyl methacrylate, tetraethylene glycol dimethacrylate and diacrylate, di- and trimethylpropane dimethacrylate (Salmon and Loan, 1972), poly-ethylene-glycol dimethacrylate (Miller, 1959).

By irradiating the polymer-polyfunctional monomer mixture the cross-linking reaction is believed to proceed the following way: (a) The mono-mer is grafted on to the PVC chains where free radicals have been

351

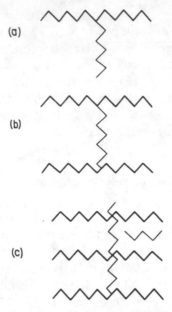

Fig. 6.29 Illustration of the process of radiation crosslinking of polyvinylchloride by poly-functional monomers: (a) grafting of the monomer onto PVC; (b) crosslinking; (c) homo-polymerization of the monomer

formed by irradiation and stabilized in the solid PVC matrix. In this way a graft copolymer is formed. (b) In the irradiation field the unreacted unsaturated bonds of the polymer would react further with the PVC or with another grafted chain to form a crosslinked network.

The process is schematically shown in Fig. 6.29. Besides grafting and crosslinking, evidently homopolymerization of the polyfunctional monomer also proceeds simultaneously. This may result in formation of separate homopolymer phases within the host polymer network. A composite material is formed this way having rather different texture and properties than those of the homopolymers. By using dielectric spectroscopy, along with mechanical thermal and chemical (extraction) measurements, important information about this structure can be gained.

In Fig. 6.30 the dielectric depolarization spectra of unplasticized PVC, that containing 16.5 % ethylene glycol dimethacrylate (EGDMA) is shown in comparison with the irradiated crosslinked compound. The depolarization spectra were measured by polarizing the samples at 150°C under an electrical field of 8 kV/cm for 30 minutes, and after cooling

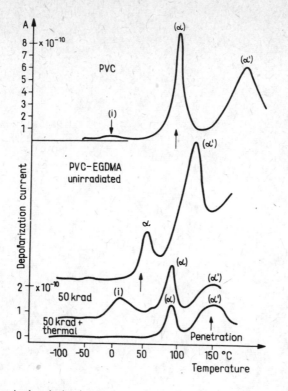

Fig. 6.30 Dielectric depolarization spectra of unplasticized PVC: of a mixture of PVC and ethylene glycol dimethacrylate 16.5 %: and of a radiation crosslinked mixture: polarization field 8 kV/cm. temperature 150°C. time 30 min: rates of heating and cooling: 2°C/min for all curves: arrows indicate transitions measured thermomechanically

down at a rate of 2°C/min the sample was heated up at the same rate while the short circuit current without external field was recorded. The positions of the depolarization peaks as well as their intensity are then comparable.

Figure 6.30 shows that the monomer EGDMA has a plasticizing effect. the glass-rubber transition (α) of the PVC-EGDMA system is much lower than that of pure PVC. This is also confirmed by dynamic mechanical and thermomechanical studies. The α' peak observed in pure PVC and in the PVC-EGDMA system as well is interpreted as being a liquid-liquid type transition. This transition is found to shift to lower temperatures by adding the monomer. Similar shifts are observed in plasticized PVC by increasing plasticizer concentration (see Chapter 5).

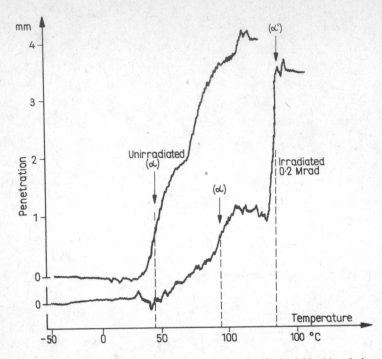

Fig. 6.31 Thermomechanical (penetration) curves of polyvinylchloride-ethylene glycol dimethacrylate (16.5%) mixture; load 34 kgf/m² speed 2° C/min.

By irradiation the α and α' peaks are both shifted to higher temperatures, the shape of the α'-peak is considerably changed. Thermomechanical measurements performed under identical conditions show that the penetration of a profile exhibits a transition in unplasticized polymer and in the PVC-EGDMA mixture at the α-transition, while in the irradiated crosslinked sample the rate of penetration is highest at the α'-transition. These thermomechanical transitions are shown in Fig. 6.30 by arrows.

In Fig. 6.31 thermomechanical curves of the unirradiated and irradiated 0.2 Mrad PVC-EGDMA mixture are shown. The penetration of a 1 mm diameter cylindrical profile was recorded as a function of the temperature at a rate of heating of 2° C/min under a load of 34 kgf/cm². The transitions measured by the dielectric depolarization method are indicated by arrows. It is seen that in the irradiated crosslinked sample the thermomechanical α-transition is much less intense than the α′-transition which seems to determine the overall mechanical properties of the sample.

By conventional dielectric spectroscopy, i.e. by recording the dielectric permittivity and loss as a function of the temperature, only the α-peaks are observed because at higher temperatures in PVC ohmic conductivity becomes very high.

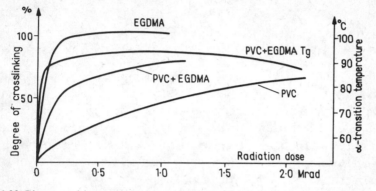

Fig. 6.32 Glass- ransition temperature and degree of crosslinking of a polyvinylchloride ethylene glycol dimethacrylate mixture as a function of the ^{60}Co γ-radiation dose

Figure 6.32 shows a comparison between the T_g (α-peak) shift and the degree of crosslinking, measured by extraction of the irradiated mixture with boiling tetrahydrofurane. It is seen that maximum degree of crosslinking is obtained at a dose of about 1 Mrad, while the glass-transition temperature is shifted from 60°C up to about 90°C at radiation doses near 50 krad. The T_g is found to decrease slightly on further increasing the irradiation dose. This shows that the PVC-submolecules involved in the microbrownian motion at the α-transition are practically unaffected by the crosslinking, i.e. the crosslinking density in the PVC-chains is small. The shift of the α-peak is mainly due to the reduction of the plasticizing effect of the monomer while it is polymerized. This interpretation is also supported by the result of the gel fraction determinations of the PVC-component by measuring chlorine content, and that of poly EGDMA. The results are also indicated in Fig. 6.32. It is seen that the T_g-shift correlates with the polymerization of the EGDMA component and not with the gel-formation of PVC.

CHAPTER 7

STUDY OF AGEING PROCESSES BY DIELECTRIC SPECTROSCOPY

The physical and chemical structure and the macroscopic properties of polymers change during processing, storage and use. This change, commonly referred to as ageing, is evidently a very complex process involving chemical reactions, physical structural changes and changes in the mechanical, thermal and electric properties of the material. Even biological problems are involved in connection with the effect of some micro-organisms on the polymeric materials.

The basic problem in studying ageing processes is that the correlation between the chemical molecular structure and the final macroscopic properties is not known. The reason lies in the peculiarity of the thermo-dynamics of macromolecular systems discussed in Chapters 1 and 2. The overall physical structure, i.e. the texture, of the polymer, which is closely related to the macroscopic properties, is determined by the molecular structure and the method of processing, hence by the thermal and thermo-mechanical history of the sample. Moreover, during storage, use or processing, two basic processes run parallel: changes of the texture involving changes in crystallinity, crystallite size, aggregate structure, packing, free volume, etc., and changes in the molecular structure. So when studying the mechanism of ageing at least three basic processes should be kept in mind: (1) changes in the chemical structure, (2) changes in the physical structure, (3) the inter-action between chemical and physical changes and the connection of these with the ultimate properties of the material.

Dielectric spectroscopy has much to contribute to understanding chemical and physical ageing processes; it is also of considerable importance because many polymers are used in the electrical industry as insulators, and changes in their dielectric properties are significant.

7.1 Typical ageing processes

Chemical ageing processes may be initiated by thermal energy, by illumination with visible or ultraviolet light or by high energy irradiation. Some are known to be initiated by mechanical action, especially during processing. Aggressive environment, gases or acids, are also highly effective in the initiation of ageing.

The *propagation* of the chemical reaction depends mainly on the physical and chemical structure of the polymer but the environmental effects are also effective. In the presence of oxygen, for example, chemical ageing processes are quite different from those in inert atmosphere or in vacuum. Even the shape and size of the sample are important, as diffusion of oxygen or other gases is a relatively slow process in solid polymers.

The *termination* of the chemical process is also mainly determined by the physical and chemical structure of the polymer. A real polymeric system usually contains stabilizers, which are low molecular weight additives acting as inhibitors to certain chemical ageing reactions.

Quite a lot of work has been done on the elucidation of these processes. Efforts have been made to eliminate the physical structural effects. Most of the kinetic studies have been performed in the molten state or in solution. Much less effort has been given to studying the change in the physical structure during the chemical process and the effect of the physical structure on the reaction. Generally the information about ageing processes gained in the fluid state cannot be extrapolated to the solid state. Consequently, ageing processes occurring during processing at high temperature are much better known than those at the temperature of utilization of the products. There is no reliable method so far for prognosis of the lifetime of polymeric materials, although this would be highly desirable.

Ageing processes involving main chain scission

An important group of ageing processes involves main chain scission. As a result of this the average molecular weight of the polymer decreases and the molecular weight distribution changes as the reaction proceeds; the polymer might be degraded down to its monomer units.

In one group of polymers, at high temperatures even in vacuum more than 90% of the degradation products is monomer. Such polymers are the polymethacrylates, polystyrene, α-methyl styrene and methacrylonitrilé. In such polymers the degradation reaction is considered to be a free radical chain reaction which is opposite to polymerization: when a polymer chain is attacked by a radical it would degrade com-

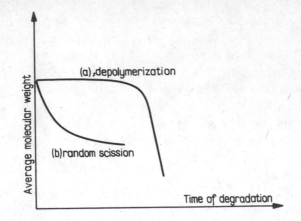

Fig. 7.1 Decrease of the average chain length molecular weight by (a) depolymerization and (b) random chain scission (after Grassie, 1956)

pletely to monomer units if no side-reactions were effective. Such a process is also referred to as depolymerization. An example of this is

$$
\sim \underset{\underset{H}{|}}{\overset{\overset{H}{|}}{C}} - \underset{\underset{R_2}{|}}{\overset{\overset{R_1}{|}}{C}} - \underset{\underset{H}{|}}{\overset{\overset{H}{|}}{C}} - \underset{\underset{R_2}{|}}{\overset{\overset{R_1}{|}}{C}} \cdot \;\; \rightarrow \;\; \sim \underset{\underset{H}{|}}{\overset{\overset{H}{|}}{C}} - \underset{\underset{R_2}{|}}{\overset{\overset{R_1}{|}}{C}} \cdot \;\; + \;\; \underset{\underset{H}{|}}{\overset{\overset{H}{|}}{C}} = \underset{\underset{R_2}{|}}{\overset{\overset{R_1}{|}}{C}}
$$

where R_1 and R_2 may be alkyl and ester groups respectively.

In depolymerizing polymers, during degradation the average molecular weight decreases as a function of reaction time (curve (a) in Fig. 7.1). It is seen that for quite a long period of time the sample contains mainly polymer of about the same molecular weight as the original one, and monomer, which is volatile at high temperatures. In depolymerizing polymers, therefore, the main transitions, especially the glass transition, are not expected to change very much during the course of degradation until the whole sample breaks down almost to the monomer units.

Depolymerization reactions, of course, cannot proceed freely through the whole chain which is attacked by the radical; it is terminated by

358

transfer of the radical to another molecule or by termination. The average chain length through which depolymerization reaction proceeds, referred to as zip-length, is defined as

$$\langle 1 \rangle = \frac{P_{prop} + P_{trans} + P_{term}}{P_{trans} + P_{term}} \tag{7.1}$$

where P_{prop} is the probability of propagation, P_{trans} is that of transfer, P_{term} is that of termination.

Table 7.1 Minimum Zip-lengths during the thermal depolymerization of various polymers. From Grassie (1956) by permission of the publishers, Butterworths, London

Polymer	Monomer Yield %	Minimum Zip-length
Polyethylene	0.025	0.01
Polyacrylonitrile	1	0.5
Polymethyl acrylate	2	1
Polyisobutene	20	3.1
Polystyrene	40	3.1
Poly p-methylstyrene	42	3.7
Poly α-methylstyrene	≫ 95	≫ 200
Polymethyl methacrylate	≫ 95	≫ 200

Table 7.1 shows the minimum zip-lengths for some polymers, after Grassie (1956). It is seen that for some polymers it is very short, which means that chain scission is practically a random process (in polyethylene for example). In such polymers the average molecular weight gradually decreases even in the first stage of degradation. This is also illustrated by curve (b) in Fig. 7.1 (random scission). The transition temperatures of randomly degrading polymers measured by dielectric or mechanical relaxation spectroscopy are expected to shift down to lower temperatures as degradation proceeds. This is, however, observable only when the average molecular weight falls below about 50,000. The reason is that the glass–rubber transition temperature is independent of the molecular weight above $M_w = 50,000$.

When a depolarizing polymer, e.g. polymethyl methacrylate, is degraded in the solid phase by ultraviolet light or by high energy irradiation, some monomer might remain trapped in the solid polymer because its diffusion in the glassy solid is very slow. In this case T_g is shifted to low temperature as a result of the plasticizing effect of the monomer. The presence of monomer in the solid can be detected by differential scanning calorimetry, dielectric spectroscopy and by NMR. From the electron spin resonance (ESR) spectra or irradiated polymethyl methacrylate, the presence of monomer radicals has also been observed (Charlesby, 1960).

Degradation is very often initiated by chemical agents coming from the environment. A classical example of this is hydrolysis of starch and cellulose derivatives. Methylated cellulose, for example, is heavily degraded to its monomer units in fuming hydrochloric acid as

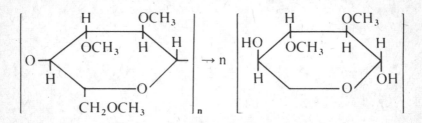

This reaction can be easily followed by measuring the OH-group concentration.

Hydrolysis is a typical random degradation process. For such processes the average chain length $\langle l(t) \rangle$ depends on the original average chain length $\langle l_0 \rangle$ as

$$\ln \left| \frac{\langle l(t) \rangle - 1}{\langle l(t) \rangle} \right| = \ln \left[\frac{\langle l_0 \rangle - 1}{\langle l_0 \rangle} \right] - kt \qquad (7.2)$$

where k is the rate constant of the reaction, t is the time.

By using statistical treatment it is possible to calculate how a given distribution of chain length (molecular weight) changes when degradation proceeds. Such a series of curves is shown in Fig. 7.2 for cellulose acetate, after Montroll (1941). The degree of degradation is defined as

$$\alpha = \frac{\langle s \rangle}{\langle l_0 \rangle - 1} \qquad (7.3)$$

where $\langle s \rangle$ is the average number of scissions per molecule.

Figure 7.2 shows that with increasing degradation the original distribution is first flattened and then sharpened again while the average is shifted to smaller chain lengths (lower average molecular weight). This

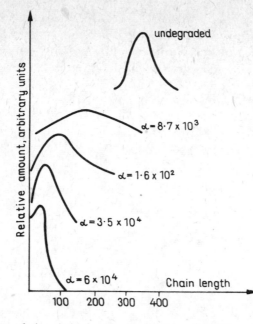

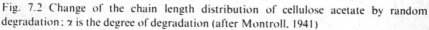

Fig. 7.2 Change of the chain length distribution of cellulose acetate by random degradation; α is the degree of degradation (after Montroll, 1941)

is a rather characteristic feature of random degradation processes. No systematic work has been done so far to follow such random degradation processes by dielectric spectroscopy, although in principle it seems possible. The change in the molecular weight distribution of the polymer during degradation should affect the dielectric relaxation time distribution, especially in the glass–rubber transition region.

Another important enviromental agent inducing degradation is oxygen. As carbon-oxygen bonds are polar, oxidation of the polymer molecules always affects the dielectric properties. (Examples of this have already been presented in section 4.1.)

Oxygen is known to accelerate random degradation processes. The general reason for this is believed to be the production of weak bonds in the polymer chains by oxidation. Especially with radiation-initiated ageing processes, very drastic effects of atmospheric oxygen are found. As an example, polytetrafluoro-ethylene (PTFE) is considered here in some detail. This polymer is practically unchanged by irradiation in vacuum with a dose of 1—10 Mrad of ^{60}Co γ-rays or of X-rays. The free radicals formed by irradiation are stable up to about 200°C. The radicals are oxidized when the sample is taken out after irradiation but this does not involve degradation. Yet if this polymer is irradiated in air, very heavy degradation occurs. On the other hand PTFE is known to be thermally extremely stable up to 250°C.

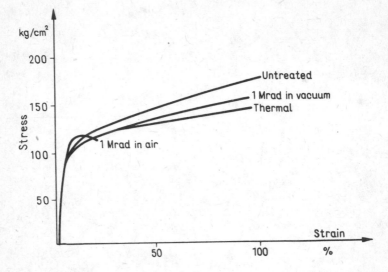

Fig. 7.3 Stress-strain curves of polytetrafluoroethylene after thermal degradation at 250°C for 24 hrs; radiation degradation in vacuum ^{60}Co 1 Mrad; and radiation degradation in air with the same dose

Figure 7.3 shows the stress-strain curves of polytetrafluoroethylene irradiated *in vacuo* and in air in comparison with the unirradiated original sample and an unirradiated one kept at 250°C for 24 hrs. It is seen that at rather low doses 1 Mrad the rupture strain decreases dramatically when the polymer is irradiated in air by ^{60}Co γ-rays, but is practically unchanged when it is irradiated in vacuum or kept at 250°C in air for a day. In both cases oxidation occurs, as on irradiation *in vacuo* stable fluorocarbon radicals are formed which are oxidized when the polymer is exposed to air. The total amount of oxidation is approximately the same for the vacuum irradiated and afterwards oxidized, and for the air-irradiated polymer, despite degradation being quite different.

362

This phenomenon is explained by the following scheme (Hedvig. 1969 a, b, c).

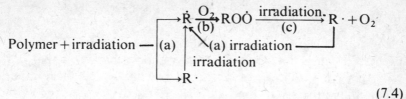

$$\text{Polymer} + \text{irradiation} \longrightarrow (a) \begin{cases} \dot{R} \xrightarrow[\text{(b)}]{O_2} RO\dot{O} \xrightarrow[\text{(c)}]{\text{irradiation}} R\cdot + O_2 \\ \quad\ \ \text{(a) irradiation} \\ \quad\ \ \text{irradiation} \\ R\cdot \end{cases}$$

(7.4)

where $R\cdot$ is a chain-end radical, in PTFE

$$R\cdot = \sim \overset{\displaystyle F}{\underset{\displaystyle F}{C}} - \overset{\displaystyle F}{\underset{\displaystyle F}{C}}\cdot$$

$\dot{R}$ is a chain radical

$$\dot{R} = \sim \overset{\displaystyle F}{\underset{\displaystyle F}{C}} - \underset{\displaystyle F}{C} - \overset{\displaystyle F}{\underset{\displaystyle F}{C}} \sim$$

and $RO\dot{O}$ is a chain peroxy radical

$$RO\dot{O} = \sim \overset{\displaystyle F}{\underset{\displaystyle F}{C}} - \overset{\displaystyle \dot{O}}{\underset{\displaystyle F}{\overset{\displaystyle |}{\underset{\displaystyle |}{C}}}} \overset{O}{} - \overset{\displaystyle F}{\underset{\displaystyle F}{C}} \sim$$

Process (a) is a fluorine abstracting reaction which takes place at low temperatures under irradiation, process (b) is oxidation which proceeds without irradiation, process (c) is a chain scission at the peroxy radical:

$$\sim \overset{F}{\underset{F}{C}} - \overset{\dot{O}}{\underset{F}{\overset{O}{C}}} - \overset{F}{\underset{F}{C}} \sim \;\rightarrow\; \sim \overset{F}{\underset{F}{C}} - \overset{F}{\underset{F}{C}}\cdot + \overset{F}{\underset{F}{C}} = \overset{F}{\underset{}{C}} - \overset{F}{\underset{F}{C}} \sim \; + O_2$$

(7.5)

363

The degradation process outlined above has been followed step by step by electron spin resonance (Hedvig, 1969a) in polytetrafluoroethylene. There is some evidence that similar processes are effective in polyethylene and in polypropylene too.

Oxidative degradation of polymers is, of course, a very complex process which is not at all fully understood. The fact is that in those polymers in which degradation is not principally a chain reaction, the random chain scissions are highly accelerated, if not induced, by the presence of oxygen. For details of oxidation processes see Grassie (1956). For results of ESR works on polymer oxidation by electron spin resonance see Hedvig and Zentai (1969); also in Kinell *et al.* (1973).

Dielectric work done in the field of oxidative degradation has been limited mainly to following the oxidation process in apolar polymers. An example will be discussed in section 7.2. Change in the dielectric transitions and relaxation time distributions during degradation has not been studied so far.

Ageing involving side group scission and conjugation of the main chain

In certain types of polymers having flexible or rigid side groups, decomposition involves mainly side group scissions and not main chain scissions. The main representatives of this group are the polyvinyl esters, which decompose into polyolefins and acids as

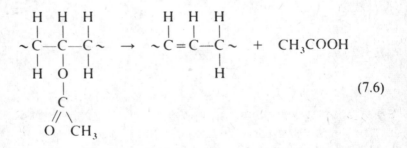

$$(7.6)$$

Such processes must strongly affect dielectric properties, as the transitions arising from the rotation of the ester side group should gradually decrease during the course of the degradation. If the acids formed remain in the system, they may exhibit individual dielectric transitions. Unfortunately this problem has not been studied so far.

Another important group of polymers which would decompose mainly by side group scission are the halogen polymers such as polyvinylchloride, polyvinyldenechloride, polyvinylfluoride. In polyvinylchloride for example the reaction is

$$
\begin{array}{ccc}
\text{H} & \text{Cl} & \text{H} \\
| & | & | \\
\sim\text{C} - \text{C} - \text{C} \sim \\
| & | & | \\
\text{H} & \text{H} & \text{H}
\end{array}
\xrightarrow[\text{radiation}]{\text{thermal}}
\begin{array}{ccc}
 & & \text{H} \\
 & & | \\
\sim\text{C} = \text{C} - \text{C} \sim \ + \ \text{HCl} \\
| & | & | \\
\text{H} & \text{H} & \text{H}
\end{array}
\qquad (7.7)
$$

The evolution of hydrochloric acid is accompanied by conjugation of the main chain. It has been shown that the initiation of this elimination-type reaction is a random process; on average, 7—8 conjugated sequences are formed in the main chain (Kelen *et al.*, 1969).

The partial conjugation of the main chain by ageing in PVC affects the d.c. and a.c. conductivities mainly, the transitions are less affected; an example will be presented in section 7.3. The process of conjugation has been followed by ultraviolet spectroscopy by Salovey *et al.* (1969, 1970). For a detailed analysis see Salovey and Bair (1970), Braun (1971).

Although in polyvinyl esters and in halogen polymers the basic chemical process of ageing is scission of side groups resulting in formation of acids and partial conjugation of main chains, in the presence of oxygen main chain scissions also occur. Figure 7.4 relating to irradiated poyvinylchloride, shows the molecular weight distribution curves measured by gel permeation chromatography (GPC) for the untreated polymer, for that irradiated *in vacuo* with ^{60}Co γ-rays and for that irradiated in air with the same dose. It is seen that the molecular weight is drastically reduced if oxygen is present during irradiation. The change of the molecular weight distribution by irradiation *in vacuo* is very probably due to the small amount of crosslinking which proceeds along the elimination reaction.

Elimination-type degradation reactions usually result in only partial conjugation of the main chain. An extensive conjugate system which would exhibit very high conductivity is not always formed. Pohl (1962) termed such polymers, in which conjugation is limited, as 'rubiconjugated'. In another class of polymers, intra- and intermolecular cyclization followed by conjugation occurs. This results in macroscopically conjugated three-dimensional networks which are highly conductive. Pohl (1962) termed such polymers as 'ekaconjugated'. An illustrative example of such polymers is polyacrylonitrile. In thermal ageing, first intramolecular cyclization occurs

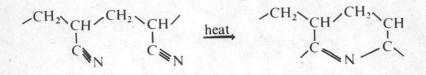

Further increase of the temperature results in interchain cyclization by forming 6-membered rings (Hedvig *et al.*, 1968)

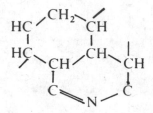

From this structure, by reduction, a highly conjugated system is formed

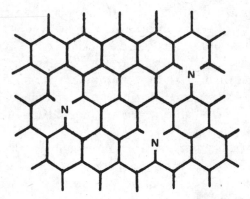

Figure 7.5 shows the decrease of the d.c. resistivity and activation energy for the conduction in comparison with the total weight loss for

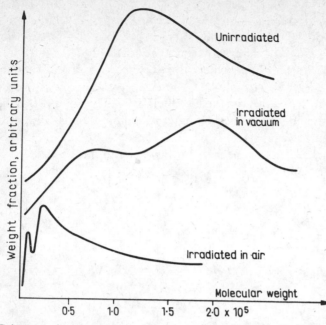

Fig. 7.4 Gel permeation chromatographic curves of polyvinylchloride irradiated by ^{60}Co γ-rays in vacuum and in air; dose: 10 Mrad

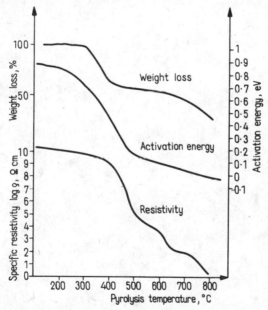

Fig. 7.5 Dependence of the weight loss, specific resistivity and activation energy of the electrical conductivity of polyacrylonitrile on the temperature of pyrolysis performed in a vacuum for 2 hours

367

polyacrylonitrile treated at different temperatures. It is seen that in the first stage of the reaction, when there is practically no weight loss, the conductivity slightly increases. This stage corresponds to intramolecular cyclization. In the stage when the weight loss drops suddenly the resistivity decreases sharply and the activation energy also decreases. At very high temperatures the conductivity reaches almost the level of the metals, and the activation energy becomes negative, indicating metallic character.

A similar cyclization mechanism has been found by Bruck (1967) in thermal decomposition of polyimides. The initial polymer is

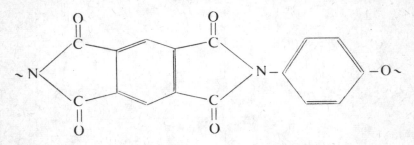

At 600°C *in vacuo* the following structure is formed

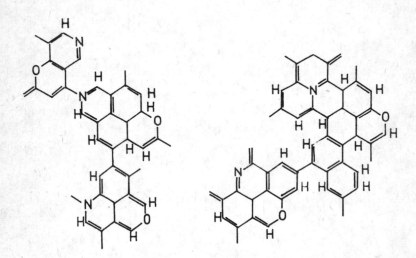

This reaction is also followed by rapid increase in the conductivity and decrease in the activation energy of conduction.

For details of highly conductive polymeric systems see Guttman and Lyons (1967), Brophy and Buttrey (1962).

Dielectric polarization in conductive polymers is usually very high: ε_0 may reach the value of 1000. This is referred to as hyperelectronic polarization (Rosen and Pohl 1966).

On the basis of this rough survey of ageing processes the following dielectric effects are principally expected to occur by ageing:

(a) Increase of the oscillator strength of the dielectric transitions in apolar polymers when ageing involves building-in polar groups in the polymer chains (oxidation, sulphuration, nitration, etc.).

(b) Increase of the overall a.c. and d.c. conductivity in cases when ageing involves chain conjugation (halogen polymers).

(c) Change of the dielectric spectra of polar polymers when ageing affects the polar group (for example, by side group scissions).

(d) Change of dielectric transitions, especially T_g, when the average molecular weight falls below about 50,000. Change in the relaxation distribution of the transition even when T_g is unchanged.

(e) As ageing might involve change in crystallinity and change in the aggregate (supermolecular) structure, the corresponding dielectric transitions are expected to change too. Especially, dielectric depolarization spectra should be sensitive to such changes.

(f) On ageing in the solid phase, polar or ionic reaction products may remain trapped in the solid, exhibiting individual dielectric transitions. Even trapped apolar reaction products might be detected by dielectric depolarization spectroscopy through the Maxwell–Wagner–Sillars (MWS) polarization.

(g) Slow changes in the texture of the polymer without involving chemical processes (physical ageing) should be detected by low frequency dielectric spectra or depolarization spectra.

In the subsequent sections some examples are presented of the dielectric technique for studying physical and chemical ageing processes.

7.2 Dielectric study of the oxidative degradation of polyethylene

As an example of oxidative degradation of nonpolar polymers, polyethylene will be discussed in some detail. It has been mentioned in section 4.1 that apolar polymers exhibit dielectric spectra as a result of the polar groups formed by oxidation. It was shown, too, that the main transitions in polyethylene correspond to the crystalline and amorphous phases as follows: T_m-melting temperature, crystalline phase; $\alpha_c(\alpha)$-

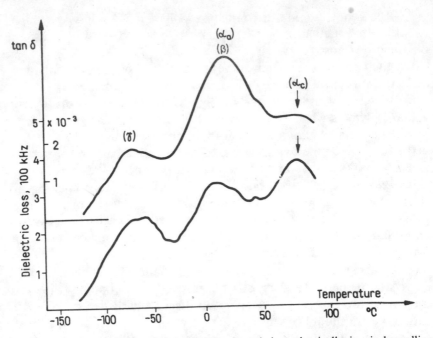

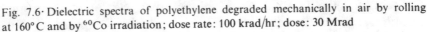

Fig. 7.6· Dielectric spectra of polyethylene degraded mechanically in air by rolling at 160°C and by ^{60}Co irradiation; dose rate: 100 krad/hr; dose: 30 Mrad

transition: crystalline phase, surface of crystallites; $\alpha_a(\beta)$-transition: amorphous phase; γ-transition: crystalline defects and amorphous phase.

This, in principle, makes it possible to localize the environment in which the oxidation occurs under given circumstances.

Figure 7.6 shows the dielectric spectrum of low-density polyethylene oxidized by ^{60}Co γ-irradiation at 25°C, recorded as a function of the temperature in comparison with that of a mechanically degraded sample. It is seen that the α_a- and γ-peaks of the radiation oxidized sample are high and the α_c-peak is low. This means that radiation oxidation proceeds mainly in the amorphous phase. This is also supported by the dielectric loss measurements of Sasakura (1963) performed as a function of the γ-irradiation dose. The $\tan\delta$ versus dose curves for polyethylenes of different density (crystallinity) are shown in Fig. 7.7. It is apparent that the sample having lowest crystallinity is oxidized to the highest degree.

Reddish and Barrie (1959) found a correlation between the carbonyl group concentration measured by infrared spectroscopy and the oscil-

370

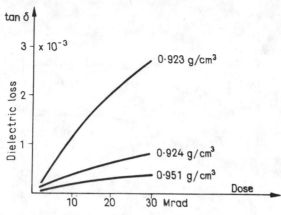

Fig. 7.7 Dielectric loss tangents of polyethylenes of different density (crystallinity), irradiated by ^{60}Co γ-rays in air

lator strength of dielectric absorption. The concentration of the carbonyl groups is given by using the Kirkwood–Fröhlich equation (1.13) as

$$N = \frac{27\,kT}{4\pi} \frac{(2\,\varepsilon_0 + \varepsilon_\infty)(\varepsilon_0 - \varepsilon_\infty)}{3\,\varepsilon_0\,g_r\,(\varepsilon_0 + 2)^2\,\mu^2} \qquad (7.8)$$

where $\mu = 2.8$ debye is the dipole moment of the carbonyl group. For the oscillator strength

$$\varepsilon_0 - \varepsilon_\infty = \frac{2}{\pi} \int \varepsilon''(\ln\omega)\,d\ln\omega = \frac{4.606}{\pi}\,A \qquad (7.9)$$

where A is the area under the $\varepsilon''(\ln\omega)$ curve. Reddish and Barrie (1959) measured ε'' by the Fourier transformation Hamon method discussed in Chapter 3 and found the $1/T$ dependence for the area A valid in a fairly wide temperature range. According to Eqns (7.8) and (7.9), considering that $3\,\varepsilon_0/2\,\varepsilon_0 + \varepsilon_\infty \approx 1$, one obtains

$$A = 2.07 \times 10^{-20}\,\frac{N}{T}\left(\frac{\varepsilon_\infty + 2}{3}\right)^2 \mu^2\,g_r \qquad (7.10)$$

This makes it possible to follow oxidation processes kinetically. Figure 7.8 shows the infrared extinction coefficient at the 5.7—5.8 μm band as a function of the irradiation dose for low density polyethylene. Irradiations were performed in air with a dose rate of 1 Mrad/hr. The

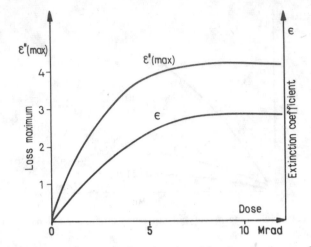

Fig. 7.8 Correlation between the infrared carbonyl band extinction coefficient ϵ and the maximum dielectric loss factor in low density polyethylene irradiated in air by ^{60}Co γ-rays

extinction coefficient ϵ is proportional to the overall carbonyl group concentration. Figure 7.8 also shows the dielectric loss maxima in the α_c-transition region. It is seen that the correlation is fair.

The absolute carbonyl group concentration values derived from the dielectric spectra are, however, much higher than those obtained by infrared spectroscopy. The difference is so high that it cannot be attributed solely to the approximate character of the Kirkwood–Fröhlich equation.

Another difficulty is that Reddish and Barrie measured the dielectric α_c-transition in which only a part of the overall carbonyl group concentration is manifested. It is expected that oxidation in low density polyethylene should be higher in the amorphous phase than at the crystallite surfaces. Correspondingly in this polymer most of the dielectric effect due to oxidation is expected in the α_a-region. It is necessary therefore to record the complete spectrum either by using the computerized Fourier transformation technique, or by the dielectric depolarization technique discussed in Chapter 3.

Figure 7.9 shows a series of dielectric depolarization spectra of low density polyethylene. All the spectra were recorded by using the same temperature program: 2° C/min rate of heating and cooling. The polarization field was 10 kV/cm, the polarization temperature 100°C, time

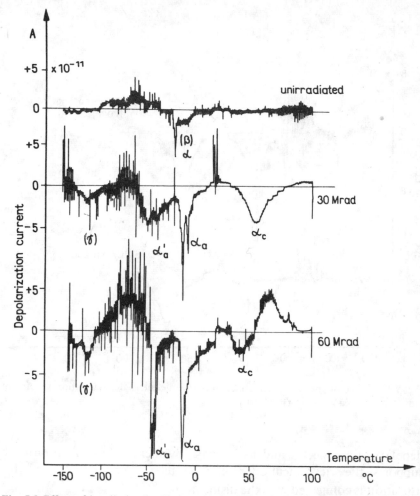

Fig. 7.9 Effect of irradiation by X-rays on the dielectric depolarization spectra of low density polyethylene; dose rate: 30 Mrad/hr; rate of heating 2°C/min.; polarization at 100°C with 10 kV/cm.

30 min. The measurement was performed on the same sample by successive irradiation with X-rays of increasing dose in air at 25°C.

It is seen that with increasing irradiation dose all the peaks increase at different rates. The highest increase is observed in the α_a-transition region, indicating that oxidation indeed proceeds preferably in the amorphous phase. The increase in the α_c-peak shows that the surfaces of the crystallites are also oxidized. The presence of the γ-peak in the

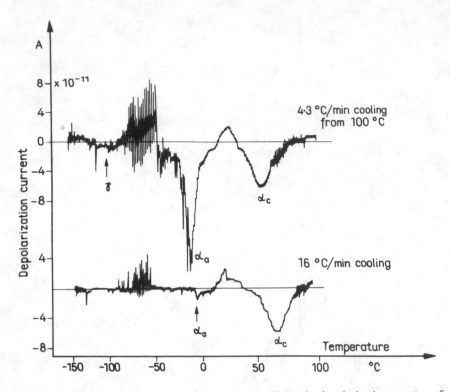

Fig. 7.10 Effect of thermal pretreatment on the dielectric depolarization spectra of radiation oxidized low density polyethylene

depolarization spectra and its increase by increasing irradiation dose supports the view mentioned in Chapter 4 Section 4.1 that this transition is connected with the dislocations.

The dielectric depolarization spectra of oxidized polyethylene are found to be highly dependent on the thermal history of the sample. In Fig. 7.10 two successive runs are represented: in the first run the rates of heating and cooling were equal, 2°C/min, in the second run the rate of cooling before recording the depolarization spectra was 10°C/min. It is seen that the α_a-peak almost completely vanishes by this treatment.

From this and from the too high absolute values of the polar group concentrations obtained from the Fourier transformed dielectric spectra, it is concluded that the low frequency dielectric properties in oxidized polyethylene are basically determined by the Maxwell–Wagner–Sillars (MWS) polarization. The oscillator strength of the

374

dielectric transition, correspondingly. is determined not only by the concentration of the carbonyl groups, but by the distribution of the oxidized regions in the polymer as well. According to this Eqn. (7.10) should contain a structure-dependent factor representing the inhomogeneity of the oxidation.

The permittivity components of a heterogeneous system in which polar heterogeneities are distributed in a nonpolar high resistivity medium are expressed by the Maxwell–Wagner–Sillars model discussed in Chapter 5 as

$$\varepsilon'(\omega, T) - \varepsilon_\infty^{(1)} = \bar{S}\, \frac{\varepsilon_\infty^{(1)}}{1 + \omega^3 \tau_{MW}^2} \qquad (7.11)$$

$$\varepsilon''(\omega, T) = \bar{S}\, \frac{\varepsilon_\infty^{(1)} - \omega^2 \tau_{MW}^2}{1 + \omega^3 \tau_{MW}^2} \qquad (7.12)$$

where

$$\tau_{MW} = \frac{\varepsilon_\infty^{(2)} \tau_2}{\Delta \varepsilon_2} \qquad (7.13)$$

is the relaxation time associated with the MWS process, τ_2 is the relaxation time associated with the polar medium (2), $\Delta \varepsilon_2$ is the corresponding oscillator strength, $\varepsilon_\infty^{(2)}$ is the corresponding unrelaxed permittivity. $\varepsilon_\infty^{(1)}$ is the unrelaxed permittivity of the non-polar medium, polyethylene; the factor $\bar{S}$ depends only on the structure, i.e. on the shape and distribution of the polar heterogeneities

$$\bar{S} = \frac{v_1}{v_2} \frac{S^2}{S - 1} \qquad (7.14)$$

where v_1 and v_2 are the volumes of the polar heterogeneities and the apolar medium respectively, S is a shape factor representing the eccentricity of the inhomogeneities. $S = 1$ for spheres, for ellipsoids S is related to the eccentricity; the more elongated the ellipsoids are, the higher is the shape factor S.

From Eqns (7.11) and (7.12) it is seen that the shape of the relaxation curve related to the MWS mechanism is somewhat different from that of the simple dipole-orientation mechanism and the oscillator strength of the transition is proportional to the structure factor S.

Consider, for example, the α_c-transition, which corresponds to the motion at the surface of the crystallites. Oxidation at the surface results in a layer of polar polymer at a surface of an apolar folded chain crystallite, as is schematically shown in Fig. 7.11. If the crystallite size and shape are constant during the course of the oxidation the area under the low frequency α_c loss peak or under the dielectric depolarization current α_c-peak would be proportional to the number of carbonyl groups built-in. If, however, the crystallite size and shape changes, the areas would change too. So by fixing the degree of oxidation it is possible

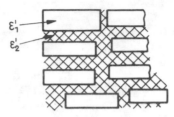

Fig. 7.11 Schematic representation of the structure of polyethylene, oxidized in the interlamellar space

to investigate the physical structural changes, and by fixing the thermal history of the sample rigorously it is possible to follow the chemical process. This goes, of course, only up to a limit until the structure crystallinity is not affected by the reaction. At a high degree of oxidation the crystallinity gradually decreases resulting in a shift of the α_c-transition to lower temperatures.

An interesting evidence that oxidation proceeds at the surface of crystallites has been found by Keller (1957). He oxidized crystalline polyethylene by fuming nitric acid and measured the molecular weight distribution of the products by gel permeation chromatography. The distribution of the oxidized sample exhibited a sharp peak at those molecular weights which corresponded to the chain folds. The chains, correspondingly, were cut at the folds by the oxidative degradation but not elsewhere.

Keeping in mind that the oxidative degradation process can be followed only by fixing the temperature history of the sample, the effect of anti-oxidants can be easily studied by low frequency dielectric spectroscopy. Figure 7.12 shows a series of Fourier-transformation spectra of unstabilized low density polyethylene in the α_c-transition region after irradiation in air with X-rays by increasing doses. Figure 7.13 shows the same sequence of spectra measured at identical

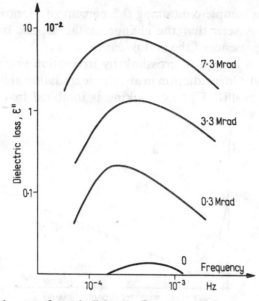

Fig. 7.12 Fourier transformed dielectric loss curves of unstabilized low density polyethylene, irradiated by X-rays in air with increasing doses. Temperature of the irradiation: 25°C; of the measurement, 50°C

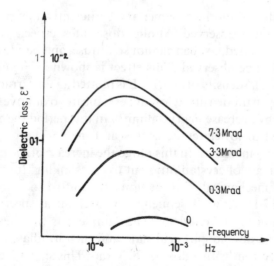

Fig. 7.13 Fourier transformed dielectric loss curves of stabilized (0.1% Nonox WSP) low density polyethylene irradiated as in Fig. 7.12

conditions for a sample containing 0.5 percent of Nonox WSP as an antioxidant. It is seen that the change in the α-peak is enormously reduced by the presence of the antioxidant.

Polyethylene is known to crosslink by irradiation *in vacuo*, see e.g. Charlesby (1960). On irradiation in air, the degradation and crosslinking processes run parallel. The crosslinking is inhibited by oxygen. So if

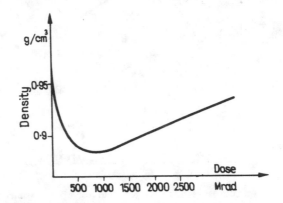

Fig. 7.14 Dependence of the density of polyethylene on the irradiation dose by accelerated 1 MeV electrons (after Heinze, 1966)

oxygen can diffuse into the polymer at a sufficiently high rate, radiation degradation only is observed. At high dose-rates, especially when thick samples are irradiated, oxygen cannot so diffuse, and beside degradation crosslinking will be observed. This effect is shown in Figure 7.14, after Heinze (1966). The density of an LDPE is plotted against irradiation dose. The sharp drop in the density at low doses indicates oxidative degradation accompanied by decrease in crystallinity. In this period the α_c-transition peak is shifted to lower temperatures and merges with α_a. At higher doses the density increases. In this range the increase of the density is not due to the increase of crystallinity, but to crosslinking. In this range the dielectric and mechanical α_c-transitions are shifted to higher temperatures. Figure 7.15 shows a sequence of mechanical spectra measured after irradiation of polyethylene. The sample was first heat-treated in order to reach a definite physical state and then irradiated with 1 MeV accelerated electrons with a dose of 20 Mrad. This shows a possibility of following textural changes in the polymer as a result of chemical processes: oxidative degradation and crosslinking.

378

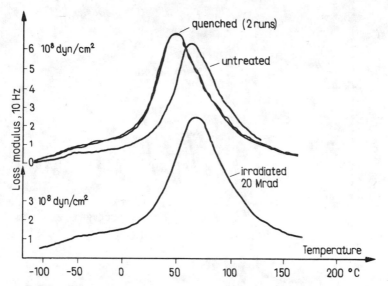

Fig. 7.15 Mechanical resonance spectra of polyethylene first quenched from 100°C and then irradiated by accelerated 1 MeV-electrons with a dose of 20 Mrad; dose rate: 200 Mrad/hr

7.3 Dielectric study of the ageing of polyvinylchloride

An example of ageing phenomena which primarily do not involve main chain scission is the thermal and radiation-induced ageing of polyvinylchloride. It has been shown in Chapters 2 and 4 that polyvinyl chloride exhibits two main dielectric transition regions: one is T_g α-transition, which is near to 80° for the unplasticized polymer, the other (β) transition is a broad band extending from $-20°$C up to about 50°C, partially merging with T_g. The dielectric effect comes from the polar $C-Cl$ bond. Correspondingly, if the elimination type degradation process discussed in section 7.1 proceeds, a decrease of the dielectric peak intensities should be expected. This is usually not observed, however, because the partial conjugation of the main chain results in a very rapid increase of the d.c. and a.c. conductivities. In the presence of a high conductivity background it is very difficult to measure small changes in the oscillator strength of the dielectric transition.

Moreover by partial conjugation of the main chain dipoles are introduced, as the dipole moment of the $-C=C$-bond is 0.69 debye. This might also contribute to the dielectric behaviour.

379

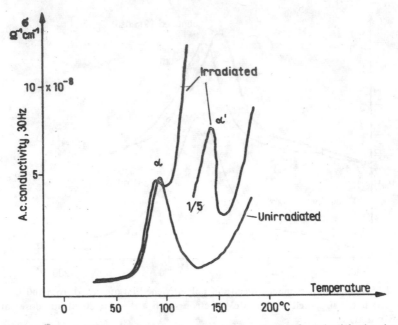

Fig. 7.16 Effect of X-irradiation on the dielectric spectrum of unplasticized polyvinyl-chloride; radiation dose 30 Mrad

Figure 7.16 shows the dielectric spectra of unplasticized PVC measured before and after X-irradiation at room temperature. It is seen that the dielectric α-peak is practically unchanged and the a.c. conductivity is appreciably raised as the irradiation dose increases. The second maximum found at about 140°C is attributed to the effect of hydrochloric acid trapped in the solid. Similar curves are obtained for samples degraded thermally *in vacuo* or in inert atmosphere. When PVC is degraded in air, first only the a.c. conductivity is raised, leaving the dielectric α-peak unchanged. At high degrees of degradation when the average molecular weight falls below 10,000 the α-peak is shifted to low temperature and the oscillator strength is decreased. This is usually rather difficult to measure dielectrically because the conductivity of such highly degraded samples is very high. The shift of T_g in such cases is easier to measure by mechanical relaxation spectroscopy.

Figure 7.17 shows a series of a.c. conductivity curves recorded during thermal ageing of PVC. These curves could be linearized in a coordinate system of lnσ versus time corresponding to unimolecular

380

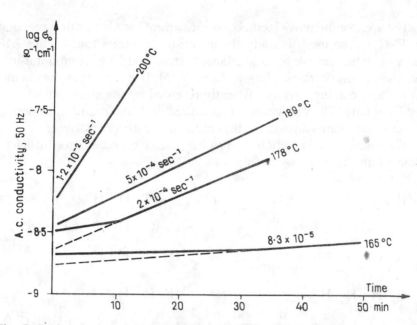

Fig. 7.17 Linearized a.c. 50 Hz conductivity curves of unstabilized unplasticized polyvinylchloride as a function of the thermal degradation time; rate constants calculated from the slopes are indicated

kinetics (Hedvig, 1971 b). The rate constants determined in this way agree satisfactorily with those obtained by measuring the amount of hydrochloric acid evolved. Figure 7.17 shows also the values of the first-order rate constants. The activation energy calculated from the temperature dependence of the rate constants also agrees well with that obtained by hydrochloric acid determination (30 kcal/mole). The breaks shown at the beginning of the linearized curves of Figure 7.17 are due to the retardation effect of the lubricant lead stearate which was used in processing the samples (Hedvig and Kisbényi, 1969).

The kinetic equation for the a.c. conductivity is

$$\frac{d\sigma(t)}{dt} = -k\sigma(t) \qquad (7.15)$$

The relative conductivity change is

$$\frac{\Delta\sigma}{\sigma_0} = \frac{\sigma(t) - \sigma(0)}{\sigma(0)} = \exp(-kt) \qquad (7.16)$$

where k is the rate constant of the process.

The a.c. conductivity method for measuring the elimination reaction in PVC can be used to study the effect of stabilizers too. Figure 7.18 shows a typical curve for a plasticized, stabilized PVC compound in a representation corresponding to Eqn. (7.16). It is seen that the change in the a.c. conductivity is efficiently retarded by the stabilizer tribasic lead sulphate. The rate constant measured after the retardation period corresponds approximately to that of the unplasticized polymer.

The elimination reaction in PVC *in vacuo* proceeds, according to Salovey and Bair (1970), as follows

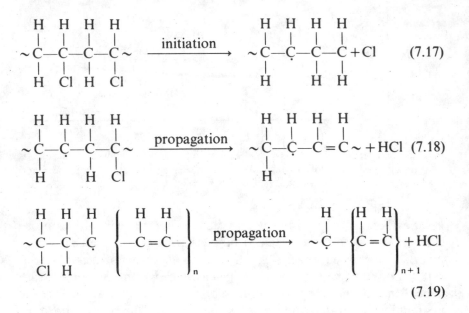

$$(7.17)$$

$$(7.18)$$

$$(7.19)$$

The total weight loss curves measured as a function of time at different temperatures can be linearized at low conversions in a $C^{-1/2}$ versus time system corresponding to 3/2 order kinetics. (Stromberg *et al.*, 1959). Here C is the ratio of the instantaneous hydrogen chloride concentration in the sample to the total Cl-concentration in the original sample.

The activation energies obtained from weight loss measurements are near to 30 kcal/mole.

In a comparison of the weight loss results with the conductivity curves one obtains

$$\ln\left(\frac{\sigma - \sigma_0}{\sigma_0}\right) \propto C^{-1/2} \qquad (7.20)$$

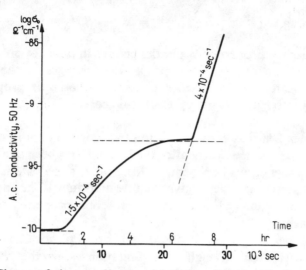

Fig. 7.18 Change of the a.c. 50 Hz conductivity of a plasticized stabilized PVC compound (tribasic lead sulphate, di-isooctyl phthalate) by thermal degradation at 185°C

As $C = \dfrac{[HCl]_t}{[HCl]_0}$ is related to the relative undegraded hydrogen chloride concentration in the sample $[HCl]_t$ the HCl evolved, i.e. the number of the degraded monomer units, is expressed as

$$N(t) = N_0 (C - 1) \qquad (7.21)$$

where N_0 is the original number of the monomer units in the sample.

There is a direct correlation between the number of degraded monomer units and the chain conjugation, i.e. number of conjugated double bonds measured by UV spectroscopy (see Salovey *et al.*, 1970).

Correspondingly, by comparison of the conductivity and weight loss measurements, it is concluded that

$$\left[\frac{1 - N(t)}{N_0} \right]^{1/2} \infty \frac{\sigma(t) - \sigma_0}{\sigma_0} \qquad (7.22)$$

where $N(t)$ is the number of degraded monomer units, i.e. the number of the double bonds.

7.4 Physical ageing

The macroscopie properties of plastics change in time not only because of chemical reactions but because of slow changes in the physical structure, the texture, and hence in such macroscopic properties as mechanical strength, density, electrical conductivity and thermal characteristics.

In Chapter 2 it was shown that the free volume of an amorphous polymer would change slowly after the material has been cooled below its glass–rubber transition temperature; and a polymeric solid is practically never in thermodynamic equilibrium. The process of restoring equilibrium changes many macroscopic properties and is also a kind of ageing.

Slow changes in the texture of a polymer or a plastic compound are difficult to measure. Although is has long been known that the thermal history has a certain effect on the macroscopic properties, even on the chemical reactions proceeding in the polymer, very little systematic work has been done in this field.

Effects of storage below and above T_g
Texture ageing

The effects of storage of amorphous polyvinylchloride at temperatures lower and higher than T_g were studied first by Illers (1969a). He found that on storing PVC just below T_g for some weeks its density gradually decreases, its differential scanning curves show typical endothermic peaks when T_g is passed through and its mechanical properties change quite drastically. All these observations are quite coherently explained by the free-volume theory discussed in Chapter 2. It has been shown there that when a polymer is cooled down from the melt and stored below the glass–rubber transition temperature, the non-equilibrium free volume decreases in time, resulting in increase of density, a process referred to as volume relaxation. According to Rusch (1968) the non-equilibrium free volume can be incorporated in the effective temperature in the Williams, Landel, Ferry equation (see Eqn. 2.15 in Chapter 2). This means that, by cooling down a polymer from the melt, its internal effective temperature will be different from that of the environment. The effective temperature is defined as

$$T_{eff} = T + \frac{w_f(T)}{\Delta\alpha} \tag{7.23}$$

384

if $T_\infty \leq T < T_g$; $w_f(T)$ is the non-equilibrium free volume, $\Delta\alpha$ is the difference of the thermal expansion coefficients below and above T_g. When the material is stored at a temperature below T_g, T_{eff} tends to equalize with the temperature of the environment T, i.e. w_f tends to zero. This process is very slow at storage temperature well below T_g, and is fast near to T_g. The time scale of this process is some weeks when the

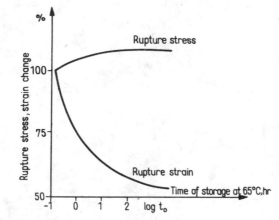

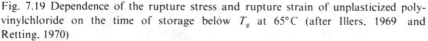

Fig. 7.19 Dependence of the rupture stress and rupture strain of unplasticized poly-vinylchloride on the time of storage below T_g at 65°C (after Illers, 1969 and Retting, 1970)

storage temperature is 5°C below T_g, and might be several years when it is 20—30°C below T_g. This is a kind of physical structural ageing: the macroscopic properties change in time very slowly without involving chemical processes.

The change of density of a polymer stored below T_g has already been demonstrated in Chapter 2 for polyvinylacetate (Fig. 2.16). This is quite common for amorphous polymers: storing them below T_g their density slowly increases. This is accompanied by the slight increase of the rupture stress and very drastic decrease of rupture strain. This is illustrated for unplasticized polyvinylchloride in Fig. 7.19. The PVC sample was stored at 65°C, i.e. about 10°C below T_g, for a considerable period of time and the density, stress-strain curve, DSC curve and some other properties were measured. It is noted that the drastic decrease in the rupture strain makes the process similar to chemical degradation: when a polymer is degraded, i.e. its average molecular weight decreases as a result of depolymerization or random chain scission, its rupture

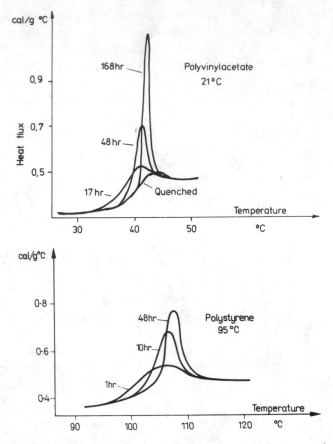

Fig. 7.20 Differential scanning calorimetric curves of polyvinylacetate and polystyrene stored for different periods of time below T_g, 21°C and 95°C respectively

strain decreases. In this case, however, there is no chemical degradation process at the temperature of storage, but the rupture strain still decreases.

Figure 7.20 shows differential scanning calorimetry curves for polystyrene and for polyvinylacetate measured after storing the samples below T_g. (For PVC see Fig. 2.10.)

It is seen that a considerable endothermic peak develops with increasing storage time, its height being proportional to the difference in the free volume corresponding to the starting structure at $T < T_g$ and that above T_g. Thus from the DSC peak it can be concluded, at least

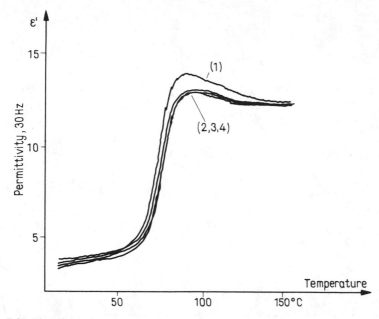

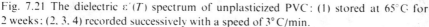

Fig. 7.21 The dielectric $\varepsilon'(T)$ spectrum of unplasticized PVC: (1) stored at 65°C for 2 weeks: (2. 3. 4) recorded successively with a speed of 3°C/min.

approximately, how far from the equilibrium structure the sample has been, before heating-up to measure the DSC curve. Such an effect is also observed in the low frequency dielectric spectra measured as a function of the temperature.

Figure 7.21 shows the dielectric $\varepsilon'(T)$ curves measured at a frequency of 30 Hz for unplasticized polyvinylchloride samples stored below T_g at different temperatures. It is seen that the oscillator strength $(\varepsilon_0 - \varepsilon_\infty)_T$ increases with increasing storage time because the relaxed permittivity increases. This effect is more marked at very low frequencies.

Figure 7.22 shows a series of Fourier transformed dielectric spectra of unplasticized polyvinylchloride after storage at 65°C for increasing periods of time. The spectra were obtained by transformation of the discharge currents recorded at 65°C.

It is seen that by increasing the storage time the spectrum band intensity decreases. This is explained by considering that polarization at such low frequencies is partially due to the MWS-effect. According to Eqns (7.11) and (7.12) and to the general discussion in section 5.3 the oscillator strength of the dielectric spectra in inhomogeneous systems

25*

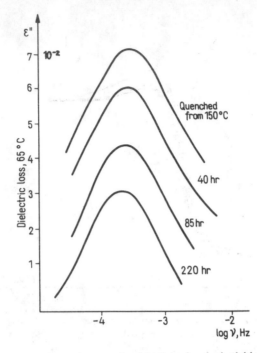

Fig. 7.22 Fourier transform spectra of unplasticized polyvinylchloride after prolonged storage at 65°C

depends on a structure factor S, which is dependent on the size and shape of the inhomogeneities. The decrease of the oscillator strength of the Fourier transformation spectrum band involves a structural change towards decreasing the structure factor S. This is not in contradiction with the increase of the 30 Hz $\varepsilon'(T)$ amplitude shown in Fig. 7.21 because in this case, during the recording of the spectra, the T_g has been passed through, while in the Fourier transformation measurement $T < T_g$. A glance at Fig. 2.13, illustrating free volume and defect concentration changes on passing through T_g, explains the problem. When the transition is measured as a function of the temperature, the temperature sweep starts from a state which is the closer to equilibrium the longer the time of storage under T_g has been. The change in the oscillator strength in this case would reflect the abrupt change of the relaxed permittivity (see Chapter 4). If, on the other hand, the transition is measured as a function of the frequency at a temperature $T < T_g$, only changes of S during the course of storage are detected.

388

The problem can be formally treated by considering the structure factor S as time and temperature dependent, with a time scale considerably larger than that of the measurement. The oscillator strength in this case is given by

$$\varepsilon_0 - \varepsilon_\infty = \varepsilon_\infty^{(1)}\, \bar{S}(t, T) \qquad (7.24)$$

where $\varepsilon_\infty \approx \varepsilon_\infty^{(1)}$ is the unrelaxed permittivity of the medium in which the inhomogeneities are present and can be considered as constant.

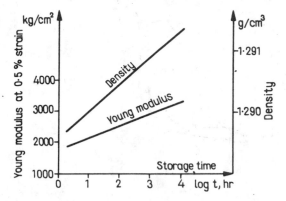

Fig. 7.23 Change of density and Young's modulus of a plasticized PVC-compound by prolonged storage at room temperature

As the area under the dielectric $\varepsilon''(\ln\omega)$ curve is proportional to $\varepsilon_0 - \varepsilon_\infty$ it is concluded that the decrease in the intensity of the Fourier transform spectra shown in Fig. 7.22 reflects the storage-time dependence of the structure factor.

The effect of storage above T_g for a plasticized PVC compound containing 27.6 % dioctyl phthalate has been studied by Juijn *et al.* (1969). The compound was stored at room temperature after processing for increasing periods, and the differential scanning calorimetric curves, Young's moduli and density were measured. The calorimetric curves show endothermic maxima above T_g, which are shifted to higher temperatures with increasing storage time. A similar effect has been observed by Illers (1969a) for unplasticized PVC stored above T_g too. The endothermic maxima are attributed to a process of aggregation.

In Fig. 7.23 the change of the density and of the Young's modulus of the plasticized PVC compound is shown as a function of the storage time at room temperature. It is seen that the process is going on continu-

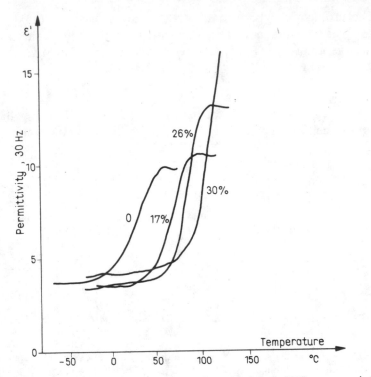

Fig. 7.24 Change of the dielectric spectra of a plasticized PVC-compound 30% diisooctyl phthalate by prolonged storage at 120°C: numbers on curves indicate percentage loss in weight

ously without showing any sign of levelling off even after half a year storage.

As aggregation is a very slow process, the endothermic peaks observed above T_g by differential scanning calorimetry are not reversible. They appear on recording the DSC curve after a certain time of storage, but vanish upon repeated heating and cooling cycles.

The high temperature (150—170°C) endothermic OSC peak may be correlated with the dielectric depolarization current peak observed in the same temperature range in plasticized PVC (cf. Fig. 5.4).

From this it is concluded that upon storage above T_g a crystallization-like process is slowly going on in the compound which makes its macroscopical properties change. This can be regarded as physical, structural, ageing. The process can be monitored by the dielectric polarization or depolarization method.

390

Loss of plasticizer

A very typical ageing process which does not involve chemical change is exudation of the plasticizer from plasticized compounds. As discussed in Chapter 5, the dielectric transitions, especially T_g, are dependent on the plasticizer content. During processing and use, the plasticizer slowly diffuses out of the polymer, resulting in shifting of T_g to higher temperature and correspondingly increasing rigidity. This process can be very easily followed by dielectric spectroscopy. Figure 7.24 shows a series of dielectric $\varepsilon'(T)$ spectra obtained from a plasticized PVC compound stored at 110°C for different periods. The original compound contained 30% di-isooctyl phthalate as plasticizer.

It is seen that by storage the T_g is shifted to higher temperatures, the transition is narrowed and the oscillator strength is increased. This is just the opposite to the effect found by measuring the dielectric spectra of compounds containing different amounts of plasticizers. The parameters of the dielectric dispersion curve can be calibrated against the plasticizer content for a given type of compound. Thus dielectric spectroscopy can be used as a rapid routine method for checking plasticizer content during processing, after prolonged storage, and after weathering of plasticized PVC-compounds.

Phase separation processes

In multi-component polymeric compounds, especially in toughened systems, the components usually form separate phases represented by separate peaks in the dielectric spectra. (Examples of this have already been presented in Chapter 5.) The presence of a separate phase exhibiting lower T_g than that of the base polymer (the rubber phase) is considered to be important in creating the toughness of the system. The shock waves of the impact are thought to be damped by this rubber phase.

Toughened systems are known to lose their toughness after prolonged storage, even if no chemical processes are involved. This process is accompanied by a gradual change of the texture: the extent of phase separation may be increased or decreased.

An example of decrease is shown in Fig. 7.25, giving the dielectric spectra of a toughened PVC blend after thermal treatment. The blend contained 40% chlorinated polyethylene as a rubber phase. After the processing, the dielectric spectra exhibited two main transitions corresponding to the PVC and rubber phases respectively. The assignment of these peaks was checked by successively increasing the amount of the

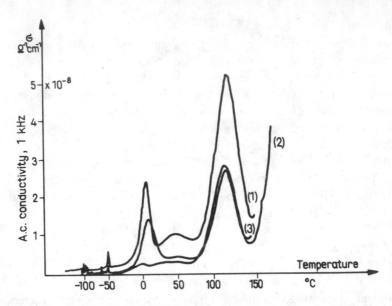

Fig. 7.25 Effect of annealing at 150°C on the dielectric spectra of a PVC-chlorinated polyethylene blend toughened PVC; annealing times in hours

rubber phase and by measuring the dielectric and mechanical spectra, the DSC and thermomechanical curves as discussed in section 5.4. These measurements showed that the rubber component in this compound had a slight effect on the T_g of the original compound by shifting it to a lower temperature.

Figure 7.25 shows that by storage at 150°C the separate rubber phase peak is gradually decreased and the PVC peak shifts further to lower temperatures. This is explained by assuming that a kind of mixing of the two components occurs by which the separate phase gradually vanishes.

The effect of thermal history on the texture of the polymer blends is extremely important. Phase separation processes may proceed rather slowly. The corresponding changes in the macroscopic properties, e.g. toughness, are often taken as being due to chemical degradation processes. To the author it seems that in most cases purely physical structural ageing is involved. By a combination of dielectric, mechanical and thermal methods such processes can be followed.

APPENDIX

RELAXATION MAPS OF POLYMERS

In this section the temperature dependences of the relaxation times associated with the main transitions in various polymers are plotted in the Arrhenius representation. These diagrams have been constructed by using dielectric and mechanical relaxation data. Shaded areas represent the deviations of data obtained by different methods for different polymers. Deviations from the straight line show that the transition is not Arrhenius-like (glass–rubber transitions). From the relaxation maps the transition frequencies corresponding to a given temperature or the transition temperatures corresponding to a given frequency are directly obtained. The transition angular frequency is related with the relaxation time as

$$\omega_0 \tau(T) = 1$$

Transition maps of the following polymers are included:

 (1) polyethylene
 (2) iso-polypropylene
 (3) polyisobutylene
 (4) polytetrafluoroethylene
 (5) polystyrene
 (6) polychlorotrifluoroethylene
 (7) polyethylene terephthalate
 (8) polyvinyl acetate
 (9) polymethyl methacrylate
(10) polyethyl methacrylate
(11) polymethyl acrylate
(12) polyethyl acrylate
(13) polypropyl acrylate
(14) polymethyl α-chloroacrylate
(15) polyethyl α-chloroacrylate

(16) polyvinyl alcohol

(17) polyvinyl acetal

(18) polyvinyl n-butyral

(19) polyvinyl n-octanal

(20) polyvinyl chloride

(21) polyvinylidene fluoride

(22) polyamide-6 (Nylon 6)

(23) polyamide-66 (Nylon 66)

(24) polyamide-610 (Nylon 610)

(25) polyethylene oxide

(26) polypropylene oxide

(27) polyoximethylene

(28) polydian isophthalate

(29) polydian terephthalate

(30) polydian carbonate

(31) polydian sebacate

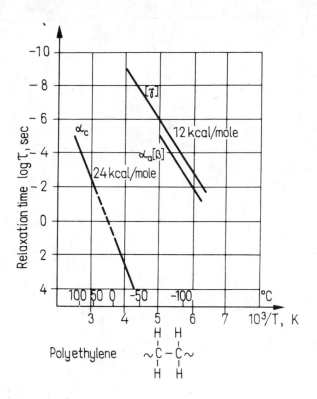

Polyethylene $\sim\overset{\displaystyle H}{\underset{\displaystyle H}{C}}-\overset{\displaystyle H}{\underset{\displaystyle H}{C}}\sim$

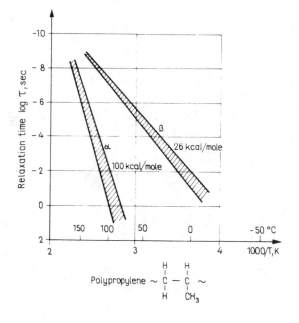

Polypropylene $\sim\overset{\displaystyle H}{\underset{\displaystyle H}{C}}-\overset{\displaystyle H}{\underset{\displaystyle CH_3}{C}}\sim$

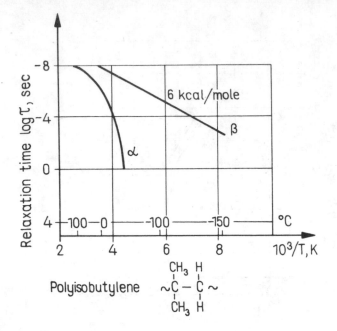

Polyisobutylene $\sim\underset{\underset{CH_3}{|}}{\overset{\overset{CH_3}{|}}{C}} - \underset{\underset{H}{|}}{\overset{\overset{H}{|}}{C}} \sim$

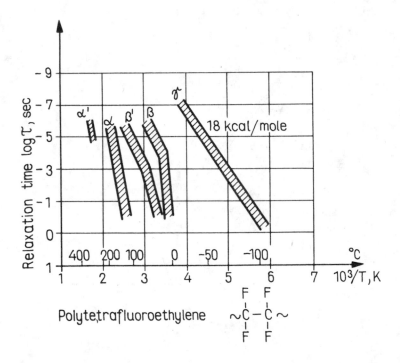

Polytetrafluoroethylene $\sim\underset{\underset{F}{|}}{\overset{\overset{F}{|}}{C}} - \underset{\underset{F}{|}}{\overset{\overset{F}{|}}{C}} \sim$

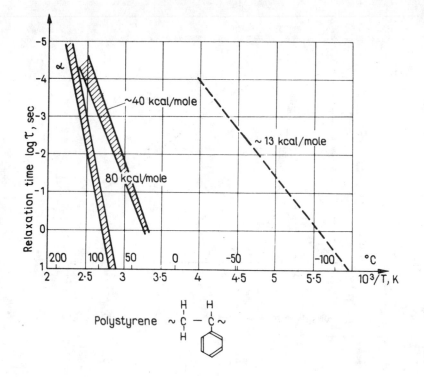

Polystyrene   ~ C – C ~

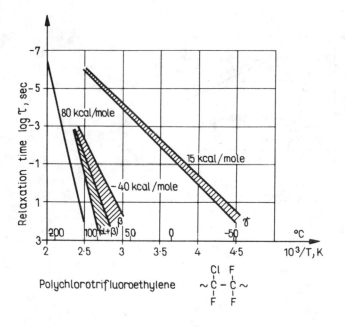

Polychlorotrifluoroethylene   ~ C – C ~

397

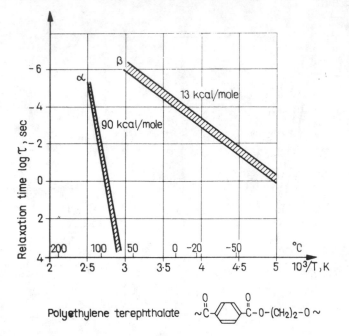

Polyethylene terephthalate

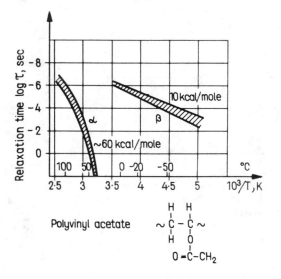

Polyvinyl acetate

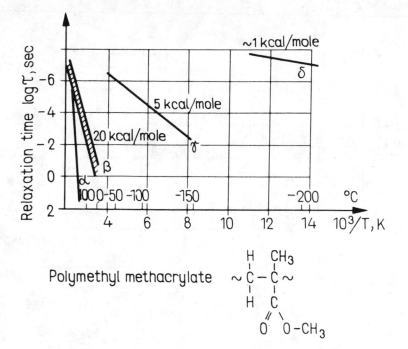

Polymethyl methacrylate

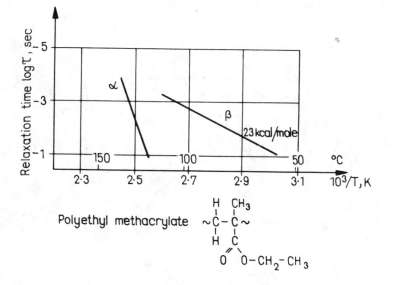

Polyethyl methacrylate

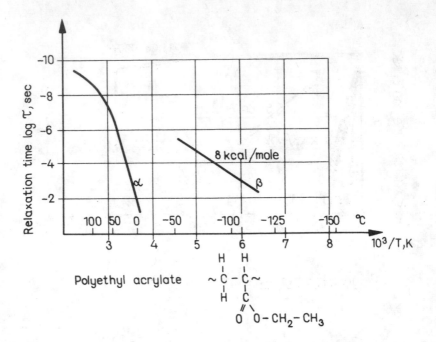

Polyethyl acrylate

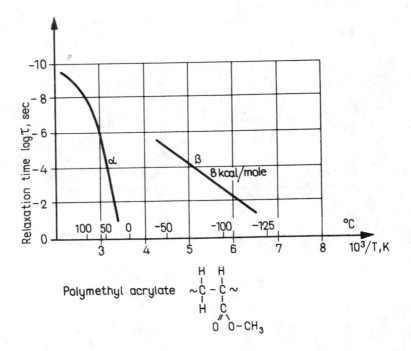

Polymethyl acrylate

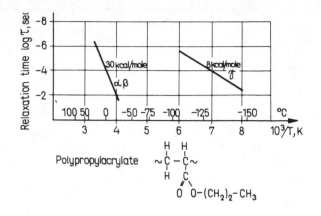

Polypropylacrylate

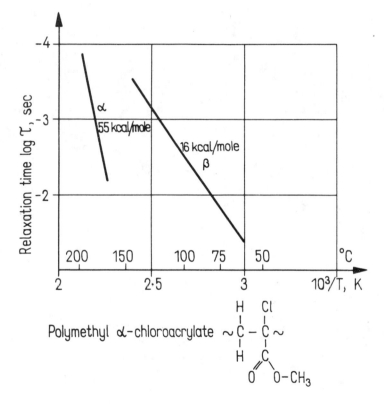

Polymethyl α-chloroacrylate

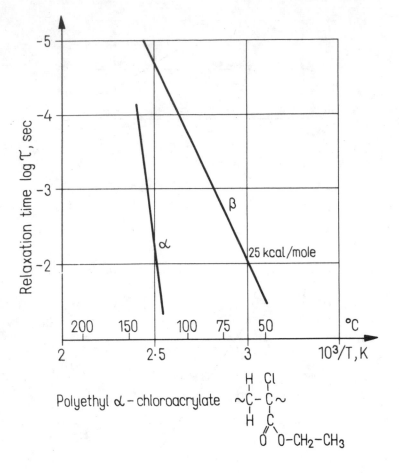

Polyethyl α – chloroacrylate

$$\sim \overset{\overset{\displaystyle H}{|}}{\underset{\underset{\displaystyle H}{|}}{C}} - \overset{\overset{\displaystyle Cl}{|}}{\underset{\underset{\displaystyle C}{|}}{C}} \sim$$
$$\overset{||}{O}\ O{-}CH_2{-}CH_3$$

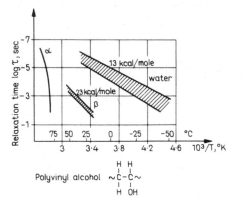

Polyvinyl alcohol $\sim \overset{\overset{\displaystyle H}{|}}{\underset{\underset{\displaystyle H}{|}}{C}} - \overset{\overset{\displaystyle H}{|}}{\underset{\underset{\displaystyle OH}{|}}{C}} \sim$

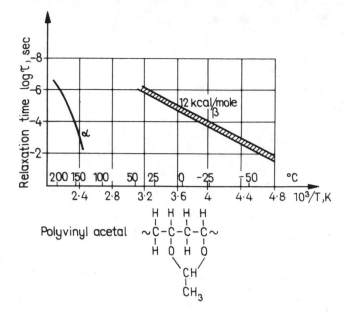

Polyvinyl acetal

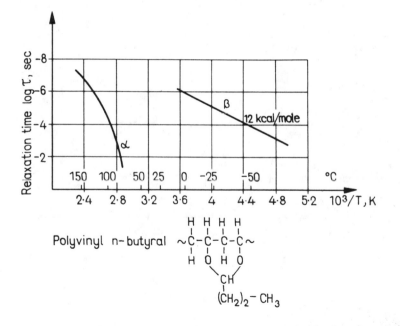

Polyvinyl n-butyral

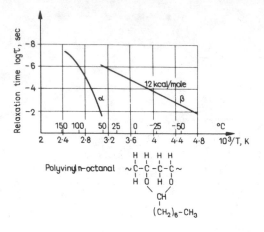

Polyvinyl n-octanal

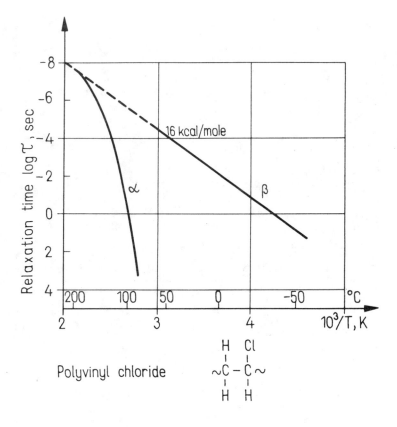

Polyvinyl chloride

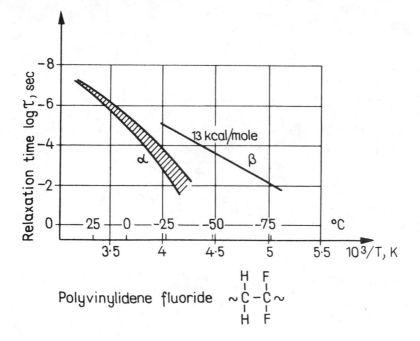

Polyvinylidene fluoride $\sim\underset{\underset{H}{|}}{\overset{\overset{H}{|}}{C}}-\underset{\underset{F}{|}}{\overset{\overset{F}{|}}{C}}\sim$

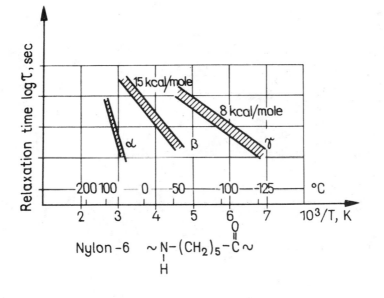

Nylon-6 $\sim\underset{\underset{H}{|}}{N}-(CH_2)_5-\overset{\overset{O}{||}}{C}\sim$

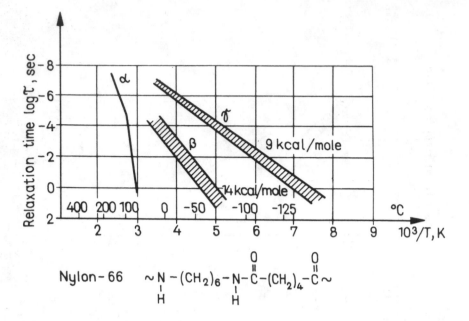

Nylon-66 $\sim N-(CH_2)_6-N-\overset{\overset{O}{\|}}{C}-(CH_2)_4-\overset{\overset{O}{\|}}{C}\sim$
              $\underset{H}{|}$ $\underset{H}{|}$

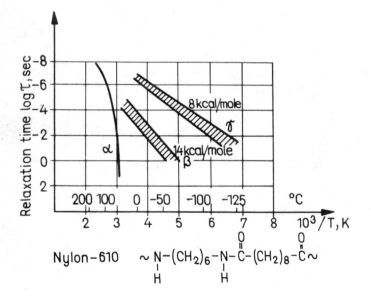

Nylon-610 $\sim N-(CH_2)_6-N-\overset{\overset{O}{\|}}{C}-(CH_2)_8-\overset{\overset{O}{\|}}{C}\sim$
               $\underset{H}{|}$ $\underset{H}{|}$

406

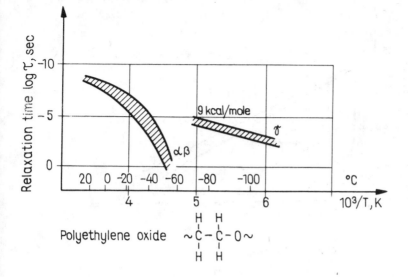

Polyethylene oxide
$$\sim \overset{\displaystyle H}{\underset{\displaystyle H}{C}} - \overset{\displaystyle H}{\underset{\displaystyle H}{C}} - O \sim$$

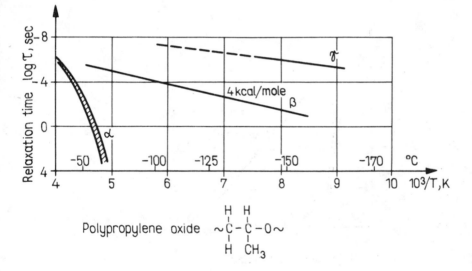

Polypropylene oxide
$$\sim \overset{\displaystyle H}{\underset{\displaystyle H}{C}} - \overset{\displaystyle H}{\underset{\displaystyle CH_3}{C}} - O \sim$$

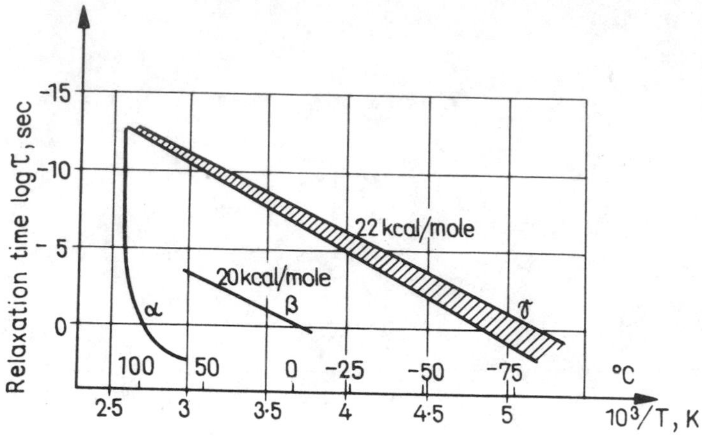

Polyoximethylene 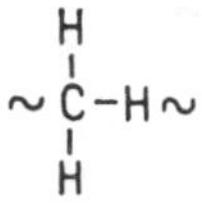

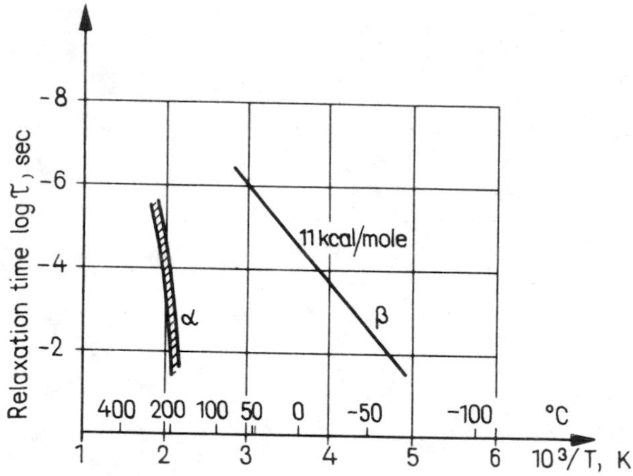

Polydian isophthalate

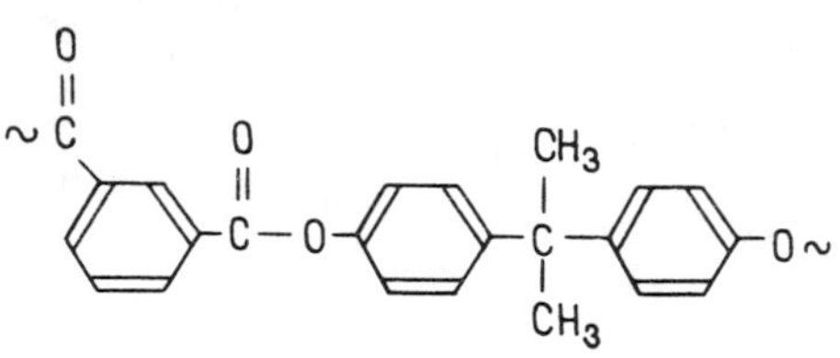

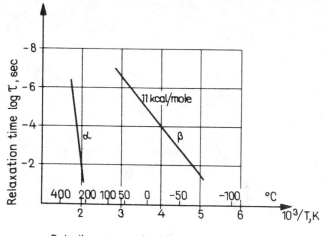

Polydian terephthalate

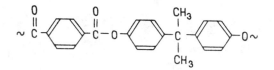

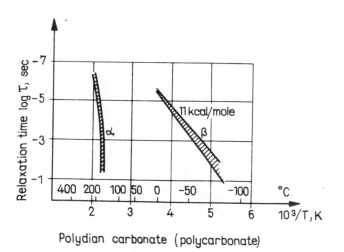

Polydian carbonate (polycarbonate)

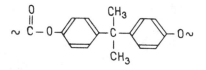

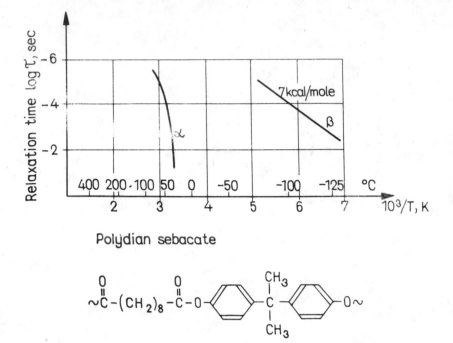

Polydian sebacate

BOOKS ON DIELECTRIC SPECTROSCOPY

J. H. VAN VLECK, *The Theory of Electric and Magnetic Susceptibilities*. University Press, Oxford, 1932.

P. DEBYE, *Polar Molecules*. Dover, New York, 1945.

G. I. SKANAVI, *The Physics of Dielectrics at Low Fields*. Gostechtheorizdat, Moscow, 1949.

C. J. E. BÖTTCHER, *Theory of Electric Polarization*. Elsevier, Amsterdam, 1952.

A. VON HIPPEL, *Dielectric Materials and Applications*. Chapman & Hall, London, 1954.

C. P. SMYTH, *Dielectric Behaviour and Structure*. McGraw-Hill, New York, 1955.

J. W. SMITH, *Electric Dipole Moments*. Butterworth, London, 1955.

H. A. STUART (ed.), *Die Physik der Hochpolymeren* (in German). Springer, Berlin, 1956.

H. FRÖHLICH, *Theory of Dielectrics* (1st edn, 1949). University Press, Oxford, 1958.

G. I. SKANAVI, *The Physics of Dielectrics at High Fields*. Gostechtheorizdat, Moscow, 1958.

V. I. ODELEVSKY *et al.*, *Physics of Dielectrics*. Izv. Akad. Nauk. SSSR, Moscow, 1960.

J. B. BIRKS, J. HART (eds), *Progress in Dielectrics*. Heywood, London; Vol. 1 1960, Vol. 2 1961.

W. HOLZMÜLLER, K. ALTENBURG, *Physik der Kunststoffe*. Akademie Verlag, Berlin, 1961.

N. G. MCCRUM, B. E. READ and G. WILLIAMS, *Anelastic and Dielectric Effects in Polymeric Solids*. Wiley, London and New York, 1967.

V. V. DANIEL, *Dielectric Relaxation*. Academic Press, New York and London, 1967.

N. E. HILL, W. E. VAUGHAN, A. H. PRYCE, M. DAVIES, *Dielectric Properties and Molecular Behaviour*. Van Nostrand, Reinhold, London, 1969.

B. I. SAZHIN (ed.), *Electrical Properties of Polymers* (in Russian). Izd. Khimia, Leningrad, 1970.

P. HEDVIG, *Electrical Conductivity and Polarization in Plastics* (in Hungarian). Akadémiai Kiadó, Budapest, 1970.

A. TAGER, *Physical Chemistry of Polymers* (revised translation from Russian). MIR Publishers, Moscow, 1972.

REFERENCES

ADACHI, K., H. SUGA, S. SEKI (1970). *Bull. chem. Soc., Japan,* **43,** 1916.

ADAMSKI, P. (1971). *Vysokomol. Soed.,* **13,** 19.

AIBASOV, A. B., KH. G. MINDIYAROV, YU. V. ZELENEV, YU. G. OGANESOV, V. G. RAEVSKY (1970) *Vysokomol. Soed.,* **12,** 10.

AKIJAMA, S. (1972). *Bull. chem. Soc., Japan,* **45,** 1381.

ALFREY, T., W. WIEDERHORN, R. STEIN, A. V. TOBOLSKY (1949). *Ind. Engng. Chem.,* **41,** 701.

ALLEN, T., J. MC. AINSH, G. M. JEFFS (1971). *Polymer* (London), **12,** 85.

ANAGNOSTOPOULOS, C. E., C. E. CORAN, A. Y. GAMRATH (1960). *J. appl. Polym. Sci.,* **4,** 181.

ANTON, A. (1968). *J. appl. Polym. Sci.,* **12,** 2417.

ANTONOV, S. N., I. M. GOURMAN, V. V. KOVRIGA, G. A. LUSHSCHEIKIN (1966). *Plast. Massy,* No 4, 38.

ARAI, K., O. YANO, Y. WADA (1968). *Rep. Progr. Polymer Phys. Japan,* **11,** 267.

ARMENIADES, C. D., E. BAER, J. K. RIEKE (1970). *J. appl. Polym. Sci.,* **14,** 2635.

BAIRD, M. E., G. T. GOLDSWORTHY, C. J. CREASEY (1971). *Polymer* (London), **12,** 159.

BARKALOV, I. M., D. P. KIRYOUCHIN, A. M. KAPLAN, V. I. GOLDANSKY (1972). *Proceedings 3rd Symposium on Radiation Chemistry* (ed. J. Dobó and P. Hedvig), Vol. 1, p 503. Publishing House, Hungarian Academy of Science, Budapest.

BARTENEV, G. M. (1951). *Doklady Akad. nauk. SSSR,* **76,** 227.

BARTENEV, G. M. (1956). *Doklady Akad. nauk. SSSR,* **110,** 805.

BARTENEV, G. M., I. A. LUKYANOV (1955). *Zh. tekh. Khim.,* **29,** 1486.

BARTENEV, G. M., I. V. RAZUMOVSKAYA, D. S. SANDITOV, I. A. LUKYANOV (1969). *J. Polym. Sci.,* **A-1,** 2147.

BARTENEV, G. M., Y. V. ZELENEV, I. P. BORODIN (1970). *J. appl. Polym. Sci.,* **14,** 393.

BAUR, H. (1970). *Kolloid Z.,* **241,** 1057.

BAUR, H. (1971a). *Z. Naturf.,* **26a,** 979.

BAUR, H. (1971b). *Kolloid Z.,* **244,** 293.

BAUR, M. F., W. H. STOCKMAYER (1965). *J. chem. Phys.,* **43,** 4319.

BEADEOT, R. A., (1966). *J. Am. chem. Soc.,* **88,** 1390.

BERGHMANS, H., G. J. SAFFORD, P. S. LEUNG (1971). *J. Polym. Sci.,* **A-2,** 1219.

BERGMANN, K., NAWOTKI, K. (1972). *BASF report,* private communications.

BEZJAK, A., Z. VEKSLI (1970). *IUPAC Symposium on Macromolecules,* Leiden.

BJERRUM, N. (1951). *K. danske Vidensk. Selsk. Mat-fys.,* **27,** No. 1.

BJORKSTEIN, J. H., H. TOVEY, B. HARKER, J. HENNING (1956). *Polyesters and Their Applications.* Chapman & Hall, London.

BLOCK, H., M. E. COLLINSON, S. M. WALKER (1973). *Polymer* (London), **14**, 68.

BLYACHMAN, E. M., T. I. BORISOVA, Z. M. LEVITZKAYA (1970). *Vysokomol. Soed.*, **A 12**, 1544.

BODOR, G. (1972). IUPAC Symposium, Helsinki, *J. Polym. Sci.* C (to be published).

BOHN, L. (1963). *Kunststoffe*, **53**, 826.

BOHN, L. (1973). *Angew. makromol. Chem.*, **29/30**, 25.

BOON, J. P., S. A. RICE (1967). *J. chem. Phys.*, **47**, 2480.

BORISOVA, T. I., V. N. CHIRKOV, V. A. SHEVELEV (1972). *Vysokomol. Soed.*, **A 14**, 1240.

BORISOVA, T. I., L. L. BURSHTEIN, G. P. MIKHAILOV (1962). *Vysokomol. Soed.*, **4**, 1479.

BORISOVA, T. I., G. P. MIKHAILOV (1959). *Vysokomol. Soed.*, **1**, 563.

BORISOVA, T. I., G. P. MIKHAILOV, M. M. KOTON (1969). *Vysokomol. Soed.*, **A 11**, 1140.

BOYD, R. H. (1959). *J. chem. Phys.*, **30**, 1276.

BOYD, R. H., G. H. PORTER (1972). *J. Polymer Sci.* **A-2, 10**. 647.

BOVEY, F. A. (1972). In *Natural and Synthetic High Polymers, NMR. Basic Principles and Progress* (ed. P. DIEHL, E. FLUCK and R. KOSFELD). Vol. 4, p 1. Springer, Berlin.

BOVEY, F. A., G. V. D. TIERS (1960). *J. Polym. Sci.*, **44**, 173.

BOYER, R. F. (1963). *Rubber Rev.*, **34**, 1303.

BOYER, R. F. (1966). *J. Polym. Sci.*, **C**, No. 14,3.

BOYER, R. F. (1968). *Polym. Engng. Sci.*, **8**, 161.

BOYER, R. F. (1973). *Macromolecules*, **6**, 288.

BOYER, R. F., R. S. SPENCER (1947). *J. Polym. Sci.*, **2**, 157.

BRAUN, D. (1971) *Pure appl. Chem.*, **26**, 173.

BROPHY, J. J., J. W. BUTTREY (ed). (1962). *Organic Semiconductors*. Macmillan, New York.

BROUCKERE, L., G. OFFERGELD (1958). *J. Polym. Sci.*, **30**, 105.

BRUCK, S. D. (1967). *J. Polym. Sci.*, **17**, 169.

BUECHE, F. (1954). *J. chem. Phys.*, **22**, 603.

BUECHE, F. (1961). *J. Polymer Sci.* **54**, 597.

CALVET, C., H. PRAT (1956). *Microcalorimétrie. Massonet, Paris.*

CARIUS, H. E. (1968). Thesis. Univ. Leipzig.

CARR, H. Y., E. M. PURCELL (1954). *Phys. Rev.*, **94**, 630.

CARTER, W. C., M. MAGAT, W. G. SCHMIEDER, C. P. SMYTH (1946). *Trans. Faraday Soc.*, **42 A**, 213.

CHARLESBY, A. (1960). *Atomic Radiation and Polymers*. Pergamon, London.

CHARLESBY, A., R. H. PARTRIDGE (1963). *Proc. Roy. Soc.*, London, **A 271**, 170, 188.

CHARLESBY, A., R. H. PARTRIDGE (1965). *Proc. Roy. Soc.*, London, **A 283**, 312, 329.

CHASSET, R., P. THIRION, V. B. NGUYEN (1959). *Revue gen. Caoutch.*, **36**, 857.

CHERNYAK, K. I., (1963). *Epoxy Resins and Their Applications*. Sudpromgiz, Moscow.

COHEN, M. H., D. TURNBULL (1959). *J. chem. Phys.*, **31**, 1164.

COHEN, M. H., D. TURNBULL (1964). *Nature*, **203**, 964.

COLE, R. H. (1965). *J. chem. Phys.*, **42**, 637.

COLE, R. H., K. S. COLE (1941). *J. chem. Phys.*, **9**, 341.

COLE, R. H., P. M. GROSS (1949). *Rev. scient. Instrum.*, **20**, 252.

CONNOR, T. M. (1970). *J. Polym. Sci.*, **A-2**, 191.

CRISSMAN, J. M., E. Passaglia (1966). *J. Polym. Sci.*, C, No. 14, 237.

CRUGNOLA, A. (1968). *Mater. Plast. Elast.*, **34**, 883.

CURTIS, A. J. 1960. In *Progress in Dielectrics* (ed. J. B. Birks and J. Hart). Vol. 2, 29.

CZVIKOVSZKY, T. (1968). Atom. Energy Rev., **6**, No. 3, 3–99.

414

DAKIN, T. W., C. N. WORKS (1947). *J. appl. Phys.*, **18**, 789.

DANIEL, V. V. (1967). *Dielectric Relaxation*. Academic Press, New York and London.

DARBY, I. R., N. W. TOUCHETTE, K. SEARS (1967). *Polym. Engng. Sci.*, **7**, No. 4, 1.

DAS, T. P. (1957). *J. chem. Phys.*, **27**, 763.

DAVIDSON, D. W., R. H. COLE, (1950). *J. chem. Phys.*, **18**, 1417.

DAVIES, R. O., G. O. JONES (1953). *Proc. R. Soc.* **A 217**, 26.

DAVIES, G. R., I. M. WARD (1969). *Polym. Letters*, **7**, 353.

DAVYDOV, A. S. (1948). *Ph. Exp. Theor. Phys.*, USSR, **18**, 210.

DAVYDOV, A. S. (1968). *Theory of Molecular Excitons* (in Russian). Izd. Nauka, Moscow.

DEBYE, P. (1921). *Ann. Physik*, **39**, 789.

DEBYE, P. (1945). *Polar Molecules*. Dover, New York.

DELMONTE, I. (1959). *J. appl. Polym. Sci.*, **2**, 108.

DEMMLER, K. (1969). *Farbe Lack*, **75**, 1051.

DICKIE, R. A. (1973). *J. appl. Polym. Sci.*, **17**, 65, 79.

DI MARZIO, E. A., J. H. GIBBS (1959). *J. Polym. Sci.*, **40**, 121.

DOMINIC, C. J. (1967). Thesis. Univ. Aston, Birmingham, England.

DOOLITTLE, A. K. (1951). *J. appl. Phys.*, **22**, 1471.

DOOLITTLE, A. K., D. B. DOOLITTLE (1957). *J. appl. Phys.*, **28**, 901.

DOTY, H. S., J. ZABLE (1946). *J. Polym. Sci.*, **1**, 90.

DRYDEN, J. S., R. MEAKINS (1957). *Proc. phys. Soc. Lond.*, **B 70**, 427.

DUNLOP, L. H., C. R. FOLTZ, A. G. MITCHELL (1972). *J. Polym. Sci., Phys.*, **10**, 2223.

DUNN, P., B. C. ENNIS (1970). *J. appl. Polym. Sci.*, **14**, 355.

ECKSTEIN, B. (1966). *Glastech. Ber.*, **39**, 455.

ECKSTEIN, B. (1967). *Phys. Status solidi*, **20**, 83.

ECKSTEIN, B. (1969). *Kolloid Z.*, **233**, 845.

EISENBERG, A., S. REICH (1969). *J. chem. Phys.*, **51**, 5706.

FARROW, G., J. McINTOSH, I. M. WARD (1960). *Macromol. Chem.*, **38**, 147.

FERRY, J. D. (1961). *Viscoelastic Properties of Polymers*. Wiley, New York.

FILIPOVICH, G., C. D. KNUTSON, D. M. SPITZER (1965). *J. Polym. Sci., Polym. Letters*, **3**, 1065.

FITZHUGH, A. F., R. N. CROZIER (1952). *J. Polym. Sci.*, **8**, 225.

FLOCKE, H. A. (1962). *Kolloid Z.*, **180**, 118.

FLOCKE, H. A. (1963). *Kolloid Z.*, **188**, 114.

FLORY, P. J. (1953). *Principles of Polymer Chemistry*. Cornell Univ. Press, New York.

FLORY, P. J. (1972). Paper presented at IUPAC Symposium, Helsinki.

FLORY, P. J., REHNER, J. J. (1943). J. chem. Phys., **11**, 521.

FLÜGGE, S. ed. (1955). In *Handbuch der Physik*.

FORSLIND, L. (1971). In *Natural and Synthetic High Polymers, NMR, Basic Principles and Progress* (ed. P. Diehl, E. Fluck and R. Kosfeld). Vol. 4, 145. Springer, Berlin.

FOX, T. G. (1956). *Bull. Am. phys. Soc.*, **1**, 123.

FOX, T. G., P. J. FLORY (1950). *J. appl. Phys.*, **21**, 581.

FRENKEL, J. (1931). *Phys. Rev.*, **37**, 1276.

FRÖHLICH, H. (1949). *Theory of Dielectrics*, 1st edn. Univ. Press, Oxford.

FRÖHLICH, H. (1958). *Theory of Dielectrics*, 2nd edn. Univ. Press, Oxford.

FUJINO, K. T., K. HORINO, K. MIYAMOTO, H. KAWAI (1962). *Rep. Prog. Polym. Phys. Japan*, **5**, 111.

FUOSS, R. M., J. G. KIRKWOOD (1941). *J. Am. chem. Soc.*, **63**, 385.

GERNGROSS, O., K. HERMANN, W. ABITZ (1930). *Z. phys. Chem.*, **B 10**, 371.

415

GIBBS, J., E. A. DI MARZIO (1958). *J. chem. Phys.*, **28**, 373, 807.

GIBBS, J. H., E. A. DI MARZIO (1959). *J. Polymer Sci.* **40**, 121.

GLARUM, S. H. (1960). *J. chem. Phys.*, **33**, 1371, 1960.

GLASSTONE, S., K. J. LAIDLER, H. EYRING (1941). *Theory of Rate Processes.* McGraw-Hill, New York.

GOLDANSKY, V. I., R. H. HERBER (editors) (1968). *Chemical Applications of Mössbauer Spectroscopy.* Academic Press, New York.

GOLDSTEIN, M. (1963). *J. chem. Phys.*, **39**, 3369.

GOLDSTEIN, M. (1969). *J. chem Phys.*, **51**, 3728.

GORDON, M., J. S. TAYLOR (1952). *J. appl. Chem.* London, **2**, 493.

GOTLIB, YU. YA., K. M. SALIKHOV (1962). *Vysokomol. Soed.*, **4**, 1163.

GOUL, V. E., L. N. TZARSKY, N. S. MAISEL, L. Z. SHENFIL, V. S. ZHUROVLEV, N. G. TZIBRYA (1968). *Electrically Conductive Polymeric Materials* (in Russian). Izd. Khimia, Moscow.

GRASSIE, N. (1956). *Chemistry of High Polymer Degradation Processes.* Butterworths, London.

GUTTMAN, F., L. E. LYONS (1967). *Organic Semiconductors.* Wiley, New York.

HABERLAND, G. D., J. B. CARMICHAEL (1966). *J. Polym. Sci.*, *C*, No. 14, 291.

HAHN, E. L. (1950). *Phys. Rev.*, **80**, 580.

HAMMER, C. F. (1971). *Macromolecules*, **4**, 69.

HAMON, B. V. (1952). *Proc. Instn. elect. Engrs*, 99, Pt IV. Monograph 27.

HANAI, T. (1961). *Kolloid Z.*, **177**, 57.

HANAI, T., N. KOIZUMI, R. GOTOH (1962). *Bull. Inst. chem. Res., Kyoto Univ.*, **40**, 240.

HARAN, E. N., H. GRINGAS, D. KATZ (1965). *J. appl. Polym. Sci.*, **9**, 3505.

HARPER, C. D., (1963). *Giessharze in der elektronischen Technik.* Carl Hanser Verlag, München.

HARTMAN, A. (1957). *Kolloid Z.*, **153**, 157.

HASHIMOTO, T., R. E. PRUDHOMME, D. A. KEEDY, R. S. STEIN (1973). *J. Polymer Sci.* **11**, 693, 709.

HAVRILIAK, S., S. NEGAMI (1966). *J. Polym. Sci.*, *C* **14**, 99.

HAVRILIAK, S. JR., N. ROMAN (1966). *Polymer* (London), **7**, 387.

HAVRILIAK, S., S. NEGAMI (1967). *Polymer* (London), **8**, 161.

HAVRILIAK, S., S. NEGAMI (1969). *Polymer* (London), **10**, 859.

HEDVIG, P. (1964). *J. Polym. Sci.*, **A-2**, 4097.

HEDVIG, P. (1967). In *Proceedings 2nd Tihany Symposium on Radiation Chemistry* (ed. J. DOBO and P. HEDVIG), 599. Publishing House, Hungarian|Academy|of Sciences, Budapest.

HEDVIG, P. (1969a). Kinetics and mechanism of polyreactions, *IUPAC Symposium, Budapest*, Vol. 5, p. 277, Hungarian Academy of Science, Budapest.

HEDVIG, P. (1969b). *J. Polym. Sci.*, **A-7**, 1145.

HEDVIG, P. (1969c). *Electrical Conductivity and Polarization in Plastics* (in Hungarian), Akadémiai Kiadó, Budapest.

HEDVIG, P. (1970a). *Müanyag és Gumi*, **7**, 283.

HEDVIG, P. (1970b). *Kémiai Közlemények*, **33**, 37.

HEDVIG, P. (1971). *J. Polym. Sci.*, *C*, No. 33, 315.

HEDVIG, P. (1972a). In *Radiation Chemistry of Macromolecules* (ed. M. Dole), Vol. 1. Academic Press, New York.

HEDVIG, P. (1972b). *International Microsymposium on Polarization and Conduction in Insulating Polymers*. Bratislava-Harmonia.

HEDVIG, P. (1972c). IUPAC Symposium, Helsinki, *J. Polym. Sci.*, C, **42**, 127.

HEDVIG, P. (1973). Dielectric depolarization spectroscopy, in *Magnetic Resonance in Chemistry and Biology* (ed. J. Herak and K.J. Adamic). Marcel Dekker, New York.

HEDVIG, P. (1975) *Experimental Quantum Chemistry*. Academic Press, New York.

HEDVIG, P., T. CZVIKOVSZKY (1972). *Angew. makromol. Chem.*, **21**, 79.

HEDVIG, P., E. FÖLDES (1974). *Angew. Makromol. Chem.* **35**, 147.

HEDVIG P. (1973) *J. Polym. Sci.*, Symposium No. 42, 1271.

HEDVIG P., GY. ZENTAI (1969). *Microwave Study of Chemical Structures and Reactions*. Iliffe, London, Akadémiai Kiadó, Budapest.

HEDVIG, P., M. KISBENYI (1969). *Angew. makromol. Chem.*, **7**, 198.

HEDVIG, P., S. KULCSÁR, L. KISS (1968). *Eur. Polym. J.* **4**, 610.

HEIJBOER, J. (1960a). *Makromol Chem.*, **35 A**, 86.

HEIJBOER, J. (1960b). *Kolloid Z.*, **171**, 7.

HEIJBOER, J. (1965). In *Physics of Non-crystalline Solids*. North Holland, Amsterdam, 237.

HEINZE, D. (1966). *Kolloid Z. u. Z. f. Polym.*, **210**, 45.

HELFAND, E. (1971). *J. chem. Phys.*, **54**, 4651.

HENDUS, H., G. SCHNELL, H. THURN, K. WOLF (1959). *Ergebn. exakt Naturw.*, **31**, 5, 220.

HESLINGS, A., A. SCHORS (1964). *J. appl. Polym. Sci.*, **8**, 1921.

HIGASI, K., H. BABA, A. REMBAUM (1965). *Quantum Organic Chemistry*. Interscience, New York.

HILL, N.E., W.E. VAUGHAN, A.H. PRYCE, M. DAVIES (1969). *Dielectric Properties and Molecular Behaviour*. Van Nostrand, Reinhold, London.

HILTNER, A., E. BAER (1972). *Polym. J.*, **3**, 378.

HIRSCHFELD, J.O., J.N. LINNETT (1950). *J. chem. Phys.*, **18**, 130.

HODGMAN, CH. D., WEAST, R.C., SELBY, S.M. (1958). *Handbook of Chemistry and Physics*. Chem. Rubber Publ. Co., Cleveland.

HOFFMAN, J.D. (1952). *J. chem. Phys.*, **20**, 541.

HOFFMAN, J.D. (1955). *J. chem. Phys.*, **23**, 1331.

HOFFMAN, J.D. (1965). *Polymer Preprints ACS*, **6**, 583.

HOFFMAN, J.D., B.M. AXILROD (1955). *J. natn. Bur. Stands*, **54**, 357.

HOFFMAN, J.D., H.G. PFEIFFER (1954). *J. chem. Phys.*, **22**, 132.

HOLZMÜLLER, W., K. ALTENBURG (1961). *Physik der Kunststoffe*. Akad. Verlag, Berlin.

HOPKINSON, J. (1877). *Phil. Trans. R. Soc.*, **167**, 599.

HORNER, F., T.A. TAYLOR, R. DUNSMUIR, J. LAMB, W. JACKSON (1946). *J. Instn. elect. Engrs*, **93**, 53.

HOSEMANN, R. (1950). *Z. Phys.*, **128**, 1, 465.

HOSEMANN, R. (1963). *J. appl. Phys.*, **34**, 25.

HUISGEN, R. (1957). *Angew. Chem.*, **69**, 341.

HYDE. P. J. (1970) *J. Inst. Elect. Engrs.* (London) 117, 1891

IKADA, E., T. WATANABE (1972). *J. Polym. Sci.*, **10**, 3457.

ILLERS, K.H. (1969a). *Makromol Chem.*, **127**, 1.

ILLERS, K.H. (1969b). *Kolloid Z. u. Z. f. Polym.*, **231**, 622.

ILLERS, K.H. (1972). *Kolloid Z. u. Z. f. Polym.*, **250**, 426.

ILLERS, K.H., H. BREUER (1963). *J. Colloid Sci.*, **18**, 1.

ISHIDA, Y. (1960). *Kolloid Z.*, **168**, 29.

ISHIDA, Y. (1969). *J. Polym. Sci.*, **A-2**, 7, 1835.

ISHIDA, Y., M. MATSUO, M. TAKAYANAGI (1965). *J. Polym. Sci.*, **B 3**, 321.

ISHINABE, T., K. ISHIKAWA (1969). *Japan. J. appl. Phys.*, **8**, 1291.

ISHINABE, T., K. ISHIKAWA (1972). *Polym. J. Japan*, **3**, 300.

JACOBS, H., E. JENCKEL (1961). *Makromol. Chem.*, **43**, 132; **47**, 72.

JASSE, B. (1968). R.G.C.P. *Plastiques*, **5**, 393.

JENCKEL, E., R. HENSCH (1953). *Kolloid Z.*, **130**, 89.

JONES, H. (1946). *Trans. IRI*, **21**, 298.

JONES, G. P. (1975). In *Magnetic Resonance in Chemistry and Biology* (ed. J. Herak, K. J. Adamic). Marcel Dekker, New York.

JORDAN, E. F., G. R. RISER, W. E. PARKER, A. N. WRIGLEY (1966). *J. Polym. Sci.*, **A-2**, 975.

JORDAN, E. F. (1971). *J. Polym. Sci.*, **A-1**, 3367.

JUDD, N. C. W. (1965). *J. appl. Polym. Sci.*, **9**, 1743.

JUHÁSZ, K. (1967). *Plaste Kautschuk*, **14**, 2.

JUIJN, I. A., J. H. GISOF, W. A. DE JOY (1969). *Kolloid Z. u. Z. f. Polym.*, **235**, 1157.

KABIN, S. P., G. P. MIKHAILOV, (1956). *Zh. tekh. Fiz.*, *SSSR*, **26**, 511.

KABIN, S. P., G. P. MIKHAILOV (1958). *Izv. Akad. nauk. SSSR, Fiz.*, **22**, 325.

KANIG, G. (1963). *Kolloid Z.*, 190, 1.

KAKIZAKI, M., J. SHIGEMI, Y. TERAMOTO, K. YANADA, T. HIDESHIMA (1969). *Rep. Prog. Polym. Phys.*, *Japan*, **12**, 387.

KAKUTANI, H., M. ASAHINA (1969). *J. Polym. Sci.*, **A-2**, 1473.

KARGIN, V. A. (1958). *J. Polym. Sci.*, 30, 247.

KARGIN, V. A., P. V. KOZLOV, R. M. ASIMOVA, L. I. ANANIEVA (1960). *Doklady Akad. Nauk. USSR*, **135**, 357.

KARGIN, V. A., G. L. SLONIMSKY (1948). *Doklady Akad. nauk. SSSR*, **62**, 239.

KARGIN, V. A., G. L. SLONIMSKY (1949). *Zh. fiz. Khim*, *SSSR*, **23**, 526.

KARGIN, V. A., G. L. SLONIMSKY, A. I. KITAIGORODSKY (1957). *Kolloid Z.*, **19**, 13.

KASHIVAGI, M., I. M. WARD (1972). *Polymer* (London), **13**, 145.

KÄSTNER, S. (1961). *Kolloid Z.*, **178**, 24, 119.

KÄSTNER, S. (1962). *Kolloid Z.*, **184**, 109; **185**, 126.

KELEN, T., G. BÁLINT, GY. GALAMBOS, F. TÜDŐS (1969). *Eur. Polym. J.*, **5**, 597.

KELLER, A. (1957). *Phil. Mag.*, **2**, 1171.

KELLER, A. (1958). In *Growth and Perfection of Crystals* (ed. R. H. Doremus). Wiley, New York.

KELLY, F. N., F. BUECHE (1961). *J. Polym. Sci.*, **50**, 549.

KHARADLY, M. M. Z., W. JACKSON (1953). *Proc. Instn. elect. Engrs*, **100** III, 199.

KINELL, PER OLOF, B. RÅNBY, V. RUNNSTRÖM-REIO (editors) (1973). ESR Applications to Polymer Research, *Nobel Symposium 22*. Almquist, Wiksell, Stockholm.

KINJO, W., T. NAKAGAWA (1973). *Polym. J.*, **4**, 143.

KIRKWOOD, J. (1939). *J. chem. Phys.*, **7**, 911.

KIRKWOOD, J. G., R. M. FUOSS (1941). *J. chem. Phys.*, **9**, 329.

KIRYOUCHIN, D. P., A. M. KAPLAN, I. M. BARKALOV, V. I. GOLDANSKY (1970). *Vysokomol. Soed.*, **10**, 491.

KISBÉNYI, M., P. HEDVIG (1969a). *Euro. Polym. J.* Suppl., 291.

KISBÉNYI, M., P. HEDVIG (1969b). Kinetics and Mechanism of Polyreactions, *IUPAC Symposium*, *Budapest*, Vol. 5, p. 163. Hungarian Academy of Science, Budapest.

KISBÉNYI, M. (1970). Thesis. Tech. Univ., Budapest.

KISBÉNYI, M. (1973). IUPAC Macromolecular Symposium, Helsinki. *J. Polym. Sci. C 2*, **42**, 647.

KNOX, R.S. (1963). *Theory of Excitons*. Academic Press, New York.

KOIZUMI, N., K. TSUNASHIMA, S. YAUO (1969). *J. Polym. Sci., Polym. Letters*, **7**, 815.

KOLESOV, S.N. (1967). *Vysokomol. Soed.*, **A 9**, 1860.

KOPPELMAN, J. (1961). *Kolloid Z.*, **175**, 97.

KOVÁCS, A.J. (1958). *J. Polym. Sci.*, **30**, 131.

KOVÁCS, A.J. (1961). *Trans. Soc. Rheol.*, **5**, 285.

KOVÁCS, A.J., R.A. STRATTON, J.O. FERRY (1962). *J. Phys. Chem.* **67**, 152.

KOVÁCS, A.J. (1966). *Rheol. Acta*, **5**, 262.

KRATOCHVIL, A. (1967). *J. Polym. Sci., C* **16**, 1257.

KRIEGBAUM, W.R., A. ROIG (1959). *J. chem. Phys.*, **31**, 544.

KRUM, F. (1959), *Kolloid Z.*, **165**, 77.

KRUM, F., F.H. MÜLLER (1959). *Kolloid Z.*, **164**, 81.

KUBO, R. (1957). *J. phys. Soc., Japan*, **12**, 570.

KUBO, R. (1958). *Lectures in Theoretical Physics*. Interscience, New York.

KUROSAKI, S., T. FURUMAYA (1960). *J. Polym. Sci.*, **43**, 137.

LABUDA, E.F., R.C. CRAW (1961). *Rev. scient. Instrum.*, **32**, 391.

LAMB, J. (1964). In *Physical Acoustics*, Vol. 2 A (ed. W.P. Mason). Academic Press, New York.

LAMBE, J., R.C. JAKLEVIC (1969). *Tunneling Phenomena in Solids*. Plenum Press, New York.

LAURIE, V.W. (1961). *J. chem. Phys.*, **34**, 291.

LEADERMAN, H. (1943). In *Elastic and Creep Properties of Filamentous Materials and Other High Polymers*. Textile Foundation, Washington DC.

LEARMONTH, G.S., G. PRITCHARD (1967). *SPE Journal*, **23**, 12, 46.

LEARMONTH, G.S., G. PRITCHARD (1969a). *J. appl. Polym. Sci.*, **13**, 2119.

LEARMONTH, G.S., G. PRITCHARD (1969b). *IEC Products Research Rev.*, **8**, 124.

LEBEDEV, V.P. (1965). *Vysokomol. Soed.*, **1**, 333.

LEKSINA, I.E., S.I. NOVIKOVA (1959). *Sov. Phys.—Solid State*, **1**, 453.

LEUCHS, O. (1956). *Kunststoffe*, **46**, 547.

LEVITZKAYA, Z.M. (1970). *Dielectric Losses and Polarization in Epoxy Resins* (ed. Yu. N. Vershinin) NIPP. Acad. Sci., SSSR, Novosibirsk.

LINDENMEYER, P.H. (1972). *Polym. J.*, **3**, 507.

LIPATOV, YU.S., F.Y. FABULYAK (1972). *J. appl. Polym. Sci.* **16**, 2131.

LOBANOV, A.M., D.M. MIRKAMILOV, P. PLATONOV (1968). *Vysokomol. Soed.* **A-10**, 1116.

LOBANOV, A.M., G.B. SPAKOVSKAYA, O.S. ROMANOVSKAYA, B.I. SAZHIN (1969). *Vysokomol. Soed.*, **11**, 755.

LUKOMSKAYA, A.I., B.A. DOGADKIN (1960). *Kolloid Z.*, **22**, 576.

MALINOVSKAYA, B.I., M.P. PLATONOV, L.A. SHIBAEV (1968). *Vysokomol. Soed.*, **B-10**, 828.

MANDELKERN, L., G.M. MARTIN, F.A. QUINN JR. (1957). *J. Res. natn. Bur. Stand.*, **58**, 137.

MANN, J., H.I. MARRINAN (1956). *J. Polym. Sci.*, **21**, 301.

MARK, N. (1969). General lecture held at the IUPAC Conference, Budapest.

MARTYNENKO, L. YA., J.B. RABINOVICH, YU. V. OVCHINNIKOV, V.A. MASLOVA (1970). *Vysokomol. Soed.*, **A-12**, 841.

MARTYNOV, M.A., K.A. BYLEGZHANINA (1972). *X-ray Spectroscopy of Polymers* (in Russian). Izv. Khimia, Leningrad.

MATSUI, M., R. MATSUI, Y. WADA (1971). *Polym. J.*, **2**, 134.

MATSUOKA, S., Y. ISHIDA (1966). *J. Polym. Sci. C*, No. 14, 247.

MAXWELL, J. C. (1892). *Electricity and Magnetism.* Clarendon, Oxford.

MCCLELLAN, M. (1963). *Tables of Experimental Dipole Moments.* Freeman, San Francisco.

MCCRONE, W. C. (1965). In *Physics and Chemistry of the Organic Solid State* (ed. D. Fox, M. M. Labes and A. Weissberger). Vol. 2. Interscience, New York.

MCCRUM, N. G. (1961). *J. Polym. Sci.*, **54**, 561.

MCCRUM, N. G. (1962). *J. Polym. Sci.*, **60**, 53.

MCCRUM, N. G., E. L. MORRIS (1964). *Proc. R. Soc.*, A **281**, 258.

MCCRUM, N. G., B. E. READ, G. WILLIAMS (1967). *Anelastic and Dielectric Effects in Polymeric Solids.* Wiley. London and New York.

MCGRAW, G. E. (1971). In *Polymer Characterization Interdisciplinary Approaches* (ed. C. D. CARVER). Plenum, New York.

MCKELVEY, J. M. (1962). *Polymer Processing.* Wiley, New York.

MCMAHON, E. G., G. R. PERKINS (1964). *IEE Trans. on Power Apparatus and Systems*, **83**, 1253.

MEERSON, S. I., I. M. ZAGRAEVSKAYA (1969). *Kolloid Z.*, **31**, 417.

MELLAN, I. (1961). *The Behavior of Plasticizers.* Pergamon, New York.

MELVEGER, A. J. (1972). *J. Polymer Sci.* A-2, **10**, 317.

MIKHAILOV, G. P. (1955). *Uszpehi Khimi*, **24**, 875.

MIKHAILOV, G. P., T. I. BORISOVA, D. A. DIMITRENKO (1956). *Zh. tekh. Fiz.*, **26**, 1924.

MIKHAILOV, G. P. (1958). *J. Polym. Sci.*, **30**, 605.

MIKHAILOV, G. P., T. I. BORISOVA (1958). *Zh. tekh. Fiz.*, **28**, 137.

MIKHAILOV, G. P., T. I. BORISOVA (1960). *Vysokomol. Soed.*, **2**, 1772.

MIKHAILOV, G. P., A. M. LOBANOV, V. A. SHEVELEV (1961). *Vysokomol. Soed.*, **3**, 794.

MIKHAILOV, G. P., T. I. BORISOVA (1964a). *Usp. fiz. Nauk.* SSSR, **83**, 63.

MIKHAILOV, G. P., T. I. BORISOVA (1964b). *Vysokomol. Soed.*, **6**, 1785.

MIKHAILOV, G. P., L. L. BURSHTEIN (1964). *Vysokomol. Soed.*, **6**, 1713.

MIKHAILOV, G. P., A. M. LOBANOV, V. A. SHEVELEV, T. P. ORLOVA (1964). *Vysokomol. Soed.*, **6**, 868.

MIKHAILOV, G. P., L. L. BURSHTEIN, L. B. KRASNEV (1965). *Vysokomol. Soed.*, **7**, 870.

MIKHAILOV, G. P., A. I. ARTYUHOV, T. I. BORISOVA (1966a). *Vysokomol. Soed.*, **8**, 1257.

MIKHAILOV, G. G. P., A. M. LOBANOV, D. M. MIRKAMILOV (1966b). *Vysokomol. Soed.*, **8**, 1351.

MIKHAILOV, G. P., A. M. LOBANOV, M. P. PLATONOV (1966c). *Vysokomol. Soed.*, **8**, 1377.

MIKHAILOV, G. P., A. I. ARTYUHOV, T. I. BORISOVA (1967a). *Vysokomol. Soed.*, B **9**, 138.

MIKHAILOV, G. P., A. I. ARTYUHOV, T. I. BORISOVA (1967b). *Vysokomol. Soed.*, A **9**, 2401.

MIKHAILOV, G. P., L. V. KRASNEV (1967). *Vysokomol. Soed.*, A **9**, 1346.

MIKHAILOV, G. P., L. L. BURSHTEIN, T. P. ANDREEVA (1967c). *Vysokomol. Soed.*, A **9**, 2693.

MIKHAILOV, G. P., L. L. BURSHTEIN, V. P. MALINOVSKAYA (1967d). *Vysokomol. Soed.*, B **9**, 162.

MIKHAILOV, G. P., L. L. BURSHTEIN, V. P. MALINOVSKAYA (1967e). *Vysokomol. Soed.*, B **9**, 162.

MIKHAILOV, G. P., A. M. LOBANOV, M. P. PLATONOV (1967f). *Vysokomol. Soed.*, A **10**, 2267.

MIKHAILOV, G. P., A. I. ARTYUHOV, T. I. BORISOVA (1968a). *Vysokomol. Soed.*, **A 10**, 1755.
MIKHAILOV, G. P., L. L. BURSHTEIN, S. I. KLENIN, V. P. MALINOVSKAYA, A. N. CHERKASOV, L. A. CHIVAEV (1968b). *Vysokomol. Soed.*, **A 10**, 556.
MIKHAILOV, G. P., A. M. LOBANOV, D. M. MIRKAMILOV (1968c). *Vysokomol. Soed.*, **A 10**, 826.
MIKHAILOV, G. P., A. I. ARTYUHOV, V. A. SHEVELEV (1969a). *Vysokomol. Soed.*, **A 11**, 553.
MIKHAILOV, G. P., L. L. BURSHTEIN, V. P. MALINOVSKAYA (1969b). *Vysokomol. Soed.*, **A 11**, 548.
MILES, J. B., JR., H. P. ROBERTSON (1932). *Phys. Rev.*, **40**, 583.
MILLER, A. A. (1959). *Ind. Engng. Chem.*, **51**, 1271.
MILLER, G. W. (1969). *Appl. Polym. Symposia*, No. 10, 35.
MINAKATA, K., M. IWASAKI (1972). *J. chem. Phys.*, **57**, 4758.
MINKIN, V. I., O. A. OSIPOV, YU. A. ZHDANOV (1968). *Dipole Moments in Organic Chemistry* (in Russian). Izv. Khimia, Leningrad.
MONTROLL, E. (1941). *J. Am. chem Soc.*, **63**, 1215.
NAGY, J. (1973). Thesis, Hungarian Ac. Sci., Budapest.
NATTA, G., P. CORRADINI (1956). *Makromol. Chem.*, **16**, 77.
NEDETZKA, T., M. REICHLE, A. MAYER, H. VOGEL (1970). *J. phys. Chem.*, **74**, 2652.
NIELSEN, L. E. (1962). *Mechanical Properties of Polymers.* Reinhold, New York.
NIKOLSKY, V. G., N. YA. BUBEN (1960). *Doklady A. N. SSSR*, **134**, 134.
NORMAN, R. H. (1953). *Proc. Instn. elect. Engrs*, **100**, II. **A**, 341.
NORTH, A. M., J. C. REID (1972). *Eur. Polym. J.* **8**, 1129.
NORTH, A. M., I. SOUTAR (1972). *J. chem. Soc., Faraday Transactions*, **1**, 68, 1101.
NOSE, T. (1971). *Polym. J.*, **2**, 124, 427, 437, 445.
NOSE, T. (1972). *Polym. J.*, **3**, 1, 196.
NOZAKI, M., K. SHIMADA, S. OKAMOTO (1970). *Japan. J. appl. Phys.*, **9**, 843.
NOZAKI, M., K. SHIMADA, S. OKAMOTO (1971). *Japan. J. appl. Phys.*, **10**, 179.
O'DWYER, J. J., E. HARTING (1967). In *Progress in Dielectrics* (ed. J. B. Birks and J. Hart), Vol. 7, 1.
OGURA, K., S. KAWAMURA, H. SOBUE (1971). *Macromolecules*, **4**, 79.
O'KONSKI, C. T. (1960). *J. phys. Chem.*, **64**, 605.
OMELCHENKO, S. I. (1962). *Epoxy resins.* Gosud. Izv. USSR, Moscow.
ONO, K., K. MURAKAMI (1971). *Rep. Prog. Phys. Japan*, **14**, 451.
O'REILLY, J. M. (1962). *J. Polym. Sci.*, **57**, 429.
O'REILLY, D. E., T. TSANG (1967). *J. chem. Phys.*, **46**, 1291.
OSAKI, S., S. UEMURA, Y. ISHIDA (1971a). *J. Polym. Sci.*, **A-2**, 9, 585.
OSAKI, S., S. UEMURA, Y. ISHIDA (1971b). *J. Polym. Sci.*, **A-2**, 9, 2099.
OSIPOV, O. A., V. I. MINKIN (1965). *Dipole Moments* (in Russian). Izv. Vys. Skola, Moscow.
PAKEN, A. M. (1962). *Epoxy Compounds and Epoxy Resins.* Goskhimizdat, Moscow.
PAULING, L. (1960). *The Nature of the Chemical Bond*, 3rd edn. Cornell Univ. Press, New York.
PAYNE, A. R. (1958). In *Rheology of Elastomers* (ed. P. Mason and N. Wookey). Pergamon Press, London.
PEREPECHKO, I. I., L. I. TREPELKOVA, L. A. BOBROVA, L. O. BUNYINA (1968). *Vysokomol. Soed.*, **B 10**, 507.

PETERLIN, A. (1965). *J. Polym. Sci.*, **9**, 61.

PETERLIN, A., H. G. ZACHMANN (1971). *J. Polym. Sci. C*, No. 34, 11.

PETERSEN, J., B. RÅNBY (1967). *Makromol. Chem.*, **102**, 83.

PETERSEN, J., B. RÅNBY (1970). *Makromol. Chem.*, **133**, 251, 263.

PEZZIN, G. (1967). *J. appl. Polym. Sci.*, **11**, 2553.

PEZZIN, G., G. AJROLDI, T. CASIRAGHI, C. GARBUGLIO (1972). *J. appl. Polym. Sci.*, **16**, 1839.

PIERCE, R. H., E. S. CLARK, J. F. WITNEY, W. D. M. BRYANT (1956). *Polymer Preprints ACS*.

PILAR, F. L. (1968). *Elementary Quantum Chemistry*. McGraw-Hill, New York.

POHL, H. A. (1962). In *Modern Aspects of the Vitreous State* (ed. J. D. Mackenzie). Butterworths, London.

POHL, G., KÄSTNER, S. (1968). *J. Polym. Sci., C*, No. 16, 4133.

POGÁNY, G. A. (1969). *Br. Polym. J.*, **1**, 177.

POGÁNY, G. A. (1970). *Polymer*, **11**, 66.

POLITZER, P., R. R. HARRIS (1970). *J. Am. chem. Soc.* **92**, 6451.

PORT, W. S. (1955). *Ind. Engng. Chem.*, **47**, 472.

PORTER, C. H., J. H. L. LAWLER, R. H. BOYD (1970). *Macromolecules*, **3**, 308.

POWLES, J. G., B. I. HUNT (1965). *Phys. Letters*, **14**, 202.

PREDĘCKI, P., W. O. STATTON (1967). *J. appl. Phys.*, **38**, 4140.

PRIGOGINE, I. (1962). *Non-equilibrium Statistical Mechanics*. Interscience, New York.

READ, B. E., G. WILLIAMS (1961). *Polymer*, **2**, 239.

REDDISH, W. (1950). *Trans. Faraday Soc.*, **46**, 459.

REDDISH, W. (1962). *Pure appl. Chem.*, **5**, 723.

REDDISH, W. (1965). *Polymer Preprints ACS*, **6**, 571.

REDDISH, W., J. T. BARRIE (1959). *Symposium über Makromoleküle, Wiesbaden*, Verlag Chemie, Weinheim.

REESE, W. (1969). *J. Macromol. Sci. Chem.*, **A 3**, 1257.

REICH, S., A. EISENBERG (1970). *J. chem. Phys.*, **53**, 2847.

REICH, S., A. EISENBERG (1972). *J. Polym. Sci.*, **A 2**, 1397.

RÉSIBOIS, P. (1964). *J. chem. Phys.*, **41**, 2979.

RETTING, W. (1969). *Angew. Makromol. Chem.* **8**, 87.

RETTING, W. (1970). *Eur. Polym. J.*, **6**, 853.

RICE, S. A., J. JORTNER (1967). In *Chemistry and Physics of the Organic Solid State* (ed. D. FOX, M. LABES and A. WEISSBERGER), Vol. 3. Interscience, New York.

ROBESON, L. M. (1969). *Polym. engng. Sci.*, **9**, 277.

ROBITSCHEK, P., A. LEWIN (1950). *Phenolic Resins*. Iliffe, London.

ROSEN, R., H. A. POHL (1966). *J. Polym. Sci.*, **A-1**, 1135.

ROUSE, P. E. (1953). *J. chem. Phys.*, **21**, 1272.

RUSCH, K. C. (1968). *J. macromol. Sci., Phys.*, **B 2**, 2, 179.

RUSCH, K. C., R. H. BECK, JR. (1969). *J. macromol. Sci.*, **B 3**, 365.

RUSCH, K. C., R. H. BECK, JR. (1970). *J. macromol. Sci., Phys.*, **B 4**, 621.

RYSHKINA, I. J., V. M. AVERIANOVA (1970). *Izv. vyssh. ucheb. Zaved. Khim.*, **13**, 59.

SACK, R. A. (1952). *Aust. J. scient Res.*, **5**, 135.

SAFFORD, G. S., A. W. NAUMANN (1967). *Adv. Polym. Sci.*, **5**, 1.

SAITO, S. (1962). *Rep. Prog. Polym. Phys. Japan*, **5**, 205.

SAITO, S. (1963). *Kolloïd Z.*, **189**, 116.

SAITO, S., H. SASABE, T. NAKAJIMA, K. YADA (1968). *J. Polym. Sci.*, **A 2**, 6, 1297.

Salmon, W. A., L. D. Loan (1972). *J. appl. Polym. Sci.*, **16**, 671.
Salovey, R., R. V. Albarino, J. P. Luongo (1970). *Macromolecules*, **3**, 314.
Salovey, R., H. E. Bair (1970). *J. appl. Polym. Sci.*, **14**, 713.
Salovey, R., R. C. Gebauer (1972). *J. Polym. Sci.*, **A-1**, 1533.
Salovey, R., J. P. Luongo, W. A. Yager (1969). *Macromolecules*, **2**, 198.
Sanditov, S., I. A. Lukyanov (1969). *J. Polym. Sci.*, **A-1**, 2147.
Sasabe, H., S. Saito (1972). *Polym. J.*, **3**, 749.
Sasabe, H., S. Saito, M. Asahina, H. Kakutani (1969). *J. Polym. Sci.*, **A-2**, 1405.
Sasabe, H., K. Sawamura, S. Saito, K. Yoola (1971). *Polym. J.*, **2**, 518.
Sasakura, H. (1963). *Japan. J. appl. Phys.*, **2**, 66.
Sauer, J. A., R. G. Saba (1969). *J. macromol. Sci., Chem.* **A 3**, 1217.
Sazhin, B. I. (1970). *Electrical Properties of Polymers* (in Russian). Izd. Khimia, Leningrad.
Sazhin, B. I., T. P. Orlova, A. M. Lobanov (1968). *Vysokomol. Soed.*, **A 10**, 1921.
Scaife, B. K. (1963). *Proc. phys. Soc. Lond.*, **81**, 124.
Schallamach, A. (1946). *Trans. Faraday Soc.*, **42 A**, 495.
Schallamach, A. (1951). *Trans. Inst. Rubber Ind.*, **27**, 40.
Schatzki, T. F. (1962). *J. Polym. Sci.*, **57**, 496.
Schatzki, T. F. (1965). *Polymer Preprints, ACS*, **6**, 646.
Schell, W. J., R. Simha, J. Aklonis (1969). *J. macromol. Sci., Chem.*, **A 3**, 1297.
Scherr, H., W. Pechhold, S. Blasenbrey (1970). *Kolloid Z. u. Z. f. Polym.*, **238**, 395.
Schmieder, K., K. Wolf (1953). *Kolloid Z.*, **134**, 149.
Scott, A. H., A. T. McPherson, H. L. Curtis (1933). J. *Res. natn. Bur. Stand.*, **11**, 373.
Scott, A. H., D. J. Scheiber, A. J. Curtis, J. I. Lauritzen, J. D. Hoffman (1962). *J. Res. natn. Bur. Stand.*, 66 A, 269.
Seanor, D. A. (1968). *J. Polym. Sci.*, **A-2**, 6, 463.
Shindo, H. (1969). *J. Sci. Hiroshima Univ.*, Ser A-II. **33**, 189.
Shindo, H., I. Murakami, H. Yamamura (1969). *J. Polym. Sci.*, **A-1**, 297.
Simha, M. (1962). *J. Polymer Sci.* **56**, 213.
Slichter, W. P. (1963). *Principles of Magnetic Resonance with Examples from Solid State Physics.* Harper & Row, New York.
Slichter, W. P. (1966). *J. Polym. Sci. C*, No. 14, 33.
Slichter, W. P., D. Ailion (1964). *Phys. Rev.*, **A 135**, 1099.
Slonim, I. Ya., A. N. Ljubimov (1970). *The N M R of Polymers.* Plenum, New York.
Slonimsky, G. L., A. A. Askadsky, A. J. Kitaigorodsky (1970). *Vysokomol. Soed.* **A, 12**, 494.
Smith, J. W. (1955). *Electric Dipole Moments,* **Butterworth,** London.
Smyth, C. P. (1955). *Dielectric Behaviour and Structure.* McGraw-Hill, New York.
Solunov, Kh., P. Hedvig (1972a). *Natura* (Univ. Plovdiv, Bulgaria), Vol. 5, 47.
Solunov, Kh., P. Hedvig (1972b). In *Proc. 3rd Symposium on Radiation Chemistry* (ed. J. Dobó and P. Hedvig), p. 899. Publishing House, Hungarian Academy of Science, Budapest.
Sperling, L. H., D. W. Taylor, M. L. Kirkpatrick, H. F. George, D. R. Bardman (1970). *J. Polym. Sci.*, **14**, 73.
Stachurski, Z. H., I. M. Ward (1968). *J. Polym. Sci.*, A-2. 6, 1817.
Stachurski, Z. H., I. M. Ward (1969a). *J. macromol. Sci. Phys.* **B 3**, 445.
Stachurski, Z. H., I. M. Ward (1969b). *J. macromol. Sci. Phys.*, **B 3**, 427.
Stafford, T. G. (1972). *Rapra Bull.*, No. 65, 66. No. 66, 102.

STEJSKAL, E. D., H. S. GUTOWSKY (1958). *J. chem. Phys.*, **28**, 388.

STOCKMAYER, W. H., M. BAUR (1964). *J. Am. chem. Soc.*, **86**, 3485.

STOCKMAYER, W. H., J. J. BURKE (1969). *Macromolecules*, **2**, 647.

STOCKMAYER, W. H., C. E. HECHT (1953). *J. chem. Phys.*, **21**, 1954.

STROMBERG, R. R., S. STRAUSS, B. G. ACHHAMMER (1959). *J. Polym. Sci.*, **35**, 355.

STROUPE, J. D., R. E. HUGHES (1958). *J. Am. chem. Soc.*, **80**, 2341.

STUART, H. A. (1959). *Ann. N. Y. Acad. Sci.*, **83**, 1.

SUVOROVA, A. I., A. A. TAGER (1966). *Polym. Sci. SSSR*, **8**, 1873.

TADOKORO, H. (1963). *Rep. Prog. Polym. Phys. Japan*, **6**, 303.

TAKAHASHI, Y. (1961). *J. appl. Polym. Sci.*, **5**, 468.

TAKEDA, M., J. IMAMURA, S. OKAMURA, T. HIGOSHIMURA (1960). *J. chem. Phys.*, **33**, 631.

TANABE, Y., J. HIROSE, Y. WADA (1970). *Polymer J.*, **1**, 107.

TANAKA, A., Y. ISHIDA (1972). *J. Polym. Sci.*, **A-2**, 10, 1029.

TARASOV, V. V. (1950). *Zh. fiz. Khim. SSSR*, **24**, 111.

THIRION, P., R. CHASSET (1951). *Trans. Inst. Rubber Ind.*, **27**, 364.

THURN, H., K. WOLF (1956). *Kolloid Z.*, **148**, 66.

THURN, H., F. WÜRSTLIN (1958). *Kolloid Z.*, **156**, 21.

TOBOLSKY, A. V. (1960). In *Properties and Structure of Polymers*. Wiley, New York.

TSUGE, K. (1964). *Japan. J. appl. Phys.*, **3**, 588.

TUCKER, J. E., W. REESE (1967). *J. chem. Phys.*, **46**, 1388.

TURLEY, S. G., H. KESKKULA (1966). *J. Polym. Sci.* C, No. 14, 69.

VAN BEEK, L. K. H. (1962). *Proc. Colloqu. A M P É R E*, **11**, 229.

VAN BEEK, L. K. H. (1964). *J. appl. Polym. Sci.*, **8**, 2843.

VAN BEEK, L. K. H. (1967). In *Progress in Dielectrics* (ed. J. B. Birks and J. Hart), Vol. 7, 69. Heywood, London.

VAN BEEK, L. K. H., J. J. HERMANS (1957). *J. Polym. Sci.*, **23**, 211.

VAN ROGGEN, A. (1972). In *International Microsymposium on Polarization and Conduction in Insulating Materials, Bratislava, Harmonia*.

VAN TURNHOUT, J. (1971). Polym. J., 2, 173.

VESELOVSKY, I. F., V. K. MATVEEV (1964). *Vysokomol. Soed.*, **6**, 1221.

VOLKENSTEIN, M. V. (1963). *Configurational Statistics of Polymeric Chains*. Interscience. New York.

VOLKENSTEIN, M. V., O. B. PTYTSIN (1956). *Zh. tekh. Fiz.*, **26**, 2204.

VON HIPPEL, A. R. (1954). In *Dielectric Materials and Applications*. Wiley, New York.

WADA, A., H. KIHARA (1972). *Polym. J.*, **3**, 492.

WANNIER, G. H. (1937). *Phys. Rev.*, **52**, 191.

WARD, I. M. (1962). *Proc. phys. Soc. Lond.*, **80**, 1176.

WARFIELD, R. W. (1958). *S P E Journal*, **14**, 39.

WARFIELD, R. W., M. C. PETREE (1959). *J. Polym. Sci.*, **37**, 305.

WARFIELD, R. W., M. C. PETREE (1962). *Makromol. Chem.*, **58**, 139.

WARING, J. R. S. (1951). *Trans. Inst. Rubber Ind.*, **27**, 16.

WARRIER, A. V. R., S. KRIMM (1970a). *J. chem. Phys.*, **52**, 4316.

WARRIER, A. V. R., S. KRIMM (1970b). *Macromolecules*, **3**, 709.

WARTMAN, L. H., W. J. FRISSEL (1956). *Plast. Technol.*, **2**, 583.

WESSLEN, B., R. W. LENZ, W. J. McKNIGHT, F. E. KARASZ (1971). *Macromolecules*, **4**, 24.

WILLIAMS, G. (1962). *Trans. Faraday Soc.*, **58**, 1041.

WILLIAMS, G. (1972). *Chem. Rev.*, **72**, 55.

WILLIAMS, G., M. COOK (1971). *Trans. Faraday Soc.*, **67**, 990.

WILLIAMS, G., D.C. WATTS (1970). *Trans. Faraday Soc.*, **66,** 80.
WILLIAMS, G., D.C. WATTS (1971a). In *Natural and Synthetic High Polymers, NMR, Basic Principles and Progress* (ed. P. Diehl, E. Fluck and R. Kosfeld), Vol. 4, p. 271. Springer Verlag, Berlin.
WILLIAMS, G., D.C. WATTS (1971b). *Trans. Farday Soc.*, **67,** 1323.
WILLIAMS, M. L., R. F. LANDEL, J. D. FERRY (1955). *J. Am. chem. Soc.*, **77,** 3701.
WOOD, L. A. (1958). *J. Polym. Sci.*, **28,** 319.
WOODWARD, A. E. (1966). *J. Polym. Sci. C,* No. 14, 89.
WRASIDLO, W. (1972). *J. Polym. Sci.*. **A-2,** 1603.
WUNDERLICH, B. (1969). *Kolloid Z. u. Z. f. Polym.*, **231,** 605.
WUNDERLICH, B., D. M. BODILY (1964). *J. Polym. Sci. C*, No. 6, 137.
WÜRSTLIN, F., H. KLEIN (1952). *Kunststoffe*, **42,** 445.
YALOF, S., W. WRASIDLO (1972). *J. appl. Polym. Sci.*, **16,** 2159.
YAMAFUJI, K., Y. ISHIDA (1962). *Kolloid Z. u. Z. f. Polym.*, **183,** 15.
YAMAMOTO, K., Y. WADA (1957). *J. phys. Soc.*, *Japan*, **12,** 374.
YANO, O., Y. WADA (1971). *J. Polym. Sci.*, **A-2,** 669.
YANO, S. (1970) *J. Polym. Sci.*, **A-2,** 1057.
YEH, G. S. Y. (1972). *J. macromol. Sci. Phys.*, **B 6,** 3, 451, 465.
ZIMM, B. H. (1956). *J. chem. Phys.*, **24,** 269.
ZIMM, B. H. (1960). In *Rheology* (ed. F. R. Eirich). Academic Press, New York.
ZIMM, B. H., G. M. ROE, L. F. EPSTEIN (1956). *J. Am. chem. Soc.*, **86,** 3485.
ZOUBAROV, D. N. (1971). *Non-equilibrium Statistical Thermodynamics* (in Russian). Izd. Nauka, Moscow.

SUBJECT INDEX

acoustic modes 111
acrylonitrile – alkyl-amide copolymers
  300
acrylonitrile – butadiene rubber 304
aerosil filler 292
activation energy 67
ageing 356
aggregate model 135
alkyl phthalates 269
amorphous structure 78
analogue automatic bridge 157
anhydride cured epoxy resins 335
anisole 280
antiplasticizing 268
arc diagrams 61, 164
automatic resonance spectrometer 167

benzene 17
Bjerrum defects 244
blends 299
bond moments 15,18
bond order 19
bridge technique 156
bulk copolymers 298
butadiene nitrile rubber 324

carbon black loaded rubber 331
catalysts 296
cavity resonators 172
cellulose acetate 360
cellulose derivatives 251
cellulose hydrate 251
  – triacetate 253
charges in electric field 35
chlorinated polyethylene 305
  – polyvinylchloride 229

chlorine-substituted polystyrenes 115
chlorobenzene 17
chloromethanes 19
Clausius – Mosotti formula 65, 201
Cohen – Turnbull theory 95
cohesion energy density 258
Cole – Cole diagram 62
compatibility 257
compliance 51
  – matrix 135
composite oscillator 180
computerized spectrometer 163
conductive fillers 293
  – polymers 367
conductivity tensor 36
conformational transitions 102
conjugation of the main chain 364
copolymers 297
Coulombic integral 16
correlation function 34
  – time 191
creep 52
crosslinked polyvinylchloride 351
crosslinking 312
cryogenic relaxation 111
crystallinity 78
crystal orbitals 27
curing 332
cyanoethyl celluloses 255

Davydov splitting 28
d.c. conductivity of a resin 344
Debye functions 122
Debye heat capacity equation 111
defect dipole 29
depolymerization 358

427